FOUNDATIONS OF MECHANICAL ENGINEERING

FOUNDATIONS OF MECHANICAL ENGINEERING

Scarlett Bennett
Zachary Banks
Natalie Powell

Kruger Brentt
Publishers

2026

Kruger Brentt Publishers UK. LTD.
Company Number 9728962

Regd. Office: 68 St Margarets Road, Edgware, Middlesex HA8 9UU

ISBN: 978-1-78715-664-7

For information on all our publications visit our website at http://krugerbrentt.com/

PREFACE

Mechanical engineering forms the backbone of modern technology and industry, encompassing the principles of mechanics, thermodynamics, materials science, and energy systems to design, analyze, and optimize machines and processes. As the field continues to evolve with innovations in automation, robotics, renewable energy, and advanced materials, a solid understanding of mechanical engineering fundamentals is essential for students, engineers, and professionals across diverse sectors.

This book, Foundations of Mechanical Engineering, is designed to provide a comprehensive introduction to the core concepts, principles, and applications of mechanical engineering. It aims to serve as a foundational guide for undergraduate students, aspiring engineers, and anyone seeking a clear understanding of mechanical systems and their practical relevance in technology and industry.

The book begins with a discussion of the basic principles of mechanics, covering statics, dynamics, and kinematics, along with their applications in analyzing forces, motion, and mechanical equilibrium. Subsequent chapters explore materials science, emphasizing the mechanical properties of materials, stress-strain relationships, and failure analysis, which are crucial for designing reliable and efficient mechanical components.

Thermodynamics and fluid mechanics are examined in detail, highlighting energy transfer, heat engines, and fluid flow systems that underpin many industrial and engineering processes. The book also covers fundamentals of machine design, manufacturing processes, and mechanical systems, emphasizing practical problem-solving and real-world applications.

Special attention is given to emerging trends and technologies in mechanical engineering, including automation, robotics, computer-aided design (CAD), and additive manufacturing, showing how traditional engineering principles are integrated with modern innovations. Real-world examples and case studies illustrate the practical applications of mechanical engineering concepts, bridging theory and practice for students and professionals alike.

Foundations of Mechanical Engineering is a comprehensive resource that equips readers with the knowledge, analytical skills, and problem-solving techniques necessary to excel in the field. By combining fundamental principles with modern applications, the book provides a solid platform for further study and professional development in mechanical engineering.

Scarlett Bennett

Zachary Banks

Natalie Powell

CONTENTS

1 Chapter

MODELING OF MECHANICAL TRANSMISSION PARAMETERS

1. INTRODUCTION: AN OVERVIEW

In mechanical engineering, transmission refers to a device that serves as an intermediary mechanism between the driving machine (e.g., engine) and the driven (consumed) machine. The function of transmission is to convey mechanical energy from the main shaft of the driving machine to the main shaft of the working machine. The choice of transmission is constrained by the cost of the entire apparatus, the operational environment, the dimensions of the device, technical regulations, and other factors. Mechanical transmissions are widely utilized in mechanical engineering and technology in general. Mechanical transmissions are systems utilized for the transmission of mechanical energy, facilitating alterations in angular velocity, force, and rotational torque. Mechanical transmissions can be categorized based on the method of transmission into: gear transmissions (sprocket pairs), belt transmissions (belt pulleys and belts), friction transmissions (friction wheels), and chain transmissions (chain pulleys and chains). (Repcic & Muminovic, 2007)

This chapter presents the findings of the research on three-dimensional (3D) geometric parameter modeling of two commonly utilized types of mechanical transmissions: gear transmissions (including various standard catalog gears such as spur gears, bevel gears, and worms) and belt transmissions (specifically belt pulleys with a cylindrical external surface, or more precisely, with a pulley rim), utilizing the CATIA V5 software system (modules: Sketcher, Part Design, Generative Shape Design, Wireframe and Surface Design, and Assembly Design).

Computer modelling relies on geometry and perspective transformations, which are not explored in detail in this chapter due to their extensive breadth.

It is recommended to parameterize mechanical transmissions to facilitate the automation of their design. The parameter modelling application enables the manipulation of the generated geometry of a 3D model via parameters included in various relationships (formulas, parameter laws, tables, etc.). All dimensions, or more specifically, the geometric variable parameters of gears and belt pulleys, can be articulated through a limited set of characteristic fixed parameters (m, z, z1, z2, and N for the chosen gear; d, Bk, dv, and s for the selected belt pulley). The geometry of the 3D mechanical transmission model is altered by variations in the values of these parameters. Designers could produce additional design solutions by the parameterization of mechanical transmission.

Due to AutoCAD's lack of support for parametric modeling and its command system, it does not facilitate straightforward modifications to completed models. Consequently, parametric-oriented software systems such as CATIA V5, SolidWorks, and Mechanical Desktop, which utilize analytical expressions for variable interconnections through parameters, are employed. CATIA V5 (Computer-Aided Three-dimensional Interactive Application) epitomizes cutting-edge technology and serves as a benchmark in the field of design (Dassault Systemes, IBM, 2011). Presently, it is the most advanced integrated CAD/CAM/CAE software system available in the market for commercial and scientific research purposes. Prominent global corporations and their subcontractors utilize them. It is especially prevalent in the automotive sector (Daimler Chrysler, VW, BMW, Audi, Renault, Peugeot, Citroen, etc.), aerospace industry (Airbus, Boeing, etc.), and the manufacturing of machinery and consumer goods. The system possesses mathematical models and programs for the presentation of graphical shapes; nevertheless, consumers lack input in this process. The solution is developed independently from the operational computer system, enabling program module architecture and user customization. The core of the system is the integrated associational data structure for parameter modeling, which allows modifications to the model to be manifested across all associated phases of product development. Consequently, the time required for the manual remodeling of models is conserved. The framework enables the parameterization of all geometric objects, including solids, surfaces, wireframe models, and constructive elements. (Karam & Kleismit, 2004; Saric et al., 2010) The entire model, or a portion thereof, can be parameterized to provide flexibility in the development of novel variant design solutions. The interdependence of intelligent elements is established for a component or assembly by parameterization. The primary feature of parameter modeling in the CATIA V5 system is its considerable flexibility, as parameters may be defined at any time, albeit it is not obligatory to do so. In addition to altering parameter values, their deletion, addition, and reconnection are also perpetually feasible. Karam and Kleismit, 2004 The complete Graphical User Interface (GUI) written in C++ is designed to resemble the tool palette and icons seen in the Windows interface. The system was primarily developed for Windows and Windows 64-bit, but it was also designed for AIX, AIX 64-bit, HP-UX, IRIX, and Solaris operating systems. To achieve optimal performance with the CATIA V5 system, it is advisable to utilize certified hardware combinations. (Certified hardware configurations for CATIA V5 systems, 2011).

Parameter modeling in the CATIA V5 system is predicated on the principle of knowledge, facilitating the creation and utilization of parameter-modeled components and assemblies. Saric et al. (2009) The creation of 3D parametric solid models is predominantly achieved by integrating the Feature-Based Design (FBD) technique with Boolean operations (Constructive Solid Geometry - CSG). (Amirouche, 2004; Shigley et al., 2004; Spotts et al., 2004) The predominant parameter types in modeling include: Real, Integer, String, Length, Angle, Mass, among others. They are categorized into two categories.:

- internal parameters which are generated during geometry creating and which define its interior features (depth, distance, activity, etc.) and
- user parameters (with one fixed or complex variables) which user specially created and which define additional information on the: *Assembly Level, Part Level* or *Feature Level.*

A parameter is a variable utilized to regulate the geometry of a component, with its value influenced by established relationships. Geometry can be controlled by the utilization of tools. color palette Diverse modalities of knowledge:

- by creating of user parameters set and by their values changing,
- by use of defined formulas and parameter laws that join parameters,
- by joining of parameters in designed tables and by selection of appropriate configured set.

The recommendation is, before components parameterising, to:

1. check the component complexity,
2. notice possible ways of component making,
3. notice dimensions which are going to change and
4. select the best way for component parameterising.

2. MODELLING OF MECHANICAL TRANSMISSION PARAMETERS

The modeling of specific mechanical transmissions was conducted using the Sketcher, Part Design, and Generative Shape Design modules of the CATIA V5 system. Knowledge of the modeling methods in the Wireframe, Surface Design, and Assembly Design modules of the CATIA V5 system is a prerequisite for this modeling approach. (Karam & Kleismit, 2004; Dassault Systemes, 2007a, 2007b; Zamani & Weaver, 2007).

Upon completion of the modeling technique, mechanical transmissions can be utilized independently within assemblies in a complex manner.

Parameter symbols and standard equations (Tables 1 and 5) utilized in the modeling of mechanical transmissions are available in the references (Repcic et al., 1998;

Repcic & Muminovic, 2007, pp. 139, 154-155, 160-161). Comprehensive elucidations on gear transmissions and belt transmissions are available in the sources (Repcic et al., 1998, pp. 54-106, 118-151).

2.1 Modelling of Transmission Gear Parameters

The following paragraph illustrates the 3D geometric parameter modeling of characteristic standard catalog gears: spur gears, bevel gears, and worms.

Gears were chosen as a representative example due to their prevalence as mechanical components and the intricate geometry of their cog profiles for modeling purposes.

Every user of a design software system is interested in the production of a complex plane curve spline that defines the geometry of the cog side profile.

The management of 3D parameterized model geometry is executed through established parameters, formulae, and parameter laws illustrated in the tree in Fig. 1. Cozzens, 2006 The study of parameters, formulas, and parameter laws within the Part documents tree is conducted using the primary selection menu.

(*Tools* → *Options* → *Part Infrastructure* → *Display*).

Table 1. Parameters and formulas

Spur gear	Bevel gear	Worm
z	z1, z2	z1=1
m		*m*=1,5
a=20 deg		
p=*m***PI*		
r=(*z***m*)/2	*r*=(z1**m*)/2	*d*=20 mm
rb=*r**cos(*a*)	*rb*=*rc**cos(*a*)	
rf=*r*-1.2**m*		
ra=r+m		*ra*=*d*/2+*ha*
rr=0.38**m*		
ha=*m*		
hf=1.2**m*		
xd=*rb**(cos(*t***PI*)+sin(*t***PI*)**t***PI*)		N=6,5
yd=*rb**(sin(*t***PI*)-cos(*t***PI*)**t***PI*)		L=(N+1)**p*
0≤*t*≤1		*gama*=atan(*m**z1/*d*)
	delta=atan(z1/z2)	*dZ*=-*L*/2
	rc=*r*/cos(*delta*)	
	lc=*rc*/sin(*delta*)	

	tc=-atan(*Relations**yd.Evaluate*(*a*/180deg)/ Relations\ *xd.Evaluate*(*a*/180deg))	
	b=0.3**rc*	
	ratio=1-*b*/lc/cos(*delta*)	
	dZ=0 mm	

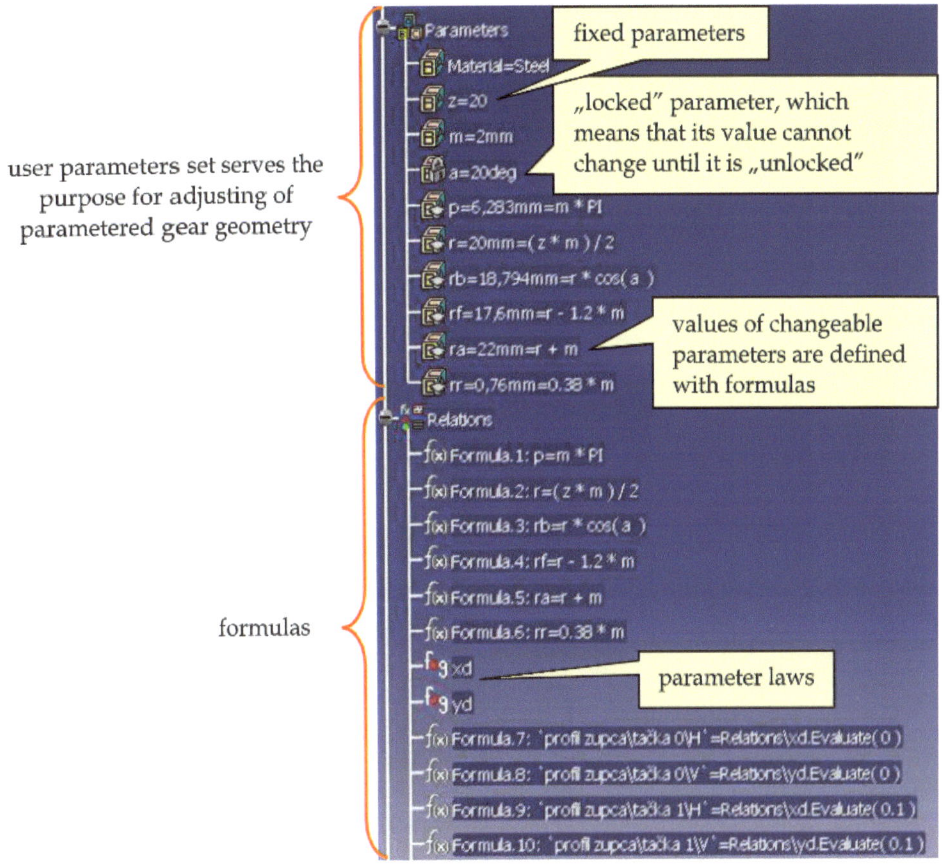

Fig. 1. Gear geometry control

2.1.1 Modeling of spur gear parameters

To establish fixed parameters (Fig. 2), we choose the Formula command from the Knowledge tools panel or the main selection menu. Subsequently, we:

1. choose desired parameter type (*Real*, *Integer*, *Length*, *Angle*) and press the button *New Parameter of type*,
2. type in a new parameter name,
3. assign a parameter value (only in the case if parameter has fixed value) and
4. press the button *Apply* to confirm a new parameter creation.

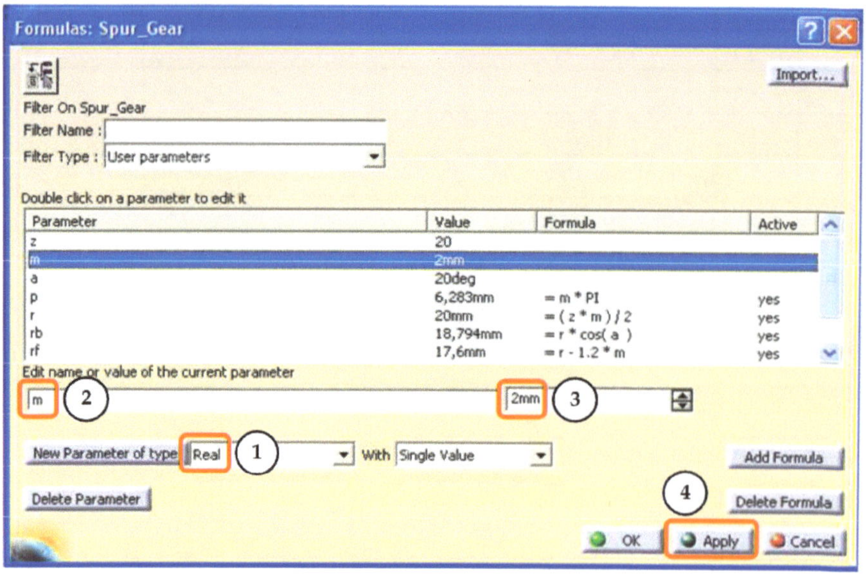

Fig. 2. Fixed parameters defining

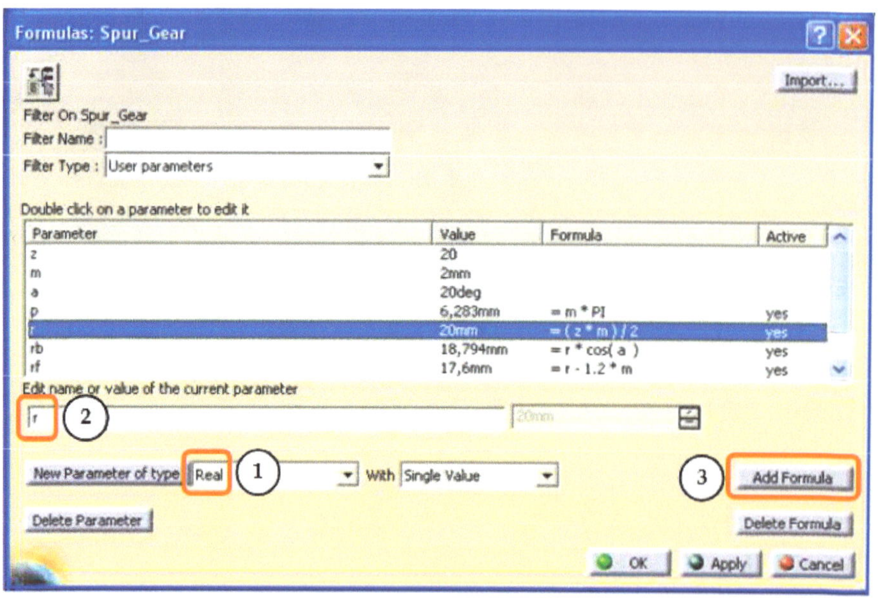

Fig. 3. Changeable parameters defining

Most geometrical gear parameters are changeable and are in the function of fixed parameters *m* and *z* (Fig. 3.). We do not need to set values for these parameters, because CATIA V5 system calculates them itself. So, instead of values setting, formulas are defined by choosing the command *Formula* (Fig. 4.). When formula has been created, it is possible to manipulate with it by the tree, similar as with any other model feature.

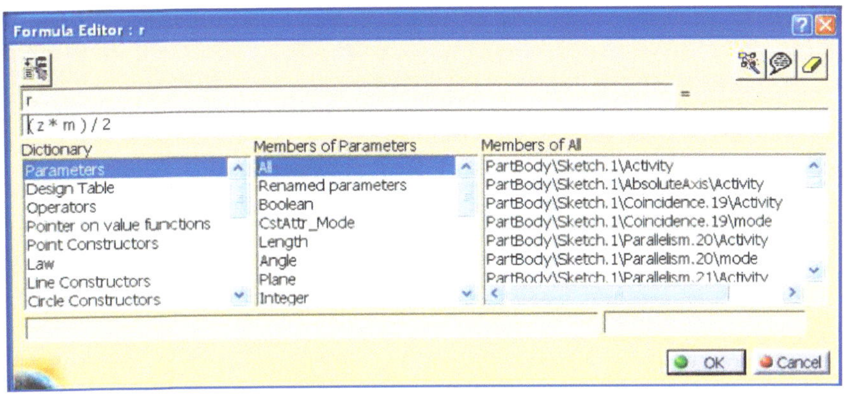

Fig. 4. Formula setting

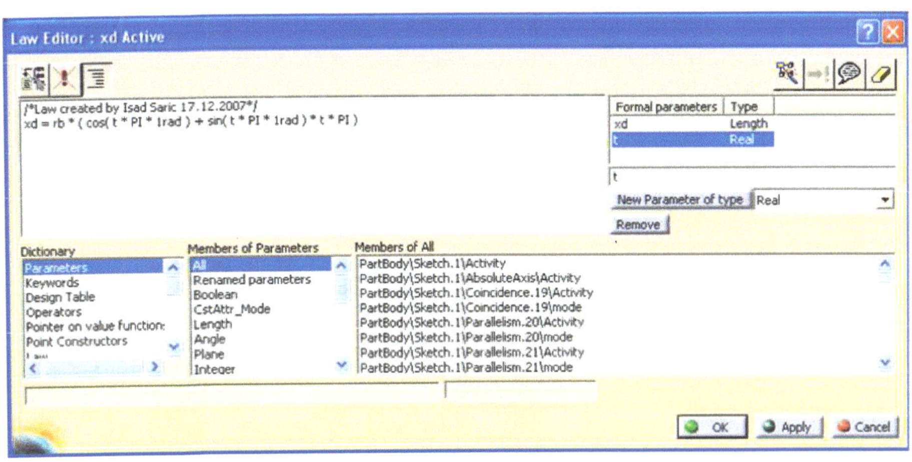

Fig. 5. Setting of parameter laws for calculation of *x* and *y* coordinates of involute points

Position of the points on involutes profile of cog side is defined in the form of parameter laws (Fig. 5.). For coordinate points of involute (x0,y0), (x1,y1), ... , (x4,y4) we most frequently define a set of parameters. To create parameter laws, we choose the command *Law* from tools palette *Knowledge*. Then, we give two laws in parameter form, which we are going to be used for calculation of *x* and *y* coordinate points of involute

$$xd = rb * (\cos(t * PI * 1rad) + \sin(t * PI * 1rad) * t * PI) \quad (1)$$

$$yd = rb * (\sin(t * PI * 1rad) - \cos(t * PI * 1rad) * t * PI) \quad (2)$$

While we use law editor, we have to take into account the following:

- trigonometric functions, specially angles, are not considered as numbers, and because of that angle constants like 1*rad* or 1deg must be used,
- *PI* is the value of the number ϖ .

For the purpose of accuracy checking of previously conducted activities, review of formulas, parameter laws and values of all defined fixed and changeable parameters is activated in the tree of *Part* document (*Tools* → *Options* → *Knowledge*).

The example of spur gear parameter modelling is shown in the next paragraph. All dimensions, or more precisely, geometric changeable parameters of spur gear are in the function of fixed parameters *m* and *z*. We can generate any spur gear by changing parameters *m* and *z*.

Table 2. Selected spur gears parameters

Part Number	m	z	d_g	d	b_z	b_g	b_k	t_k
G2-20	2	20	30	15	20	35	5	2,35
G3-40	3	40	60	25	30	50	8	3,34
G4-60	4	60	120	30	40	60	10	3,34

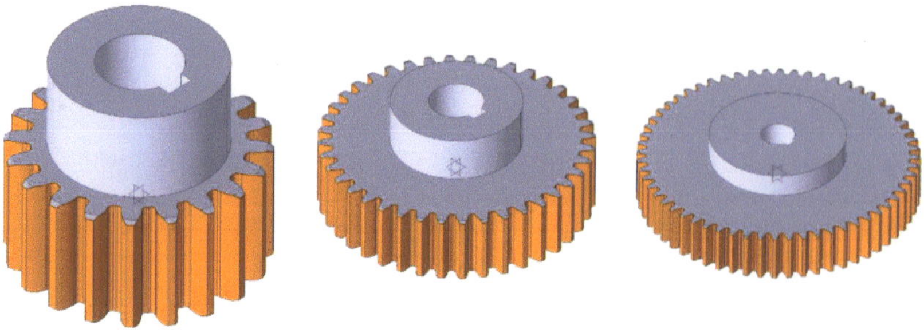

1° G2-20 (m=2 mm, z=20) 2° G3-40 (m=3 mm, z=40) 3° G4-60 (m=4 mm, z=60)

Fig. 6. Different spur gears are the result of parameter modelling

Fig. 6. shows three different standard catalogue spur gears made from the same CATIA V5 file, by changing parameters *m* and *z*. (Saric et al., 2009, 2010)

2.1.2 Modelling parameters of bevel gears

The subsequent paragraph illustrates the parameter modeling of bevel gears. All dimensions, or more specifically, the geometric variable parameters of bevel gears, depend on the fixed parameters m, z1, and z2. Any bevel gear can be produced by altering the parameters m, z1, and z2.

Table 3. Selected bevel gears parameters

Part Number	m	z_1	z_2	d_g	d	b_z	b_g	Connection between hub and shaft
B2-25	2	25	25	40	12	10,6	25,52	M5
DB3-15	3	15	30	36	18	17	36,26	M6
FB4-15	4	15	60	48	20	34	59,9	M8

Fig. 7. shows three different standard catalogue bevel gears made from the same CATIA V5 file, by changing parameters *m*, z_1 and z_2. (Saric et al., 2009, 2010)

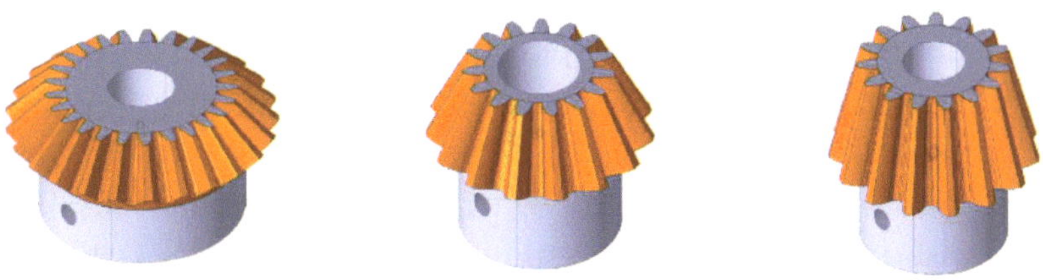

1° B2-25 (m=2 mm, z_1=25, z_2=25) — 2° DB3-15 (m=3 mm, z_1=15, z_2=30) — 3° FB4-15 (m=4 mm, z_1=15, z_2=60)

Fig. 7. Different bevel gears are the result of parameter modelling

2.1.3 Worms parameter modelling

The example of worm parameter modelling is shown in the next paragraph. All dimensions, or more precisely, geometric changeable parameters of worm are in the function of fixed parameters *m*, z_1 and *N*. We can generate any worm by changing parameters *m*, z_1 and *N*.

Table 4. Selected worms parameters

Part Number	m	z_1	d_g	d	L	L_g	Connection between hub and shaft
W1,5-1	1,5	1	23	10	35	45	M5
W2,5-2	2,5	2	35	15	45	60	M6
W3-3	3	3	41	20	55	70	M8

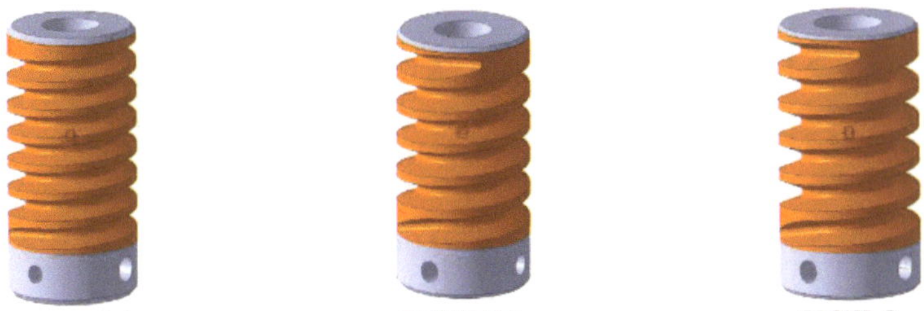

1° W1,5-1 (m=1,5 mm, z_1=1, N=6,5) — 2° W2,5-2 (m=2,5 mm, z_1=2, N=5) — 3° W3-3 (m=3 mm, z_1=3, N=5)

Fig. 8. Different worms are the result of parameter modelling

Fig. 8. shows three different standard catalogue worms made from the same CATIA V5 file, by changing parameters *m*, z_1 and *N*. (Saric et al., 2009, 2010)

2.2 Modeling of belt transmission parameters

This application encompasses a broad spectrum of the business due to the frequent necessity of belt transmission. The belt transmission design process typically involves estimating the required driving power, selecting the appropriate belt pulley, determining the belt's length and breadth, and calculating the safety factor, among other considerations. The quality of the final design can be assessed based on efficiency, compactness, and serviceability. If an engineer does not utilize parameter modeling, he or she must endure a laborious design phase, relying on lessons learned from prior errors to achieve standardized components such as belt pulleys and belts, integrated into the desired construction. This technique is automated by parameter modeling. During this procedure, attributes such as the recorded distance between belt pulleys and the length of the belts are also generated. These features also record linkages, belt angular velocities, and exit angular velocity. The outcomes for specified belt lengths can be derived from the feasibility study. A limited number of independent feasibility evaluations on various belt lengths are juxtaposed with requirements for compactness. Consequently, several belt transmission designs can be evaluated, allowing for the identification of the optimal final construction solution.

The subsequent paragraph illustrates the modeling of belt pulley parameters. The belt pulley K, seen in Fig. 9, has multiple mutually welded components: hub G, pulley rim V, plate P, and twelve lateral ribs BR. All dimensions, or more specifically, the geometric variable parameters of the belt pulley, are dependent on the fixed parameters d, Bk, dv, and s. We may produce any belt pulley with a cylindrical exterior by modifying the parameters d, Bk, dv, and s.

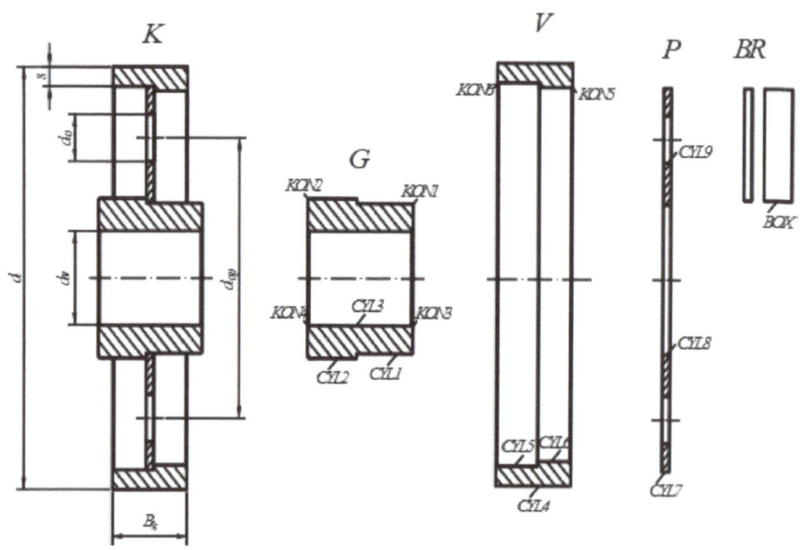

Fig. 9. Modelling of belt pulley parts with cylindrical external surface

The dimensions of the hub are contingent upon the diameter of the shaft, denoted as dv, upon which the hub is mounted. The shaft diameter serves as the input value from which the other hub dimensions are derived.

The hub shape can be derived from the addition and subtraction of cylinders and cones seen in Fig. 9.

$$G = CYL1 + CYL2 - CYL3 - KON\,1 - KON\,2 - KON\,3 - KON\,4 \tag{3}$$

Pulley rim of belt pulley depends from diameter of belt pulley d, pulley rim width B_k, diameter of shaft d_v and minimal pulley rim thickness s.

$$V = CYL4 - CYL5 - CYL6 - KON\,5 - KON6 \tag{4}$$

Plate dimensions depend from diameter of belt pulley d, minimal pulley rim thickness s and diameter of shaft d_v.

$$P = CYL7 - CYL8 - 6 \cdot CYL9 \tag{5}$$

Side ribs are side set rectangular plates which can be shown by primitive in the form of prism.

$$BR = BOX \tag{6}$$

Whole belt pulley is obtained by adding of formed forms.

$$K = G + V + P + 6 \cdot BR \tag{7}$$

Table 5. Parameters and formulas

Belt pulley with cylindrical external surface	
CYL1: $D=1{,}6 \cdot d_v$, $H=0{,}75 \cdot d_v$	CYL9: $H= 0{,}1 \cdot d_v$
CYL2: $D=1{,}7 \cdot d_v$, $H=0{,}65 \cdot d_v+2$ mm	KON1: $D=1{,}6 \cdot d_v$, $H=1$ mm, angle 45°
CYL3: $D=d_v$, $H=1{,}4 \cdot d_v+2$ mm	KON2: $D=1{,}7 \cdot d_v$, $H=1$ mm, angle 45°
CYL4: $D=d$, $H=B_k$	KON3: $D=d_v$, $H=1$ mm, angle 45°
CYL5: $D=d-2 \cdot s$, $H=B_k/2+0{,}05 \cdot d_v+1$ mm	KON4: $D=d_v$, $H=1$ mm, angle 45°
CYL6: $D=d-2 \cdot s-0{,}1 \cdot d_v$, $H=B_k/2-0{,}05 \cdot d_v-1$ mm	KON5: $D=d-2 \cdot s-0{,}1 \cdot d_v$, $H=1$ mm, angle 45°
CYL7: $D=d-2 \cdot s$, $H=0{,}1 \cdot d_v$	KON6: $D=d-2 \cdot s$, $H=1$ mm, angle 45°
CYL8: $D=1{,}6 \cdot d_v$, $H=0{,}1 \cdot d_v$	BOX: $A=[(d-2 \cdot s)-1{,}8 \cdot d_v]/2$, $B=0{,}35 \cdot B_k$, $C=0{,}1 \cdot d_v$

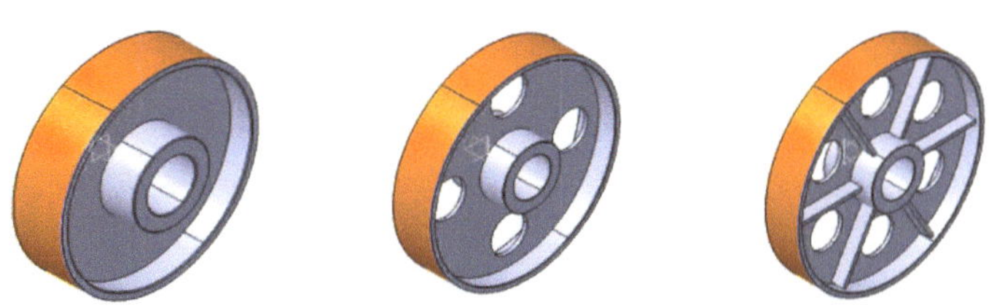

1° K200 (d=200 mm, B_k=50 mm, d_v=50 mm, s=4 mm)

2° K315 (d=315 mm, B_k=63 mm, d_v=60 mm, s=4,5 mm, d_{op}=204 mm, d_o=60 mm)

3° K400 (d=400 mm, B_k=71 mm, d_v=70 mm, s=5 mm, d_{op}=250 mm, d_o=70 mm)

Fig. 10. Different belt pulleys with cylindrical external surface are the results of parameter modelling

Fig. 10. shows three different standard belt pulleys with cylindrical external surface made from the same CATIA V5 file, by changing parameters d, B_k, d_v and s. (Saric et al., 2009).

Use of side ribs that are posed between holes on the plate is recommended during modelling of belt pulleys with longer diameter (Fig. 10.).

Rotary parts of belt pulley shown in the Fig. 9., can be modelled in a much more easier way. More complex contours, instead of their forming by adding and subtraction, they can be formed by rotation. In the first case, computer is loaded with data about points inside primitive which, in total sum, do not belong inside volume of component. In the second case, rotary contour (bolded line in the Fig. 11.) is first defined, and, then, primitive of desired shape is obtained by rotation around rotate axis.

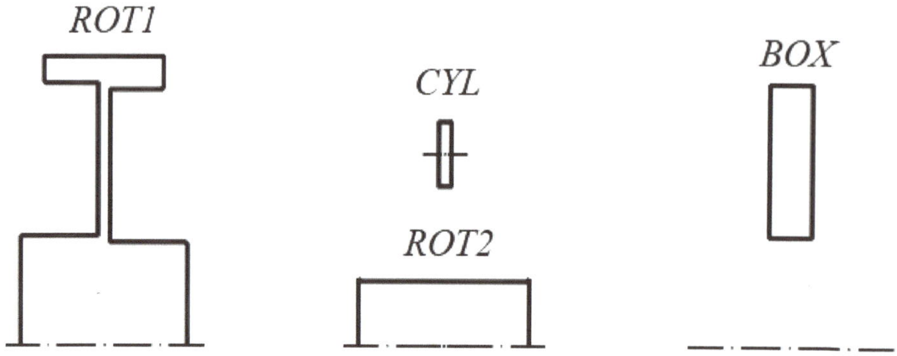

Fig. 11. Modelling of rotary forms

For primitives, shown in the Fig. 11., final form is obtained after the following operations

$$K = (ROT\ 1 - ROT\ 2 - 6 \cdot CYL) \cup 6 \cdot BOX \tag{8}$$

3. CONCISE SUMMARY

The designer must be deeply involved in shaping the component's form. Consequently, after the algorithm for modeling the component shape is established, it is stored in computer memory and utilized when modeling the same or analogous shape with comparable dimensions. Saric et al. (2009)

Components unsuitable for interactive modeling are represented by parameters. During the geometric mechanical transmission modeling in the CATIA V5 system, it is unnecessary to generate shapes directly; rather, parameters can be incorporated inside geometric and/or dimensional limitations. Modifying distinctive fixed parameters yields a three-dimensional solid model of mechanical transmission. In this manner, the designer can produce more alternative design samples, directing attention towards functional features of the design, without a particular emphasis on the intricacies of element shapes. Saric et al. (2010)

To achieve the final aim and expedite product market entry, the time allocated for product development is identified as a critical aspect for maximizing profit. The duration required for the design process of mechanical transmissions can be diminished by up to 50% with the application of parameter modeling, particularly during the preparatory phase. (Fig. 12.).

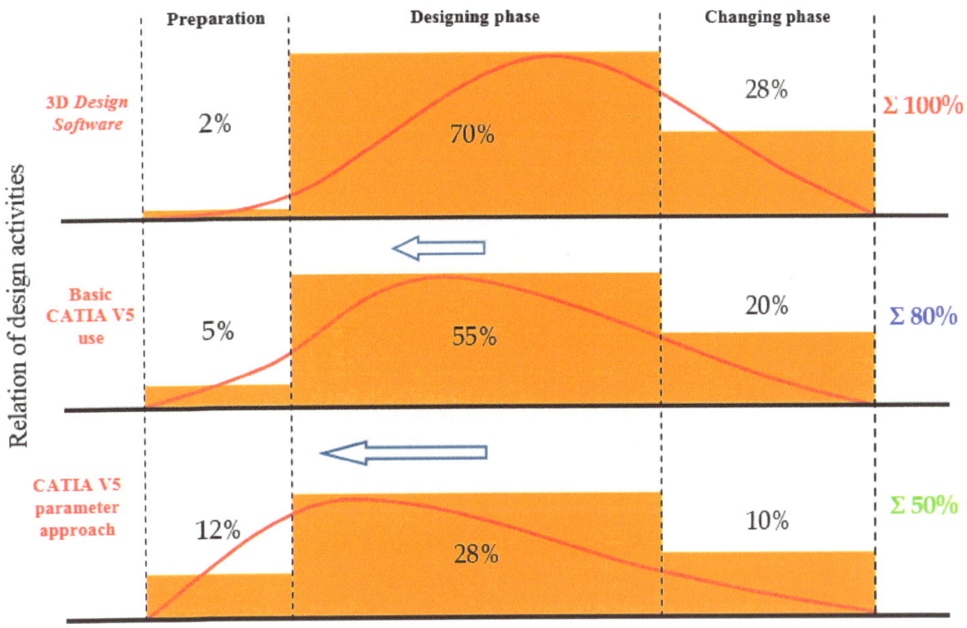

Fig. 12. Relation of design activities and reducing of time spent for design by parameter modelling

These are the advantages of parameter modelling use:

- possibility to make family of parts with the same shape based on one created model,
- forming of libraries basis of standard mechanical elements which take up computer memory, similar to the classic approach of 3D geometric modelling, is not necessary,
- use of parameters enables global modification of whole assembly (automatic reconfiguration),
- development of the product is faster, etc. (Saric et al., 2009)

The CATIA V5 technology enables the construction of geometric associations defined by relationships created between parameters. Consequently, component parameterization must necessarily be implemented in conjunction with contemporary geometric modeling techniques. Direct financial impacts are evident in the lowering of production costs, which enhances productivity. Consequently, profits are greater and product prices are reduced. Saric et al. (2010)

The 3D model acquired from the CATIA V5 system serves as the foundation for technical documentation, stress and deformation analysis via Finite Element Method (FEM), generation of NC/CNC programs for part production on machines (CAM/NC), and Rapid Prototyping (RP), among other applications.

4. NOMENCLATURE

A	mm	side rib length
a	°	line of contact angle
B	mm	side rib width
B_k	mm	pulley rim width
BR	-	side rib
b	mm	cog width
b_g	mm	hub width
b_k	mm	keyway width
b_z	mm	gear width
C	mm	side rib thickness
CYL	-	cylinder
D	mm	appropriate diameter of belt pulley components
d	mm	gear pitch circle; diameter of the hole for shaft; diameter of belt pulley
d_g	mm	interior diameter of hub
d_o	mm	diameter of the hole on the plate
d_{op}	mm	diameter on which holes on plate are set
d_v	mm	diameter of shaft
delta	°	a half of an angle of front cone
dZ	mm	translation of geometry over z axis; translation of worm surface over z axis
G	-	hub
gama	°	angle of helix
H	mm	appropriate length of belt pulley components
ha	mm	addendum part of cog height
hf	mm	root part of cog height
K	-	belt pulley
KON	-	cone

L	mm	helix length
L_g	mm	hub length
lc	mm	cone axis length
m	mm	module
N	-	number of helix
P	-	plate
PI	-	value of number ϖ
p	mm	step on pitch circle
ROT	-	rotation
r	mm	pitch radius
ra	mm	addendum radius
ratio	-	factor of scaling exterior to interior cog profile
rb	mm	basic radius
rc	mm	length of generating line of back (additional) cone
rf	mm	root radius
rr	mm	radius of profile root radius
s	mm	minimal pulley rim thickness
t	-	involutes function parameter
t_k	mm	keyway depth
tc	°	cutting angle used for contact point putting in zx plane
V	-	pulley rim
xd	mm	x coordinate of involutes cog profile generated on the base of parameter t
yd	mm	y coordinate of involutes cog profile generated on the base of parameter t
(x,y)	mm	coordinates of involute points
z	-	cog number
z_1	-	cog number of driver gear; number of turn of a worm
z_2	-	cog number of following gear

REFERENCES

1. Amirouche, F. (2004). *Principles of Computer-Aided Design and Manufacturing* (2nd edition), Prentice Hall, ISBN 0-13-064631-8, Upper Saddle River, New Jersey
2. Certified hardware configurations for CATIA V5 systems. (May 2011). Available from: <http://www.3ds.com/support/certified-hardware/overview/>
3. Cozzens, R. (2006). *Advanced CATIA V5 Workbook: Knowledgeware and Workbenches Release 16*. Schroff Development Corporation (SDC Publications), ISBN 978-1-58503-321-8, Southern Utah University
4. Dassault Systemes. (2007a). *CATIA Solutions Version V5 Release 18 English Documentation*
5. Dassault Systemes. (2007b). *CATIA Web-based Learning Solutions Version V5 Release 18 Windows*

6. Dassault Systemes - PLM solutions, 3D CAD and simulation software. (May 2011). Available from: <http://www.3ds.com/home/>
7. International Business Machines Corp. (IBM). (May 2011). Available from: <http://www.ibm.com/us/en/>
8. Karam, F. & Kleismit, C. (2004). *Using Catia V5*, Computer library, ISBN 86-7310-307-X, Cacak
9. Repcic, N. et al. (1998). *Mechanical Elements II Part*, Svjetlost, ISBN 9958-10-157-7 (ISBN 9958- 11-075-X), Sarajevo
10. Repcic, N. & Muminovic, A. (2007). *Mechanical Elements II*, Faculty of Mechanical Engineering, Sarajevo
11. Saric, I., Repcic, N. & Muminovic, A. (2006). 3D Geometric parameter modelling of belt transmissions and transmissions gear. *Technics Technologies Education Management - TTEM*, Vol. 4, No. 2, (2009), pp. 181-188, ISSN 1840-1503
12. Saric, I., Repcic, N. & Muminovic, A. (1996). Parameter Modelling of Gears, *Proceedings of the 14th International Research/Expert Conference „Trends in the Development of Machinery and Associated Technology - TMT 2010"*, pp. 557-560, ISSN 1840-4944, Mediterranean Cruise, September 11-18, 2010
13. Shigley, J.E., Mischke, C.R. & Budynas, R.G. (2004). *Mechanical Engineering Design* (7th edition), McGraw-Hill, ISBN 007-252036-1, New York
14. Spotts, M.F., Shoup, T.E. & Hornberger, L.E. (2004). *Design of Machine Elements* (8th edition), Prentice Hall, ISBN 0-13-126955-0, Upper Saddle River, New Jersey
15. Zamani, N.G. & Weaver, J.M. (2007). *Catia V5: Tutorials Mechanism Design & Animation*, Computer library, ISBN 978-86-7310-381-5, Cacak

2

Chapter

SIMULATION MODELS OF GEARBOXES INCORPORATING GEAR AND BEARING DEFECTS

1. INTRODUCTION: AN OVERVIEW

Simulation serves as a potent instrument for comprehending the intricate interactions of transmission components within dynamic environments. The vibro-dynamics simulation of defective gears and rolling element bearings enables analysts to examine the impact of compromised components in a controlled environment, facilitating data collection without incurring the costs associated with actual failures or the expenses of experiments necessitating numerous seeded fault specimens. Fault simulation can supply the data necessary for training neural network-based diagnostic and prognostic algorithms.

2. ELEMENTS IN GEARBOX SIMULATION

2.1 Transmission error

Gears inherently generate vibrations due to the significant pressure exerted between the interlocking teeth during power transmission. The meshing of gears entails alterations in the amplitude, position, and direction of substantial concentrated loads exerted on the interfacing gear teeth, thus inducing vibrations. Prolonged exposure to noise and vibration is a prevalent cause of operational weariness, communication challenges, and health risks. The mitigation of noise and vibration in working machinery has been a significant priority for enhancing safety and efficiency in machine operations.

The design and development of quieter, more dependable, and more economical gears have been a prominent research focus for decades in the automotive and aerospace sectors. The vibration of gears, which is directly associated with the noise and vibration of geared machines, is primarily influenced by the tooth meshing and shaft revolution frequencies, along with their harmonics and sidebands, resulting from low shaft frequency modulation of the higher tooth-mesh frequency components. The contribution from the gear meshing components predominantly influences the overall composition of the recorded gearbox vibration spectrum (refer to Figure 2.1.1).

Transmission Error (TE) is a critical and foundational notion essential for comprehending vibrations in gears. The term 'Transmission Error' was initially created by Professor S. L. Harris of Lancaster University, UK, and his PhD student, R.G. Munro. They recognized that the excitation forces inducing gear vibrations were contingent upon the tooth meshing inaccuracies resulting from manufacturing and the flexural deformation of the teeth under load [1].

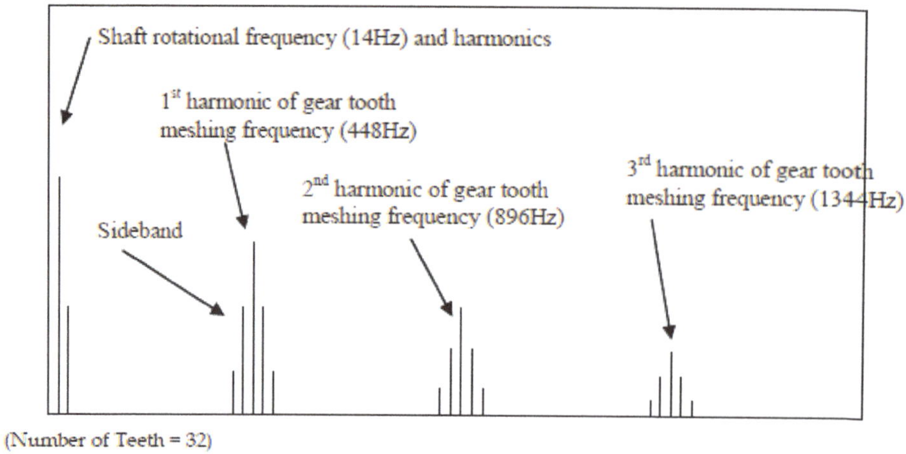

Fig. 2.1.1. Typical spectrum composition of gear vibration signal.

TE is defined as the deviation of the angular position of the driven gear from its theoretical position calculated from the gearing ratio and the angular position of the pinion (Equation- 2.1.1). The concept of TE is illustrated in Figure-2.1.2.

$$TE = \left(\theta_{gear} - \frac{R_{pinion}}{R_{gear}} \theta_{pinion} \right) \quad (2.1.1)$$

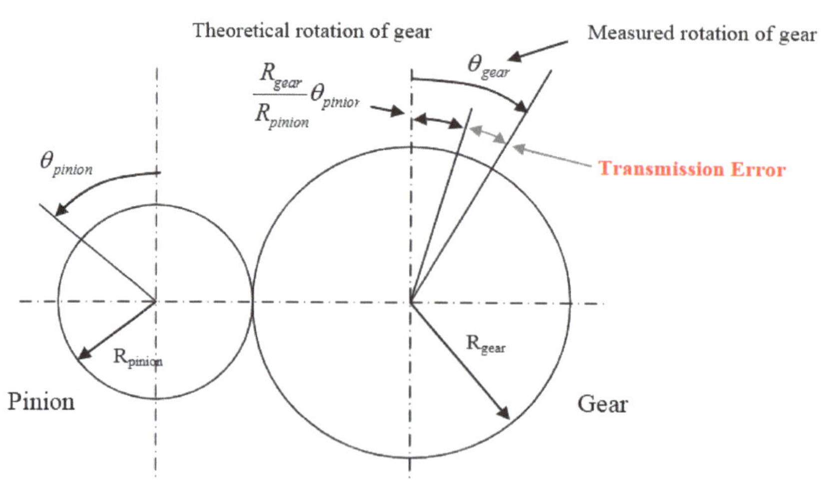

Fig. 2.1.2. Definition of Transmission Error.

The TE's robust correlation with gear noise and vibrations rendered it particularly intriguing for gear engineers and researchers. TE can be quantified using many sorts of devices. Commonly employed techniques include magnetic signal methods, strain gauges on the drive shaft, torsional vibration transducers, tachometers, tangential accelerometers, and rotary encoder systems. Smith [2] identifies three primary origins of TE: 1) Geometric inaccuracies of gears, 2) Elastic deformation of gears and related components, and 3) Mounting mistakes. Figure 2.1.3 depicts the correlation between TE and its origins.

Transmission errors manifest in three forms: 1) Geometric, 2) Static, and 3) Dynamic. Geometric TE (GTE) is assessed at low velocities and in an unloaded condition. It is frequently employed to assess the impact of production inaccuracies. Static TE (STE) is also assessed in low-speed situations when in a loaded state. STE encompasses the influence of elastic deflection of the gears along with the geometrical inaccuracies. Dynamic TE (DTE) encompasses the influence of inertia in addition to all error factors accounted for in GTE and STE. Comprehending the TE and the behavior of machine elements in the geared transmission system facilitates the creation of realistic gear rotor dynamics models..

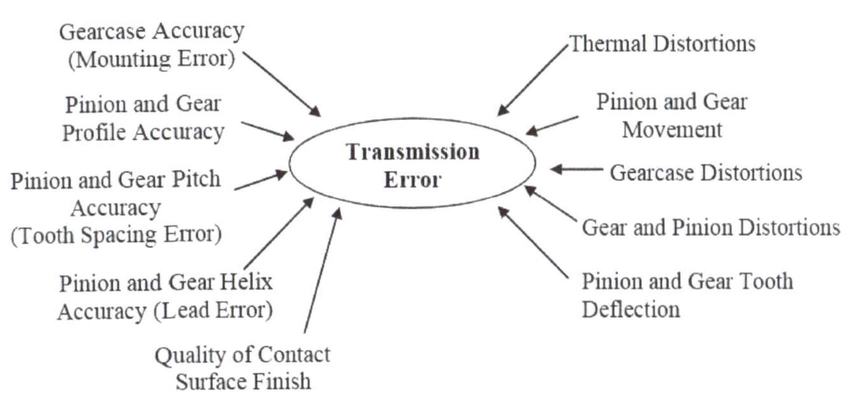

Fig. 2.1.3. Sources of Transmission Error.

2.2 Impact of gear geometry inaccuracies on transmission mistake

A typical GTE of a spur gear is shown in Figure-2.2.1. It shows a long periodic wave (gear shaft rotation) and short regular waves occurring at tooth-mesh frequency. The long wave is often known as: Long Term Component: LTC, while the short waves are known as: Short Term Component: STC.

The LTC is typically caused by the eccentricity of the gear about its rotational centre. An example is given in Figure-2.2.2 to illustrate how these eccentricities can be introduced into the gears by manufacturing errors; it shows the error due to a result of the difference between the hobbing and the shaving centres.

The effect of errors associated with gear teeth appears in the STC as a localized event. The parabolic-curve-like effect of tooth tip relief is shown in Figure-2.2.3(a). The STC is caused mainly by gear tooth profile errors and base pitch spacing error between the teeth. The effect of individual tooth profile errors on the GTE is illustrated in Figure-2.2.3(a). The GTE of a meshing tooth pair is obtained by adding their individual profile errors. The STC of gear GTE is synthesised by superposing the tooth pair GTEs separated by tooth base pitch angles (Figure-2.2.3 (b)).

Another common gear geometric error is tooth spacing or pitch errors, shown in Figure- 2.2.4. The tooth spacing error appears in GTE as vertical raise or fall in the magnitude of a tooth profile error.

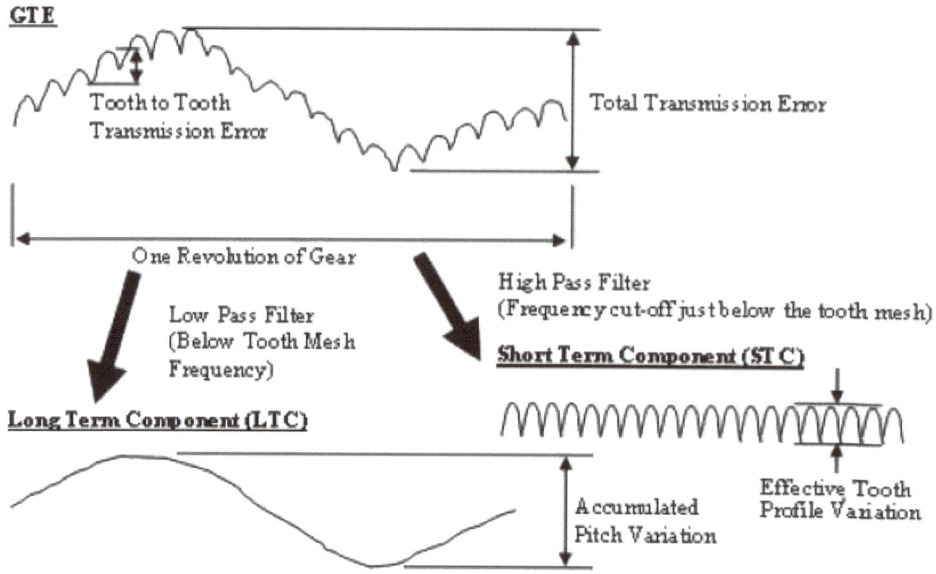

Fig. 2.2.1. A typical Geometrical Transmission Error.

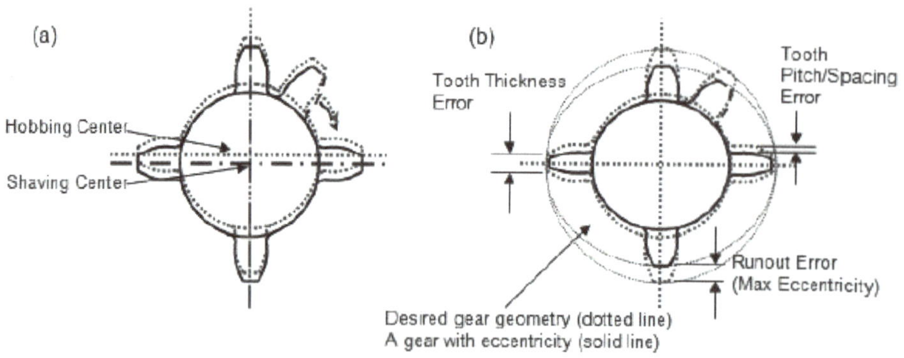

Fig. 2.2.2. (a) Eccentricity in a gear caused by manufacturing errors, (b) Resulting errors in gear geometry.

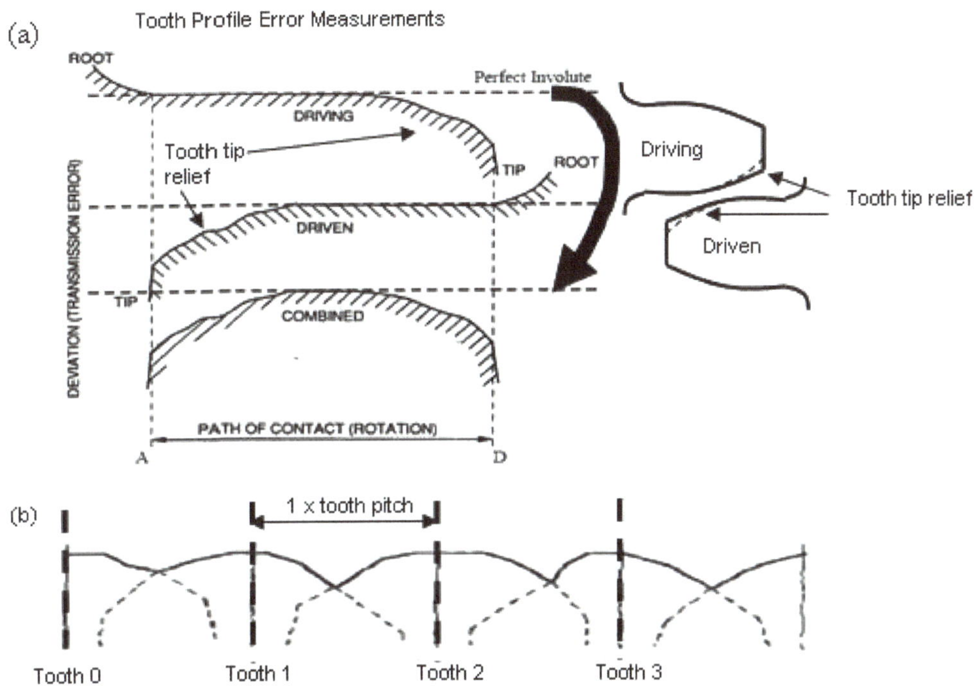

Fig. 2.2.3. (a) GTE of a meshing tooth pair (b) Resulting Short Term Component of GTE.

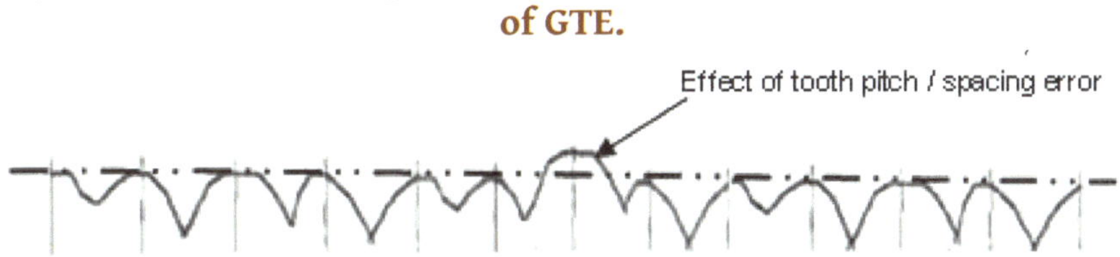

Fig. 2.2.4. Effect of spacing error appearing in short term component of GTE.

2.3 Effect of load on transmission error

Elastic deflections in gears are an additional source of TE. Despite gears often being rigid and engineered to support substantial loads, their deflection under load is significant. The standard deflection of gear teeth occurs in the range of micrometers (μm). The impact of deflection on TE may surpass the influence of gear geometry, contingent upon the load carried by the gears.

A pertinent load-deflection metric indicates that a load of 14N per 1mm of tooth face width yields a deflection of 1μm for a steel gear; thus, stiffness is calculated as 14E109 N/m/m for a tooth pair engaging at the pitch line. The stiffness of a tooth pair is notably independent of its size (or tooth module) [3]. The deflection of gear teeth displaces them from their theoretical positions, resulting in a persistent tooth pitch inaccuracy, as seen in Figure 2.3.1 (a). The impact of gear deflection is manifested in the TE (STE) as a displacement of the GTE: Figure-2.3.1 (b).

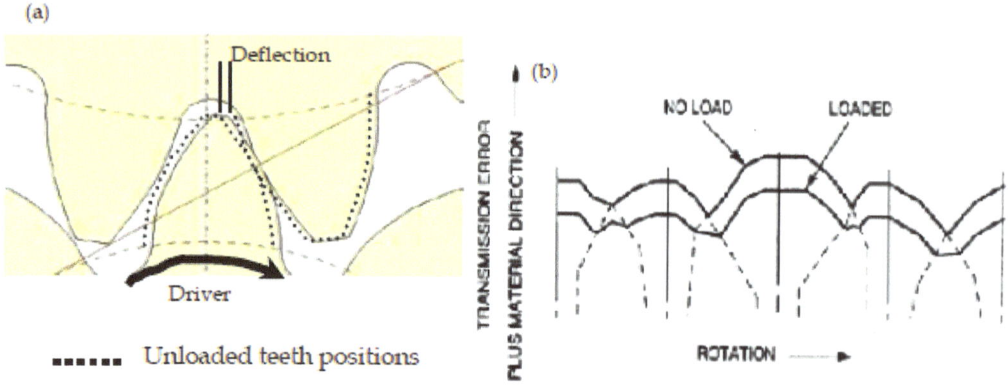

Fig. 2.3.1. (a) Deflection of gear tooth pair under load, (b) Effect of load on transmission error (TE).

Consider the more general situation where the deflection in loaded gears affects the TE significantly. Note that the following discussion uses typical spur gears (contact ratio = 1.5) with little profile modification to illustrate the effect of load on TE. Figure-2.3.2 illustrates the STE caused by the deflection of meshing gear teeth. The tooth profile chart shows a flat line indicating the ideal involute profile of the tooth:

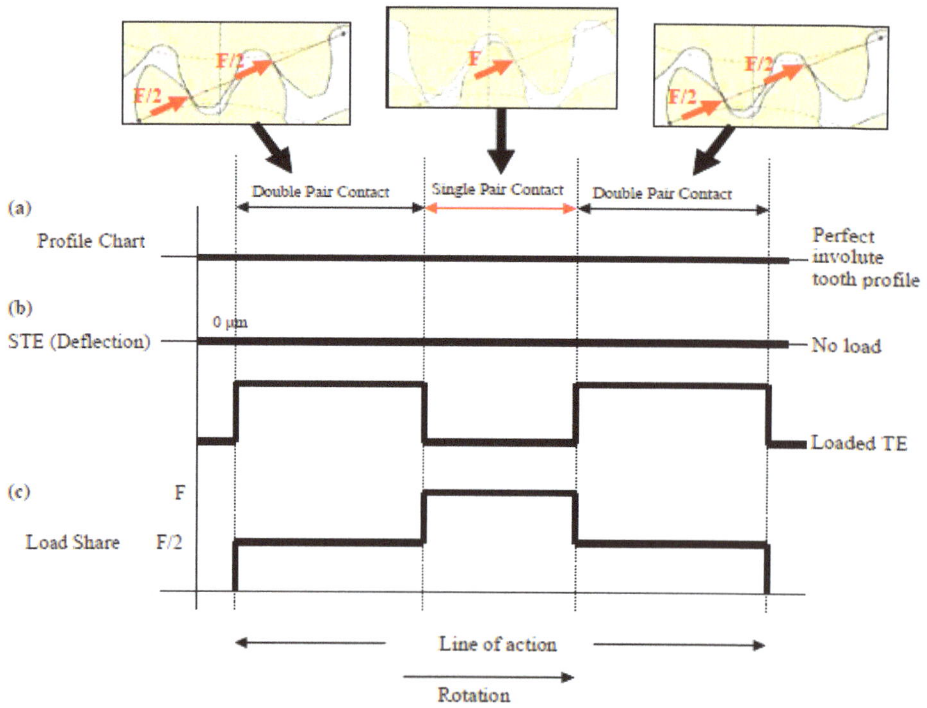

Fig. 2.3.2. Effect of Load on TE, (a) Tooth Profile Chart, (b) Static Transmission Error, (c) Loading acting on a tooth

Figure-2.3.2 (a). The impact of mesh stiffness fluctuations resulting from alterations in the quantity of meshing teeth pairs is displayed as steps in the STE plot: Figure 2.3.2 (b). The degree of deflection escalates when a solitary pair of teeth bears the weight and diminishes when the load is distributed among an additional pair. Figure 2.3.2 (c) illustrates the proportion of force transmitted by a tooth during the meshing cycle.

A 1963 work by S.L. Harris, R. Wylie Gregory, and R.G. Munro shown that transmission error can be mitigated through the application of suitable corrections to the involute gear profile [4, 5]. Figure 2.3.3 of the Harris map illustrates that any gear can be engineered to achieve a steady-state error (STE) with zero variation, characterized by a flat STE with a constant offset value, for a specific load. This technique's fundamental concept is that the geometry of gear teeth can be engineered to negate the impact of tooth deflection under a specified load.

Furthermore, the variance of torque error can be minimized by augmenting the contact ratio of the gear pair. In other words, engineer the gears to distribute the load among a larger quantity of tooth pairs..

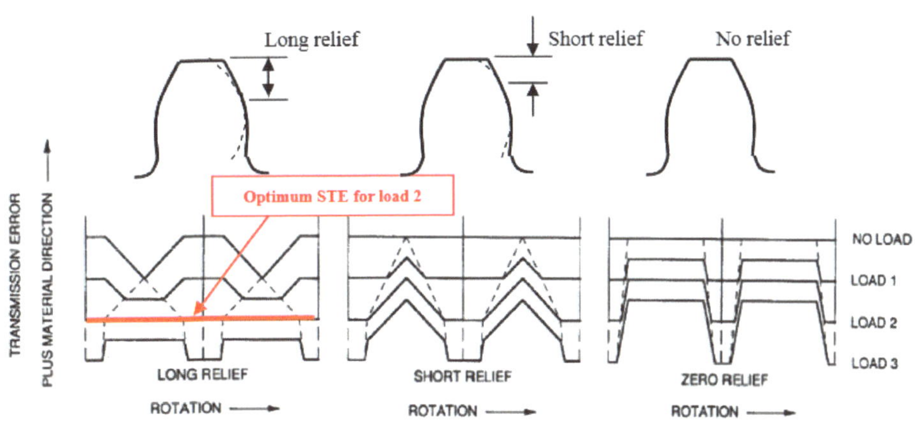

Fig. 2.3.3. Optimum tooth profile modification of a spur gear.

2.4 Modeling gear dynamics

A systematic design methodology is utilized to perform STE analysis, guaranteeing the smooth interlocking of gears under load situations. This section clarified how the strong correlation between the TE and gear vibration makes the TE an essential metric for forecasting the acoustic performance of gear drives. However, a precise depiction of the gear's dynamic attributes cannot be achieved without modeling the dynamics of the entire gear drive system. Addressing engineering difficulties often requires the mathematical representation of a physical system. A properly validated model improves understanding of the problem and provides relevant information for engineers to make informed and prudent decisions.

Ozguven and Houser [6] provide a thorough overview of the evolution of gear dynamic model development. They have examined 188 literary works pertaining to gear dynamic modeling available until 1988.. In Table-2.4.1, different types of gear dynamics models were classified into five groups according to their objectives and functionality. Traditionally lumped parameter modelling (LPM) has been a common technique that has been used to study the dynamics of gears. Wang [7] introduced a simple LPM to rationalize the dynamic factor calculation by the laws of mechanics.

Table 2.4.1. Classification of Gear Dynamic Models. (Ozguven & Houser [6])

Group-1: Simple Dynamic Factor Models
Most early models belong to this group. The model was used to study gear dynamic load and to determine the value of dynamic factor that can be used in gear root stress formulae. Empirical, semi-empirical models and dynamic models constructed specifically for determination of dynamic factor are included in this group.
Group-2: Models with Tooth Compliance
Models that consider tooth stiffness as the only potential energy storing element in the system. Flexibility of shafts, bearings etc is neglected. Typically, these models are single DOF spring-mass systems. Some of the models from this group are classified in group-1 if they are designed solely for determining the dynamic factor.
Group-3: Models for Gear Dynamics
A model that considers tooth compliance and the flexibility of the relevant components. Typically these models include torsional flexibility of shafts and lateral flexibility of bearings and shafts along the line of action.
Group-4: Models for Geared Rotor Dynamics
This group of models consider transverse vibrations of gear carrying shafts as well as the lateral component *(NOTE: Transverse: along the Plane of Action, Lateral: Normal to the Plane of Action)*. Movement of the gears is considered in two mutually perpendicular directions to simulate, for example, whirling.
Group-5: Models for Torsional Vibrations
The models in the third and fourth groups consider the flexibility of the gear teeth by including a constant or time varying mesh stiffness. The models belonging to this group differentiate themselves from the third and fourth groups by having rigid gears mounted on flexible shafts. The flexibility at the gearmesh is neglected. These models are used in studying pure (low frequency) torsional vibration problems.

He proposed a model that relates the GTE and the resulting dynamic loading. A large number of gear dynamic models that are being used widely today are based on this work. The result of an additional literature survey on more recently published materials by Bartelmus [8], Lin & Parker [9, 10], Gao & Randall [11, 12], Amabili & Rivola [13], Howard et al [14], Velex & Maatar [15], Blankenship & Singh [17], Kahraman & Blankenship [18] show that the fundamentals of the modelling technique in gear simulations have not changed and the LMP method still serves as an efficient technique to model the wide range of gear dynamics behaviour. More advanced LPM models incorporate extra functions to simulate specialized phenomena. For example, the model presented by P. Velex and M. Maatar [15] uses the individual gear tooth profiles as input and calculates the GTE directly from the gear tooth profile. Using this method they simulated how the change in contact behaviour of meshing gears due to misalignment affects the resulting TE.

FEA has become one of the most powerful simulation techniques applied to broad range of modern Engineering practices today. There have been several groups of researchers who attempted to develop detailed FEA based gear models, but they were troubled by the challenges in efficiently modelling the rolling Hertzian contact on the meshing surfaces of gear teeth. Hertzian contact occurs between the meshing gear teeth which causes large concentrated forces to act in very small area. It requires very fine FE mesh to accurately model this load distribution over the contact area. In a conventional finite element method, a fully representative dynamic model of a gear requires this fine mesh over each gear tooth flank and this makes the size of the FE model prohibitively large.

Researchers from Ohio State University have developed an efficient method to overcome the Hertzian contact problem in the 1990s' [16]. They proposed an elegant solution by modelling the contact by an analytical technique and relating the resulting force distribution to a coarsely meshed FE model. This technique has proven so efficient that they were capable of simulating the dynamics of spur and planetary gears by [19, 20] (see Figure-2.4.1). For more details see the CALYX user's manuals [21, 22].

For the purpose of studies, which require a holistic understanding of gear dynamics, a lumped parameter type model appears to provide the most accessible and computationally economical means to conduct simulation studies.

A simple single stage gear model is used to explain the basic concept of gear dynamic simulation techniques used in this chapter. A symbolic representation of a single stage gear system is illustrated in Figure-2.4.2. A pair of meshing gears is modelled by rigid disks representing their mass/moment of inertia. The discs are linked by line elements that represent the stiffness and the damping (representing the combined effect of friction and fluid film damping) of the gear mesh. Each gear has three translational degrees of freedom (one in a direction parallel to the gear's line of action, defining all interaction between the gears) and three rotational degree of freedoms (DOFs). The stiffness elements attached to the centre of the disks represent the effect of gear shafts and supporting mounts. NOTE: Symbols for the torsional stiffnesses are not shown to avoid congestion.

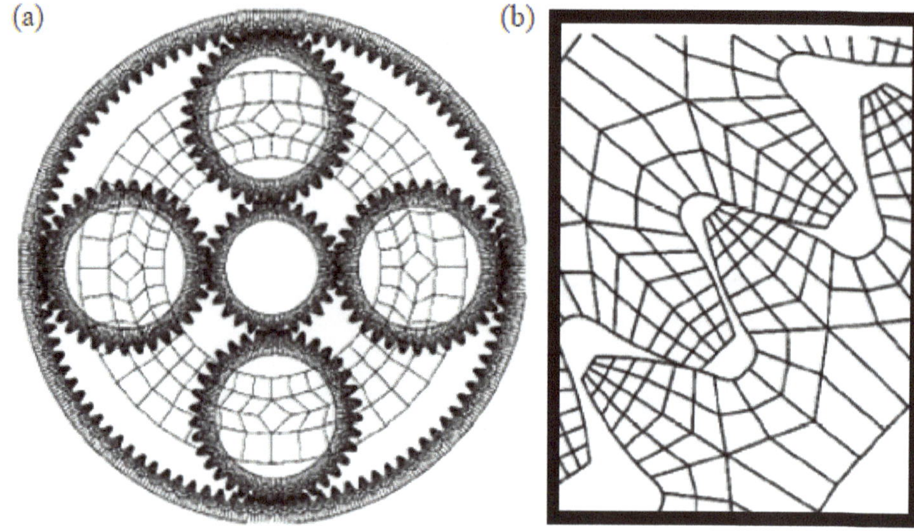

Fig. 2.4.1. (a) Parker's planetary gear model and (b) FE mesh of gear tooth. Contacts at the meshing teeth are treated analytically. It does not require dense FE mesh. (Courtesy of Parker et al. [20])

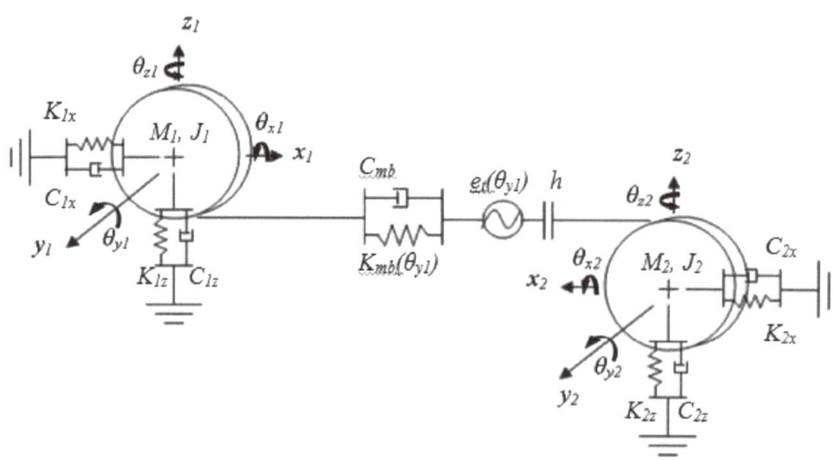

x_i, y_i, z_i	Translation at i_{th} Degrees of Freedom.
θ_{xi}, θ_{yi}, θ_{zi}	Rotation about a translational axis at i_{th} Degrees of Freedom.
C, C_{mb}	Damping matrices. The subscript '*mb*' refers to damping at the gearmesh. Typically for C_{mb}, $\zeta = 3 \sim 7\%$.
K, K_{mb}	Linear stiffness elements. The subscript '*mb*' refers to stiffness at the gearmesh.
h	An 'on/off' switch governing the contact state of the meshing gear teeth.
$\underline{e_t}$	A vector representing the combined effect of tooth topography deviations and misalignment of the gear pair.
i	Index: i=1,2, 3 ...etc.

Fig. 2.4.2. A Typical Lumped Parameter Model of Meshing Gears.

The linear spring elements representing the Rolling Element Bearings (REB) are a reasonable simplification of the system that is well documented in many papers on gear simulation. For the purpose of explaining the core elements of the gear simulation model, the detail of REB as well as the casing was omitted from this section; more comprehensive model of a gearbox, with REB and casing, will be presented later in section 2.5.

Vibration of the gears is simulated in the model as a system responding to the excitation caused by a varying TE, 'e_t' and mesh stiffness 'K_{mb}'. The dominant force exciting the gears is assumed to act in a direction along the plane of action (PoA). The angular position dependent variables 'e_t' and 'K_{mb}' are expressed as functions of the pinion pitch angle (θ_{y1}) and their values are estimated by using static simulation. Examples of similar techniques are given by Gao & Randall [11, 12], Du [23] and Endo and Randall [61].

Equations of motion derived from the LPM are written in matrix format as shown in Equation-2.4.1. The equation is rearranged to the form shown in the Euqation-2.4.2; the effect of TE is expressed as a time varying excitation in the equation source. The dynamic response of the system is simulated by numerically solving the second order term (accelerations) for each step of incremented time. The effect of the mesh stiffness variation is implemented in the model by updating its value for each time increment.

$$M\ddot{\underline{x}} + C(\dot{\underline{x}} - \dot{\underline{e}}_t) + K(\underline{x} - \underline{e}_t(\theta)) = F_s \tag{2.4.1}$$

$$M\ddot{\underline{x}} + C\dot{\underline{x}} + K\underline{x} = F_s + hC_{mb}(\theta)\dot{\underline{e}}_t + hK_{mb}(\theta)\underline{e}_t \tag{2.4.2}$$

where,

$\underline{x}, x, x$	Vectors of translational and rotational displacement, velocity and acceleration.
θ	Angular position of pinion.
K, K_{mb}	Stiffness matrices (where K includes the contribution from K_{mb}). The subscript '*mb*' refers to stiffness at the gearmesh.
C, C_{mb}	Damping matrices (C including contribution from C_{mb}). The subscript '*mb*' refers to damping at the gearmesh.
h	An 'on/off' switch governing the contact state of the meshing gear teeth.
F	Static force vector.
et, et	A vector representing the combined effect of tooth topography deviations.

2.5 Modeling rolling element bearings and gearbox housings

Simplified models of gear shaft supports, such as representing the effects of rolling element bearings and casing as simple springs with constant stiffness, can serve as helpful tools for various practical applications. Nevertheless, comprehensive representations of these components are crucial for achieving more thorough and precise simulation modeling.

To achieve a comprehensive and realistic model of the gearbox system, it is essential to include thorough representations of the REBs and the gearbox casing to accurately capture the interactions among the gears, the REBs, and the impacts of the transfer path and dynamic response of the casing.

Comprehending the interplay between the supporting structure and the rotating elements of a transmission system has posed significant challenges in the development of intricate gearbox simulation models. The characteristics of the structure supporting REBs and a shaft substantially affect the system's dynamic response. A more comprehensive representation of the REBs and gearbox casing enhances the precision of the effect transmission path that distorts the diagnostic information arising from problems in the gears and REBs. Many applications of machine health monitoring require that the technology be minimally obtrusive to machine operation. This requirement frequently necessitates the placement of sensors and/or transducers at easily accessible locations on the machine, such as the exposed surface of the gearbox casing, the machine skid, or an accessible structural frame to which the machine is affixed.

The capacity to precisely describe and replicate transmission path effects enables more realistic and effective training of diagnostic algorithms based on artificial intelligence.

2.5.1 Modeling of rolling element bearings

A number of models of REBs exist in literatures [24, 25, 26, 27] and are widely employed to study the dynamics and the effect of faults in REBs. The authors have adopted the 2 DoF model originally developed by Fukata [27] in to the LPM of the gearbox. Figure-2.5.1 (a) illustrates the main components of the rolling element bearing model and shows the load zone associated with the distribution of radial loads in the REB as it supports the shaft. Figure-2.5.1 (b) explains the essentials of the bearing model as presented by [28]. The two degrees of freedom. The REB model delineates the load-deflection relationships, disregarding the influence of mass and the inertia of the rolling elements. The two degrees of freedom (xs, ys) pertain to the inner race (shaft). The contact forces are aggregated for each rolling element to determine the total forces acting on the shaft.

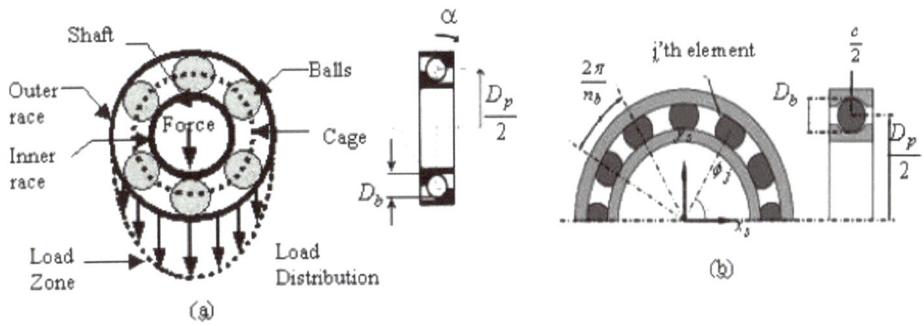

Fig. 2.5.1. (a) Rolling element bearing components and load distribution; (b) Two degree of freedom model. [28]

The overall contact deformation (under compression) for the j'th -rolling element δj is a function of the inner race displacement relative to the outer race in the *x* and *y* directions ($(xs - xp)$, $(ys - yp)$), the element position Φj (time varying) and the clearance (c). This is given by:

$$\delta j = (xs - xp)\cos\Phi j + (ys - yp)\sin\Phi j - c - \beta jCd \quad (j = 1,2..) \tag{2.5.1}$$

Accounting for the fact that compression occurs only for positive values of δj , Γj (contact state of δj the rolling elements) is introduced as:

$$\gamma_j = \begin{cases} 1, & if\ \delta_j > 0 \\ 0, & otherwise \end{cases} \tag{2.5.2}$$

The angular positions of the rolling elements Φ_j are functions of time increment *dt*, the previous cage position Φ_o and the cage speed ωc (can be calculated from the REB geometry and the shaft speed ωs assuming no slippage) are given as:

$$\phi_j = \frac{2\pi(j-1)}{n_b} + \omega_c dt + \phi_o \quad \text{with} \quad \omega_c = (1 - \frac{D_b}{D_p})\frac{\omega_s}{2} \tag{2.5.3}$$

The ball raceway contact force *f* is calculated by using traditional Hertzian theory (non- linear stiffness) from:

$$f = kb\,\delta^n \tag{2.5.4}$$

The load deflection factor *kb* depends on the geometry of contacting bodies, the elasticity of the material, and exponent n. The value of n=1.5 for ball bearings and n=1.1 for roller bearings. Using Equation-2.5.4 and summing the contact forces in the *x* and *y* directions for a ball bearing with *nb* balls, the total force exerted by the bearings to the supporting structure can be calculated as follows:

$$f_x = k_b \sum_{j=1}^{n_b} \gamma_j \delta_j^{1.5} \cos\phi_j \quad \text{and} \quad f_y = k_b \sum_{i=1}^{n_b} \gamma_i \delta_j^{1.5} \sin\phi_j \tag{2.5.5}$$

The stiffness of the given REB model is non-linear, and is time varying as it depends on the positions of the rolling elements that determine the contact condition. The effect of slippage was introduced to the model by adding random jitters of 0.01-0.02 radians to the nominal position of the cage at each step.

2.5.2 Gearbox casing model – Component mode synthesis method

Lumped parameter modelling (LPM) is an efficient means to express the internal dynamics of transmission systems; masses and inertias of key components such as gears, shafts and bearings can be lumped at appropriate locations to construct a model. The advantage of the LPM is that it provides a method to construct an effective dynamic model with relatively small number of degrees-of-freedoms (DOF), which facilitates computationally economical method to study the behaviour of gears and bearings in the presence of nonlinearities and geometrical faults [32, 33, 34, 35].

One of the limitations of the LPM method is that it does not account for the interaction between the shaft and the supporting structure; i.e. casing flexibility, which can be an important consideration in light weight gearboxes, that are common aircraft applications. Not having to include the appropriate effect of transmission path also results in poor comparison between the simulated and measured vibration signals.

Finite Element Analysis (FEA) is an efficient and well accepted technique to characterize a dynamic response of a structure such as gearbox casings. However, the use of FEA results in a large number of DOF, which could cause some challenges when attempt to solve a vibro- dynamic model of a combined casing and the LMP of gearbox internal components. Solving a large number of DOFs is time consuming even with the powerful computers available today and it could cause a number of computational problems, especially when attempting to simulate a dynamic response of gear and bearing faults which involves nonlinearities.

To overcome this shortcoming, a number of reduction techniques [36, 37] have been proposed to reduce the size of mass and stiffness matrix of FEA models. The simplified gearbox casing model derived from the reduction technique is used to capture the key characteristics of dynamic response of the casing structure and can be combined with the LPM models of gears and REBs.

The Craig-Bampton method [37] is a dynamic reduction method for reducing the size of the finite element models. In this method, the motion of the whole structure is represented as a combination of boundary points (so called master degree of freedom) and the modes of the structure, assuming the master degrees of freedom are held fixed. Unlike the Guyan reduction [38], which only deals with the reduction of stiffness matrix, the Craig-Bumpton method accounts for both the mass and the stiffness. Furthermore, it enables defining the frequency range of interest by identifying the modes of interest and including these as a part of the transformation matrix. The decomposition of the model into both physical DOFs (master DOFs) and modal coordinates allows the flexibility of connecting the finite elements to other substructures, while achieving a reasonably good result within a required frequency range. The Craig-Bumpton method is a very convenient method for modelling a geared transmission system as the input (excitation) to the system is not defined as forces, but as geometric mismatches at the connection points (i.e. gear transmission error and bearing geometric error). The following summary of the Craig-Bampton method is given based on the references [39-41].

In the Craig-Bampton reduction method, the equation of motion (dynamic equilibrium) of each superelement (substructure), without considering the effect of damping, can be expressed as in Equation-2.5.6:

$$[M]\{\ddot{u}\}+[k]\{u\}=\{F\} \tag{2.5.6}$$

Where $[M]$ is the mass matrix, $[k]$ is the stiffness matrix, $\{F\}$ is the nodal forces, $\{u\}$ and $\{u\}$ are the nodal displacements and accelerations respectively. The key to reducing the substructure is to split the degrees of freedom into masters $\{u_m\}$ (at the connecting nodes) and slaves $\{u_s\}$ (at the internal nodes). The mass, the stiffness and the force matrices are re- arranged accordingly as follows:

$$\overbrace{\begin{bmatrix} M_{mm} & M_{ms} \\ M_{sm} & M_{ss} \end{bmatrix}}^{M}\begin{Bmatrix} \ddot{u}_m \\ \ddot{u}_s \end{Bmatrix}+\overbrace{\begin{bmatrix} k_{mm} & k_{ms} \\ k_{sm} & k_{ss} \end{bmatrix}}^{k}\begin{Bmatrix} u_m \\ u_s \end{Bmatrix}=\begin{Bmatrix} F_m \\ 0 \end{Bmatrix} \tag{2.5.7}$$

The subscript m denotes master, s denotes slave. Furthermore, the slave degrees of freedom (internals) can be written by using generalized coordinates (modal coordinates by (q) using the fixed interface method, i.e. using the mode shapes of the superelement by fixing the master degrees of freedom nodes (connecting/ boundary nodes). The transformation matrix (T) is the one that achieves the following:

$$\begin{Bmatrix} u_m \\ u_s \end{Bmatrix}=T\begin{Bmatrix} u_m \\ q \end{Bmatrix} \tag{2.5.8}$$

For the fixed interface method, the transformation matrix (T) can be expressed as shown in Equation-2.5.9:

$$T=\begin{bmatrix} I & 0 \\ G_{sm} & \phi_s \end{bmatrix} \tag{2.5.9}$$

where,

$$G_{sm} = -k_{ss}^{-1} k_{sm} \tag{2.5.10}$$

and Φ_s is the modal matrix of the internal DOF with the interfaces fixed.

By applying this transformation, the number of DOFs of the component will be reduced. The new reduced mass and stiffness matrices can be extracted using Equations 2.5.11 & 2.5.12 respectively:

$$Mreduced = T^t MT \tag{2.5.11}$$

and

$$kreduced = Tt\ kT \tag{2.5.12}$$

Thus Equation-2.5.7 can be re-written in the new reduced form using the reduced mass and stiffness matrices as well as the modal coordinates as follows:

$$\overbrace{\begin{bmatrix} M_{bb} & M_{bq} \\ M_{qb} & M_{qq} \end{bmatrix}}^{M_{reduced}} \begin{Bmatrix} \ddot{u}_m \\ \ddot{q} \end{Bmatrix} + \overbrace{\begin{bmatrix} k_{bb} & 0 \\ 0 & k_{qq} \end{bmatrix}}^{k_{reduced}} \begin{Bmatrix} u_m \\ q \end{Bmatrix} = \overbrace{\begin{Bmatrix} F_m \\ 0 \end{Bmatrix}}^{F_{reduced}} \tag{2.5.13}$$

Where *Mbb* is the boundary mass matrix i.e. total mass properties translated to the boundary points. *kbb* is the interface stiffness matrix, i.e. stiffness associated with displacing one boundary DOF while the others are held fixed. The is the transpose of *Mbq*). *Mbq* is the component matrix (*Mqb*

If the mode shapes have been mass normalized (typically they are) then:

$$k_{qq} = \begin{bmatrix} \backslash & & 0 \\ & \lambda_i & \\ 0 & & \backslash \end{bmatrix} \tag{2.5.14}$$

where Λi is the eigenvalues; $\Lambda i = ki / mi = \omega^2 i$, and,

$$M_{qq} = \begin{bmatrix} \backslash & & 0 \\ & I & \\ 0 & & \backslash \end{bmatrix} \tag{2.5.15}$$

Finally the dynamic equation of motion (including damping) using the Craig-Bampton transform can be written as:

$$\begin{bmatrix} M_{bb} & M_{bq} \\ M_{qb} & I \end{bmatrix} \begin{Bmatrix} \ddot{u}_m \\ \ddot{q} \end{Bmatrix} + \begin{bmatrix} 0 & 0 \\ 0 & 2\zeta\omega \end{bmatrix} \begin{Bmatrix} \dot{u}_m \\ \dot{q} \end{Bmatrix} + \begin{bmatrix} k_{bb} & 0 \\ 0 & \omega^2 \end{bmatrix} \begin{Bmatrix} u_m \\ q \end{Bmatrix} = \begin{Bmatrix} F_m \\ 0 \end{Bmatrix} \tag{2.5.16}$$

where $2\zeta\omega$ = modal damping (ζ = fraction of critical damping)

For more detailed explanation of the techniques related to the modelling of rolling element bearings and the application of component mode synthesis (CMS) techniques, refer to the works by Sawalhi, Deshpande and Randall [41].

2.6 Solving the gear dynamic simulation models

A block diagram summarizing the time integral solution of a typical dynamic model with some time varying parameters is shown in Figure-2.6.1. There are a range of numerical algorithms available today to give solution of the dynamic models: direct time integration, harmonic balancing and shooting techniques, to name some commonly recognized methods.

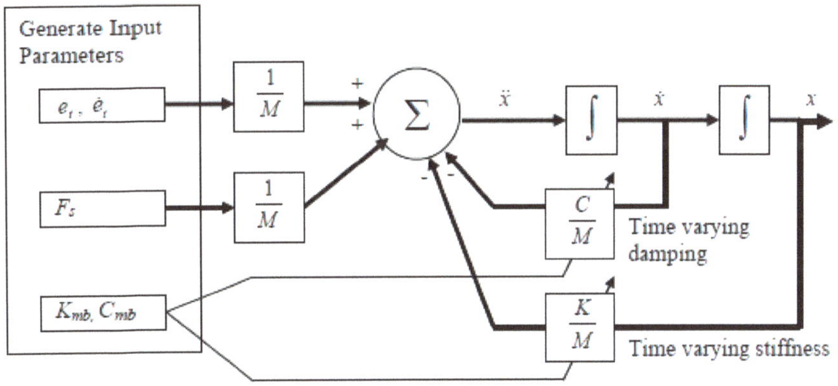

(NOTE: '∫' stands for integration over a single step of incremented time)

Fig. 2.6.1. Dynamic Simulation Process

Sometimes the solution for gears requires simulation of highly non-linear events, for example, rattling and knocking in gears, which involve modelling of the contact loss. The works presented by R. Singh [42], Kahraman & Singh [43], Kahraman & Blankenship [18] and Parker & Lin [9, 10] show some examples of the "stiff" problems involving non-linearity due to contact loss and clearances.

The solution for these Vibro-Impact problems presents difficulties involving ill-conditioning and numerical "stiffness". In [42] Singh explains that ill-conditioning of a numerical solution occurs when there is a component with a large frequency ratio: ratio of gear mesh frequency to the natural frequency of the component.

The numerical stiffness in the gear dynamic simulation becomes a problem when gears lose contact. The relationship between the elastic force, relative deflection and gear mesh stiffness is illustrated in Figure-2.6.2. The gradient of the curve represent the gear mesh stiffness.

Contact loss between the gears occurs when the force between the gears becomes zero. The gears are then unconstrained and free to move within the backlash tolerance. The presence of a discontinuity becomes obvious when the derivative of the curve in Figure-2.6.2 (i.e. mesh stiffness) is plotted against the relative deflection. The discontinuity in the stiffness introduces instability in the numerical prediction.

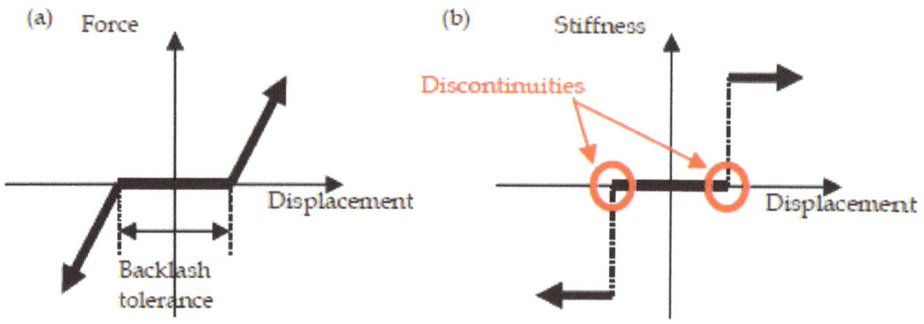

Fig. 2.6.2. Non-linearity due to contact loss in meshing gears; a) Force vs. Displacement, b) derivative of former, i.e. Stiffness vs. Displacement

In more formalized terms the "stiffness" of a problem is defined by local Eigen-values of the Jacobian matrix. Consider an equation of motion expressed in simple first order vector form '***f*(x, t)**', (Equation-2.6.1). Typically, the solution of an equation of motion is obtained by linearizing it about an operating point, say '$\mathbf{x_0}$', (Equation-2.6.2). Usually, most of the higher order terms above the 1st derivative are ignored for linearization, which leaves the following expression (Equation-2.6.3). The differential term '***J***' is called the Jacobian matrix (or Jacobian in short). The problems involving gear contact losses are "Stiff" problems because of the discontinuity in system derivatives (Jacobians). For a more detailed discussion on this topic refer to the work presented by Singh [42].

$$\underline{f}(x,t) = {}^{dx}\!/_{dt} \tag{2.6.1}$$

$$\underline{f}(x,t) \cong \underline{f}(x_0,t) + \left({}^{d\underline{f}}\!/_{dx} \right)_{x_0} (x - x_0) \tag{2.6.2}$$

$$\underline{f}(x,t) \cong \underline{f}(x_0,t) + \underline{J}(x - x_0) \tag{2.6.3}$$

3. MODELLING GEARBOX FAULTS

The study of gear faults has long been an important topic of research for the development of gear diagnostic techniques based on vibration signal analysis. Understanding how different types of gear tooth faults affect the dynamics of gears is useful to characterise and predict the symptoms of the damage appearing in vibration signals [44, 45]. The strong link between the TE and the vibration of the gears was explained earlier. The effect of different types of gear tooth faults on TE can be studied by using the static simulation models. The result of static simulation can be then used to determine how different types of gear faults can be modelled into the dynamic simulation.

Gears can fail for a broad range of reasons. Finding a root cause of damage is an important part of developing a preventative measure to stop the fault from recurring. Analysis of gear failure involves a lot of detective works to link the failed gear and the cause of the damage. Comprehensive guidelines for gear failure analysis can be found in Alban [46], DeLange [47] and DANA [48]. AGMA (American Gear Manufacturers Association) recognizes four types of gear failure mode and a fifth category which includes everything else: Wear, Surface Fatigue, Plastic Flow, Breakage and associated gear failures [49].

The effect of gear tooth fillet cracks (TFC) and spalls on gear transmission error was studied in detail by using a static simulation models (FEA and LTCA (HyGears [50])). A pair of meshing gears were modelled and analysed in step incremented non-linear static environment. Note: the transmission error obtained from the static simulation models are referred to as Motion Errors (ME) here forth by following the HyGears convention.

It was explained earlier that the interaction between two meshing gears can be expressed in the dynamic model as time-varying stiffness, damping and gear tooth topological error elements linking the two lumped mass moments of inertia. The effect of gear tooth faults can be implemented into the dynamic simulation model as changes to these parameters. The understanding gained from the detailed simulation model studies of TFCs and spalls on gear motion has lead to the method of modelling the effect of the faults in dynamic model. The relevance between the types of gear faults to the selected parameters will be explained through subsequent sections.

Further to the simulation of gear tooth faults, this chapter also briefly touches on the modelling of spalls in rolling element bearings (REB), which is also a common type of faults in geared transmission systems.

3.1 Modelling the effect of a fatigue crack in tooth fillet area of a gear

Classical tooth root fillet fatigue fracture is the most common cause of gear tooth breakages (Figure-3.1.1). Stress raisers, such as micro cracks from the heat treatment, hob tears, inclusions and grinding burns are common causes that initiate the cracks. The cracks occurring in the gear tooth fillet region progressively grow until the whole tooth or part of it breaks away. The breakage of a tooth is a serious failure. Not only the broken part fails, but serious damage may occur to the other gears as a result of a broken tooth passing though the transmission [48].

Fig. 3.1.1. A spur gear missing two teeth. They broke away due to the propagation of fatigue cracks. (Courtesy of DANA [48])

The research conducted for NASA by Lewicki [51] sets a clear guideline for predicting the trajectory of cracks occurring at the gear tooth fillet. Lewicki predicted crack propagation paths of spur gears with a variety of gear tooth and rim configurations, including the effect of: rim and web thickness, initial crack locations and gear tooth geometry factors (Diametral pitch, number of teeth, pitch radius and tooth pressure angle). A summary of the results is presented in Figure-3.1.2.

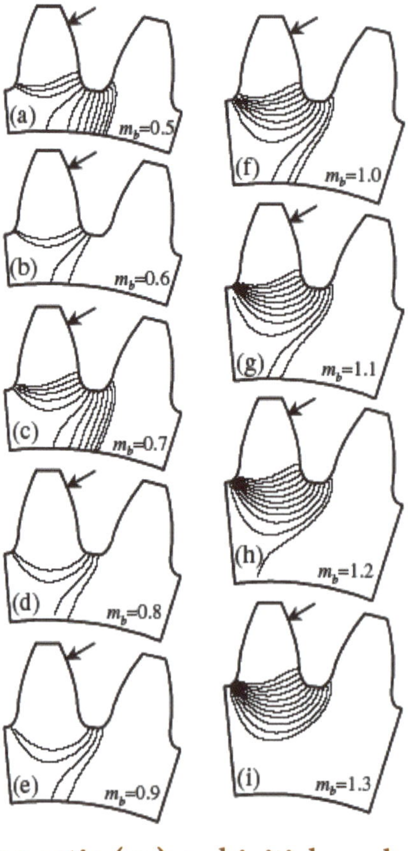

Fig. 3.1.2. Effect of backup ratio (m_b) and initial crack location on propagation path. (Lewicki [51])

A set of spur gears later used in the validation of the simulation result have a "backup ratio" (rim thickness divided by tooth height) greater than m_b=1.3. Therefore based on the Lewicki's prediction the cracks occurring in the tooth fillet region are most likely to propagate in the trajectory shown in Pattern Figure-3.1.2 (i); roughly 30~45° in to the tooth relative to the radial line path through the symmetric axis of the spur gear tooth profile.

Full 3D modelling of a propagating gear tooth crack is one of the actively researched areas. Some examples of simulation studies using the Boundary Element Method (BEM) are given in [52, 53, 54, 55]. The simulation studies using 3D models show complex behaviour crack growth from the small crack seeded at the middle of the gear tooth fillet. An example is shown in Figure-3.1.3 from "Modelling of 3D cracks in split spur gear", by Lewicki [52]. The crack front expands rapidly across the width of the gear tooth as it progresses into the thickness of the tooth. The tooth fillet crack (TFC) model used in this work assumes 2D conditions which approximates the weakening of the cracked gear tooth when the crack is extended across the whole width of the tooth face.

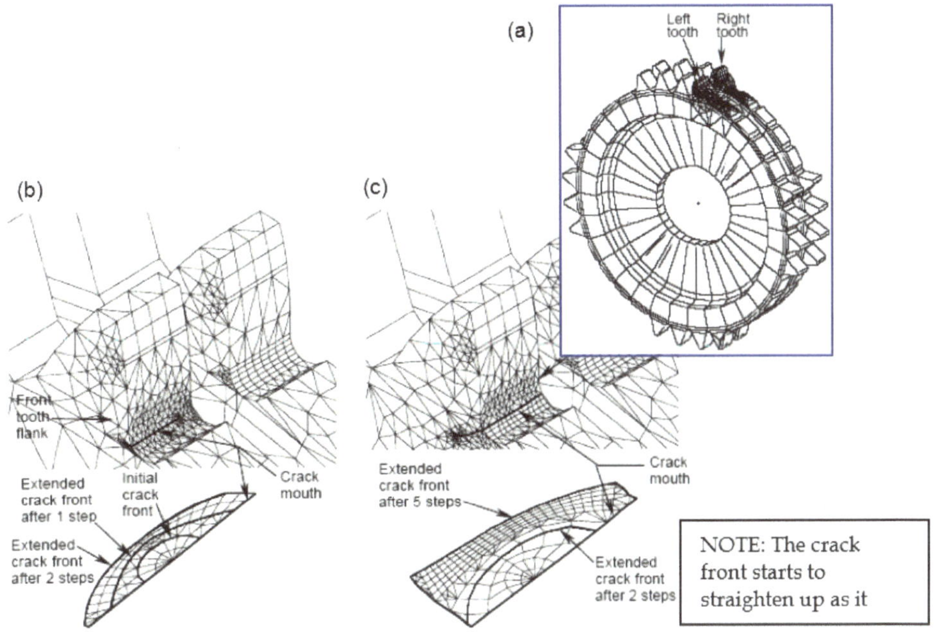

Fig. 3.1.3. 3D Crack propagation model. a) Boundary Element Model of Split Spur Gear, b) Close up of view of the gear teeth and crack section at earlier stage of development and c) more progressed crack. (Lewicki [53])

The motion error (ME), as obtained from the finite element (FE) model of a gear pair (32x32 teeth) presented in Figure-3.1.4, is shown in Figure-3.1.5. The MEs of the gears at different amount of loadings (25, 50, 100 & 200Nm) are compared in the same plot. The magnitude of MEs increases with the larger loads as the deflection of the meshing teeth become greater with the larger loads. The change in the amount of ME is roughly in linear relationship with the load, which reflects the linear elastic behaviour of the gear tooth deflection. Note the square pattern of the ME, which resulted from the time (or angular position) dependent variation of gear mesh stiffness, due to the alternating single and double tooth engagement.

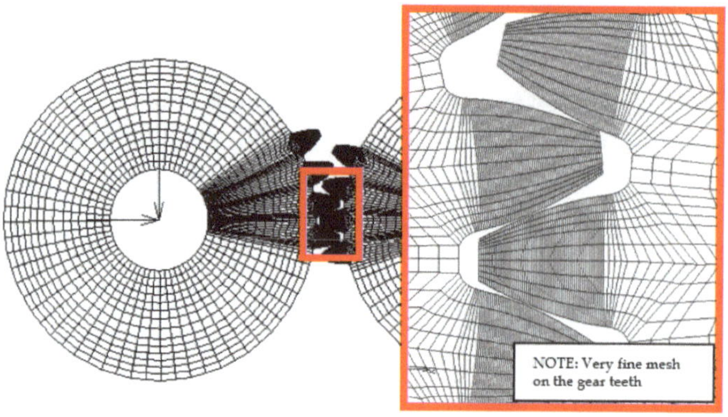

Fig. 3.1.4. Spur Gears (32x32) and its FE mesh (L), detailed view of the mesh around the gear teeth (R). [56]

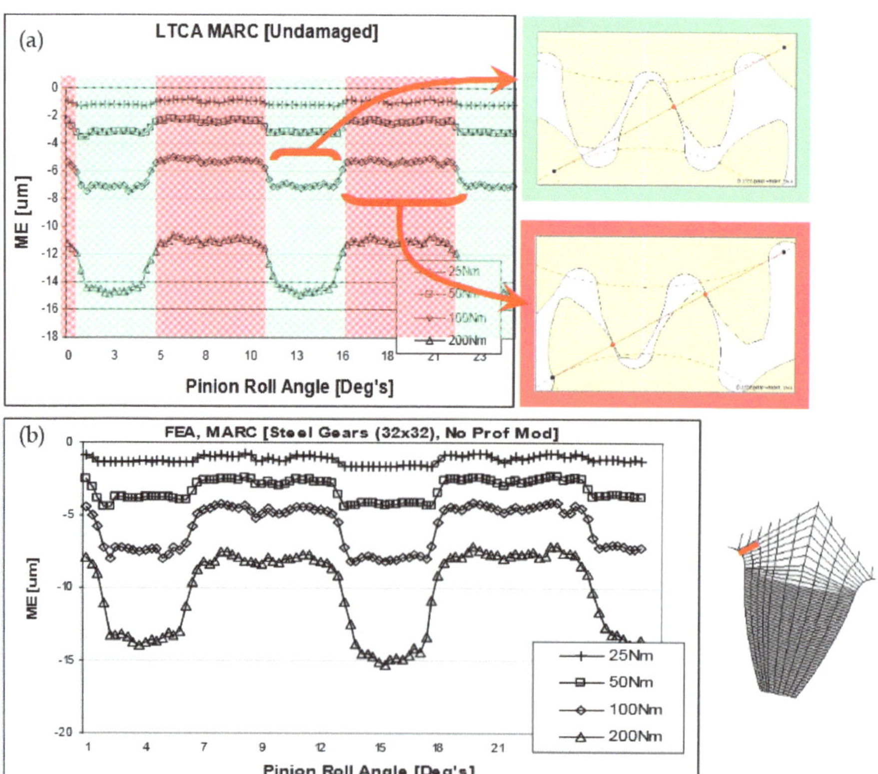

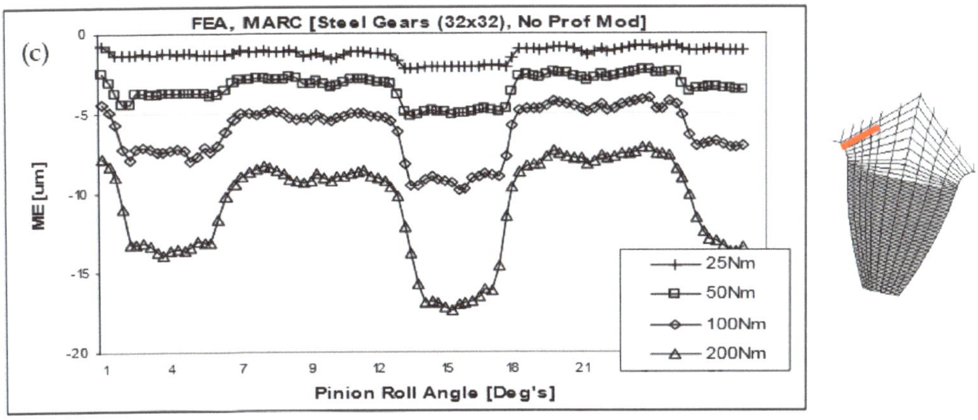

Fig. 3.1.5. Motion Errors of (32x32) teeth gear pairs; (a) Undamaged gear set; (b) TFC (L=1.18mm); (c) TFC (L=2.36mm)

The plots presented in Figures 3.1.5 (b) & (c) show the MEs of the gears with tooth fillet cracks of three different sizes. The localized increase in the amount of ME over a period of ME pattern is a direct consequence of the reduced gear tooth stiffness caused by the TFC.

The plots shown in Figure-3.1.6 (a) ~ (c) are the residual ME (RME) obtained by taking the difference between the MEs of uncracked gears and the ones with TFCs. The RMEs show a "double stepped" pattern that reflects the tooth meshing pattern of the gears, where smaller step with less deflection occurs as the crack tooth enters the mesh and share the load with the adjacent tooth; the larger second step follows when the cracked tooth alone carries the load.

The RMEs of the TFCs show linearly proportional relationship between the amount of loading and the change in RME for a given crack size. The linear relationship between the loading and the amount of tooth deflection on a cracked gear tooth indicates that the effect of TFC can be modelled effectively as a localized change in the gear mesh stiffness.

The plots in Figure-3.1.7 show the transmission errors (TEs) measured from a pair of plastic gears with a root fillet cut (Figure-3.1.7 (a)), which the cut replicates a tooth fillet crack. The TEs of Figure 3.1.7 (c1-c3 and d1-d3) are compared to the simulated patterns of MEs (figure 3.1.7 (b)). Composite TEs (CTEs) and the zoomed view of the CTEs are shown in Figure-3.1.7 (c1 & d1) and (c2 & d2) respectively. The CTE combines the both long and short term components of the TE (LTC and STC) presented earlier in section 2.2 (figure 2.2.1). The resemblance between the simulated MEs and measured the TEs confirms the validity of the simulated effect of TFC. The STCs (c3 & d3) were obtained by high pass filtering the CTE. The pattern in the STCs shows clear resemblance to the simulated TFC effect (Figure3.1.7 (b)).

The simulation model used in this study does not consider the effect of plasticity. This assumption can be justified for a gear tooth with small cracks where localised effect of plasticity at the crack tip has small influence on the overall deflection of the gear tooth, which is most likely the case for the ideal fault detection scenario.

More recent work published by Mark [57, 60] explains that the plasticity can become a significant factor when work hardening effect can cause a permanent deformation of the cracked tooth. In this case, the meshing pattern of the gears changes more definitively by the geometrical error introduced in the gears by the bent tooth. In some cases the bent tooth result in rather complex meshing behaviour that involves tooth impacting. Further explanation on this topic is available from the works published by Mark [57, 60].

Given the constraints of the aforementioned simulated TFC model, modeling the TFC as a localized alteration in gear mesh stiffness is deemed appropriate for a minor fracture developing in the gear tooth fillet region. This model provides a viable way for building a dynamic simulation of a geared transmission system with an emerging TFC.

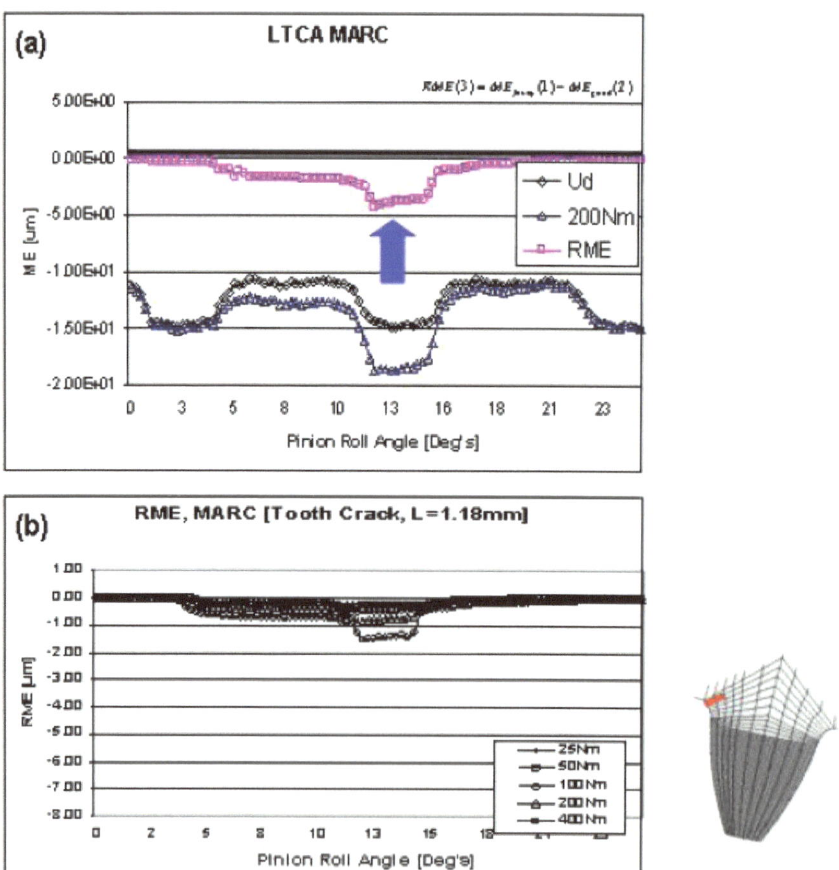

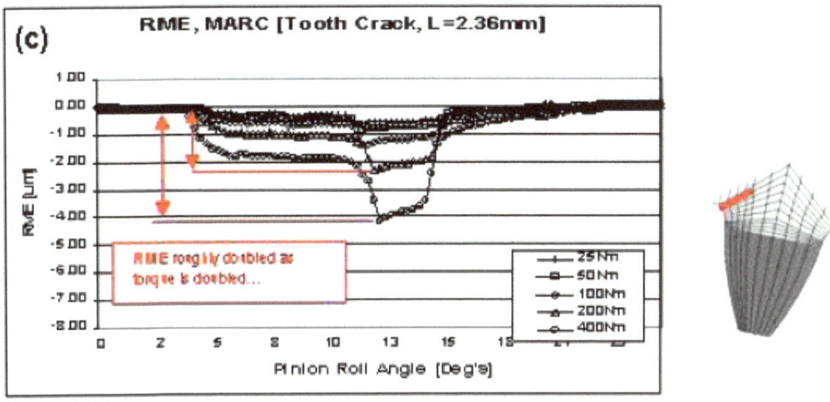

Fig. 3.1.6. Comparison of RMEs of gears with TFCs; (a) Illustrated definition of RME; (b) RMEs of TFC sizes L=1.18mm; (c) RMEs of TFC sizes L=2.36mm.

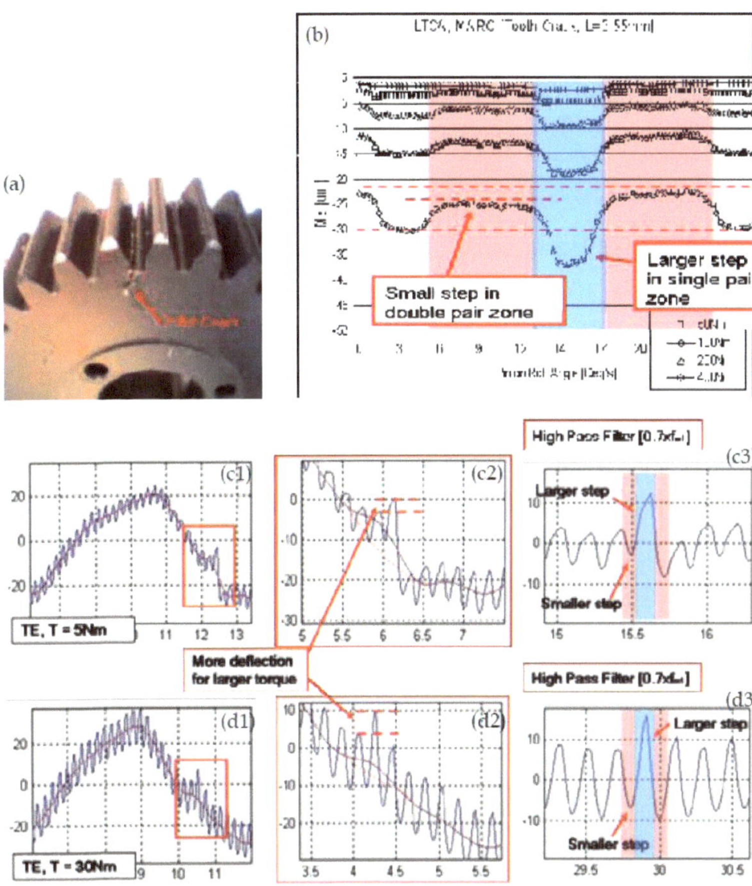

Fig. 3.1.7. Comparison of simulated vs. measured transmission errors; (a) Picture of a gear with a seeded TFC; (b) MEs from a FE model. (c1) ~ (c3) TEs of the gears with TFC loaded with 5Nm; (d1) ~ (d3) TEs of the gear with TFC loaded with 30Nm; CTE (c1, d1), Zoomed views (c2, d2) and STC (c3, d3).

3.2 Modelling the effect of a spall on a tooth face of a gear

Symptoms of surface fatigue vary, but they can generally be noticed by the appearance of cavities and craters formed by removal of surface material. The damage may start small or large and may grow or remain small. In some cases gears cure themselves as they wear off the damage: Initial pitting [47]. The terms "Spalling" and "Pitting" are often used indiscriminately to describe contact fatigue damages. Figure-3.2.1 shows some examples of spalls and pitting occurring on a gear tooth.

(a) (b)

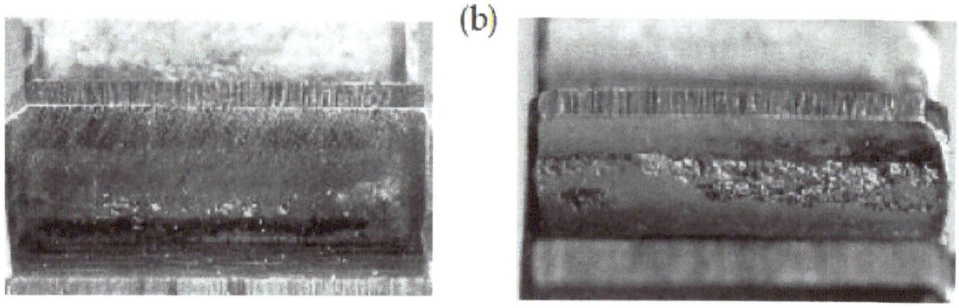

Fig. 3.2.1. Examples of (a) progressive pitting and (b) severe spalling damages on gear teeth. (DANA [48])

This work follows the definition of contact surface damage given by Tallian [58]: Spalling designated only as macro-scale contact fatigue, reserving the term pitting for the formation of pores and craters by processes other than fatigue cracking.

There are three distinctive phases in the development of surface fatigue damage:

1. Initial Phase: Bulk changes in the material structure take place around the highly stressed area under the contact path. Change in hardness, residual stress and some microscopic changes in the grain structure of the metal.
2. A Long Stable Phase: Microscopic flow occurs in the highly stressed area changing the material structure and residual stress conditions at the microscopic level. The change in the structure brought out by the microscopic flow can be observed by eyes in the illuminated etched areas.
3. Macroscopic Cracking: This is the last failure phase instituting the crack growth.

Spalls have a distinctive appearance that is characterised by how they were formed. A fully developed spall typically has its diameter much larger than its depth. The bottom of the spall has a series of serrations caused by propagating fatigue cracks running transverse to the direction of rolling contact. The bottom of the spall parallels the contact surface roughly at the depth of maximum unidirectional shear stress in Hertzian contact.

The sidewalls and the wall at the exiting side of the spall (as in the exiting of rolling contact) are often radially curved as they are formed by material breaking away from the fatigued area. The entrance wall of the spall is characterised by how it was initiated. Tallian [58] explains that shallow angled entry (less than 30° inclination to the contact surface) occurs when the spall is initiated by cracks on the surface. Spalls with steep entry (more than 45°) occur when the spall is initiated by subsurface cracks.

Surface originated spalls are caused by pre-existing surface damage (nicks and scratches). It is also known that lubrication fluid could accelerate the crack propagation when contact occurs in such a way that fluid is trapped in the cracks and squeezed at extremely high pressure as the contacting gear teeth rolls over it.

The subsurface originated spalls are caused by presence of inclusions (hard particles and impurities in the metal) and shearing occurs between the hard and the soft metal layers formed by case hardening. A spall caused by the initial breakage of the gear tooth surface continues to expand by forming subsequent cracks further down the rolling direction. Figure-3.2.2 illustrates the formation and expansion of spall damage by Ding & Rieger [59].

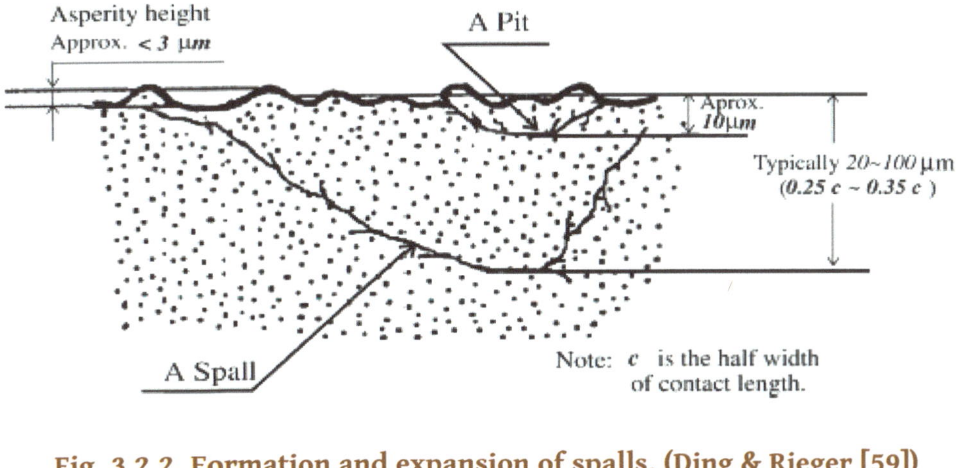

Fig. 3.2.2. Formation and expansion of spalls. (Ding & Rieger [59])

The result of contact stress analysis from HyGears [50] shows the occurrence of high stress concentration at the entrance and exit walls of the spall (Figure-3.2.3). The high stress concentration at these edges implies the likelihood of damage propagation in that direction. The result of the simulation corresponds to Tallian's [58] observation that spalls tend to expand in the direction of rolling contact. In the HyGears simulation the spall was modelled as a rectangular shaped recess on the tooth surface in the middle of the pitch line.

Typical spur gear tooth surfaces are formed by two curvatures: profile and lead curvatures. The 2D models are limited to simulating the effect of the spall crater on the gear tooth profile only. 3D simulation is required to comprehend the complete effect of spalls on the contact surface of the gears. A study was carried out to investigate the effect of spalls on the ME using a 3D gear tooth model [62]. The result of the study showed that the effect of the spalls on the ME is completely dominated by the displacement caused by the topological error of the gear tooth surface caused by the fault.

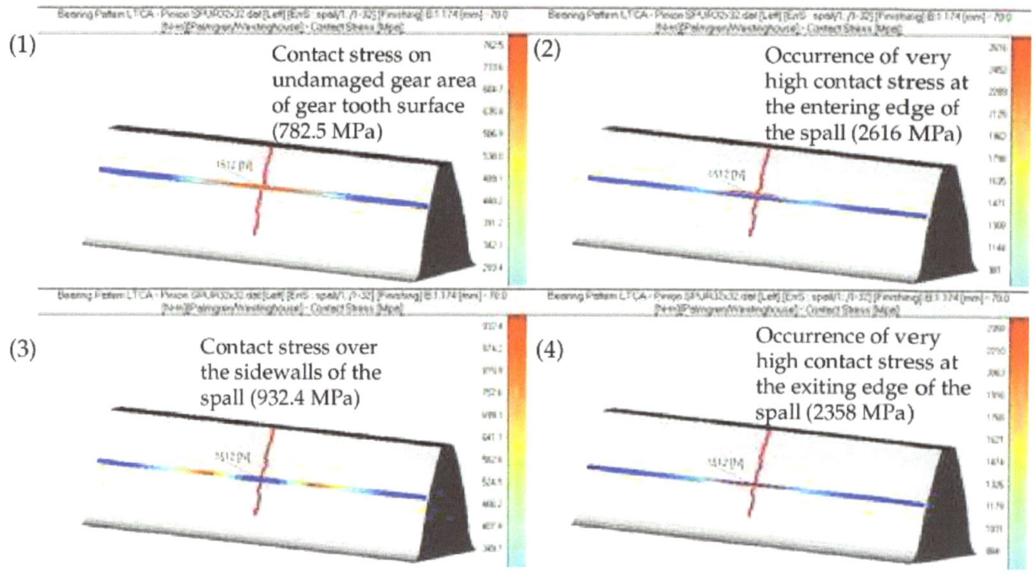

Fig. 3.2.3. HyGears analysis of contact stresses around the spall. [62]

Formation of spalls is a complicated process and it is one of the active research topics which have been studied for some time. Although there are several items in the literature that describe the properties and process of formation of spalls, no literature was sighted which defines the specific definition of shapes and sizes of spalls occurring on spur gear teeth.

Papers presented by Badaoui et al. [63, 64] and Mahfoudh et al. [65] show a successful example of simulating the effect of a spall on spur gear tooth by modelling the fault as prismatic slot cut into the gear tooth surface. Their results were validated by experiment. The model of the spall used in this work follows the same simplification of the general shape of the spalling fault. The simulation of the spall is bounded by these following assumptions:

A spall is most likely to initiate at the centre of the pitchline on a gear tooth where maximum contact stress is expected to occur in the meshing spur gear teeth as the gears carry the load by only one pair of teeth.

The spall expand in size in the direction of rolling contact until it reaches the end of the single tooth pair contact zone as shown in Figure-3.2.4.

When the spall reaches the end of the single tooth pair contact zone, the spall will then expand across the tooth face following the position of the high contact stresses. The contact stress patterns shown in Figure-3.2.3 strongly support this tendency of spall growth.

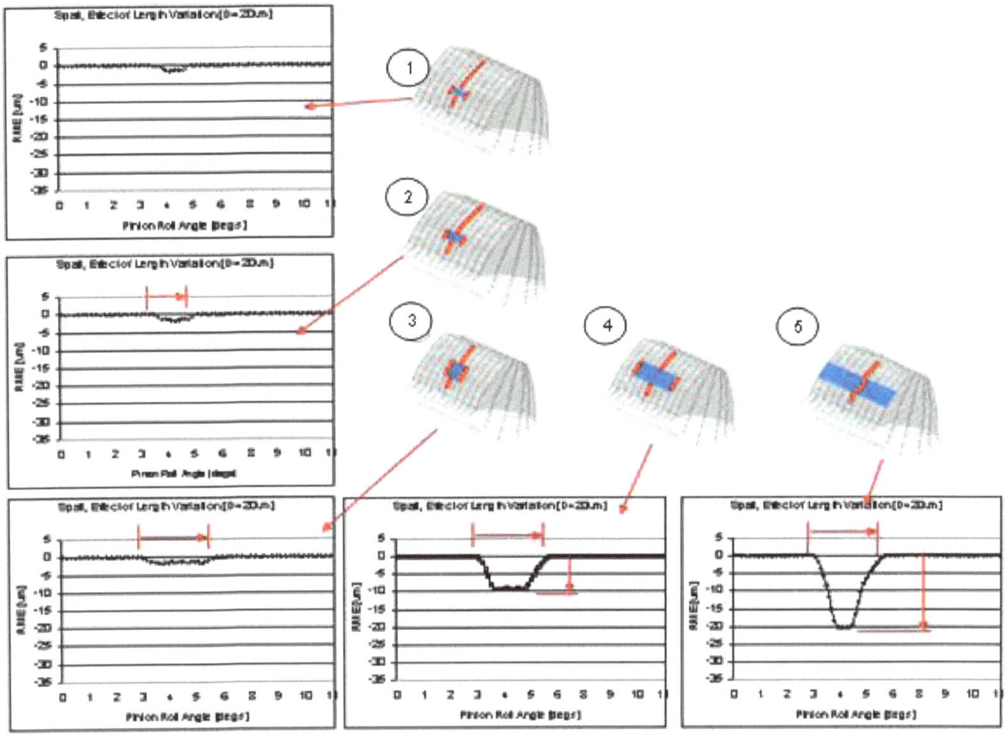

Fig. 3.2.4. A model of spall growth pattern and the resulting RME. [62]

The models of spalls were developed by following the set of assumptions described above. Resulting plots of the RMEs are presented in Figure-3.2.4 along with the shape illustration of spalls on a gear tooth. Note that the RMEs of the spalls form bucket shapes and their length and the depth are determined by the length and width of the spall; i.e. the shape of RME and the progression of the spall is directly related. This information can be used in diagnosis and prognosis of the fault.

The plots in Figure-3.2.5 show TEs measured from a pair of plastic gears with a spall (Figure-3.2.5 (a)). The change in the pattern of the TE due to the spall is comparable to the simulated pattern of the MEs (Figure-3.2.5 (b)). Composite TEs (CTEs) and the zoomed views of the CTEs are shown in Figure-3.2.5 (c1~c3). The resemblance between the simulated MEs and measured TEs confirms the validity of the simulated effect of TFC. The STCs (d1~d3) were obtained by high pass filtering the CTE. The pattern in the STCs shows clear correlation between the simulated and measured patterns of spall motion errors.

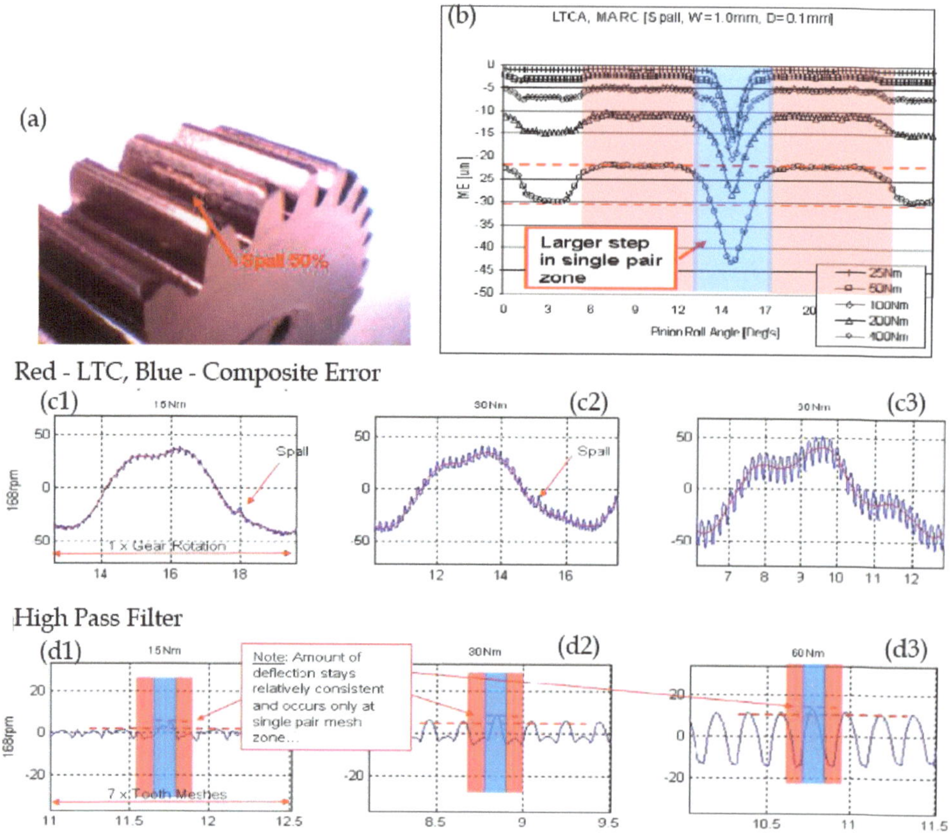

Fig. 3.2.5. Comparison of simulated vs. measured transmission errors; (a) Picture of a gear with a seeded Spall; (b) MEs from a FE model; (c1) ~ (c3) CTEs of the gear with the spall; (d1) ~ (d3) STCs of the gear with the spall; Applied torque: 15Nm (c1 & d1), 30Nm (c2, d2), 60Nm (c3, d3).

3.3 Simulating the effect of TFCs and spalls in a gear dynamics model

The FEA model based study of TFCs and spalls has lead to identifying some useful properties of the faults that can be used to model the effect of the faults in the lamped parameter type gear dynamics models.

The Residual Motion Errors (RME) of TFCs have shown double stepped patterns that were load dependent. The change in the amount of deflection in the gear mesh (i.e. ME) with a cracked tooth is influenced by the size of the crack and also by the amount of loading on the gears. The simulation result showed that the linearly proportional relationship between the torque applied to the gears and the resulting RME value. The bucket shaped RMEs of spalls were not affected by the loading condition but purely driven by the change in the contact path patterns of the meshing teeth, due to the geometrical deviation of the gear tooth surface caused by a spall. It was also understood from the simulation studies that the size and the shape of a spall affect the length and the depth of the bucket.

Based on the observation above, the effect of TFC was modelled as locally reduced tooth meshing stiffness and the spalls as direct displacement due to the topological alteration of the gear tooth surface. In the gear dynamic model is shown in Figure-3.3.1, the effect of a TFC was implemented as a reduction in stiffness "K_m" over one gear mesh cycle. The change in the value of K_m was calculated from the FEA model mapped into an angular position dependent function in the gear dynamic model. A spall was implemented as a localized displacement mapped on the "e_t". An illustration of a TFC and a spall models in a gear rotor dynamic model is shown in Figure-3.3.1 [61].

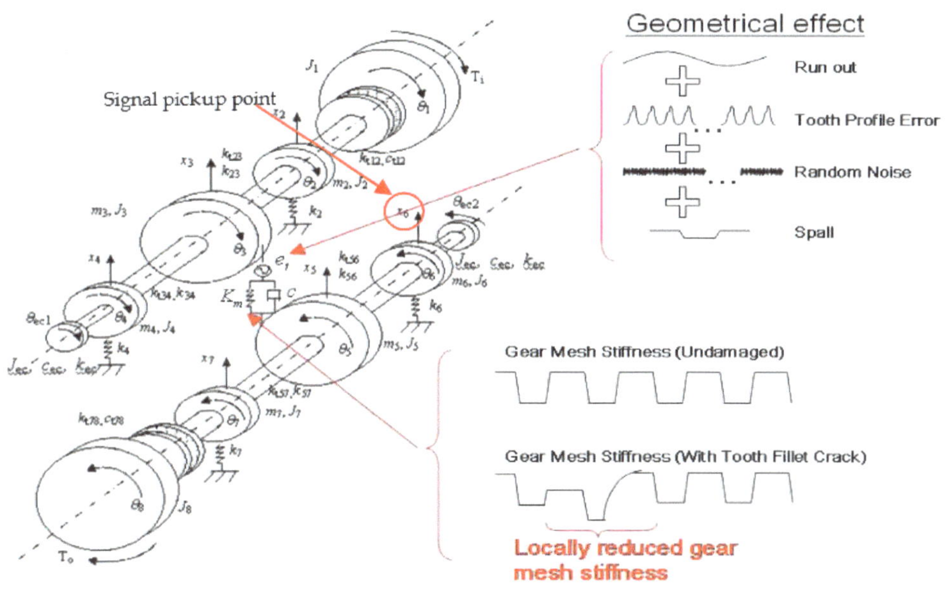

Fig. 3.3.1. Modelling of Gear Tooth Faults in Dynamic Model. [61]

Figures 3.3.2 (a1~a3) and 3.3.3 (a1~a3) show the simulated vibration signal (acceleration) from the LPM shown in the Figure-3.3.1. The signals were measured from the free end of the driven shaft (see Figure-3.3.1). The residual signals shown in (b1~b3) of Figures 3.3.2 and 3.3.3 were obtained by subtracting the simulated vibration of undamaged gearbox from the damaged one. Two identical simulation models, one with a gear tooth fault (TFC or spall) and the other undamaged, were run in parallel and the difference between the two model out puts were taken to separate the effect of the gear faults. The impact like effect of the gear tooth faults is seen in both the gears with a TFC and a spall.

A comparison of the simulated signals shows that the magnitude of the TFC impulses is affected by the amount of torque applied to the gears, while the spall impulses are not. This response is consistent with the observation made in the static simulation of the gears in mesh.

Careful observation of the residual signal also reveals that the fault information is not only buried in the dominant effect of gearmesh, but also somewhat distorted by the effect of transmission path from the gearmesh to the point where the signal was measured. The effect of transmission path appears in the residual signal As the transient "tail" effect convolved over the impulse due to the gear fault (see Equation-3.3.1 and Figure-3.3.4 for illustration).

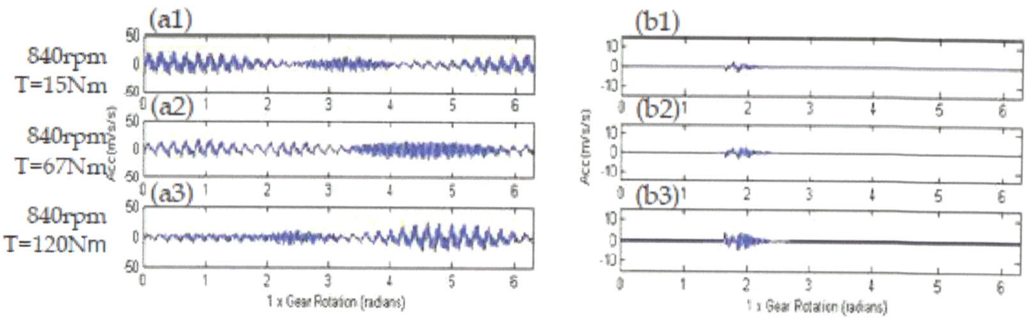

Fig. 3.3.2. Simulated gearbox vibration signal with the effect of a TFC.

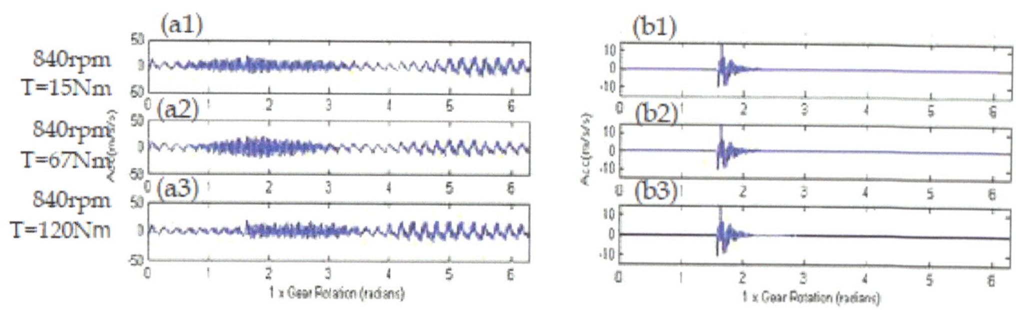

Fig. 3.3.3. Simulated gearbox vibration signal with the effect of a spall.

The residual signal of the TFC and the spall provide a useful means to understand the nature of diagnostic information of the faults. In machine condition monitoring signal processing techniques are often developed to detect and quantify the symptoms of a damage buried in a background noise. By being able to see the "clear" effect of a fault, the most effective signal processing technique can be applied to target and monitor the symptoms of the damage. The idea of using the simulated fault signals to design and improve the fault detection and machine condition monitoring techniques has been put to effective uses by Randall, Sawalhi and Endo [34, 35, 62, 66].

$$y = (e + w + n) * h \text{ (Note: * represents convolution)} \tag{3.3.1}$$

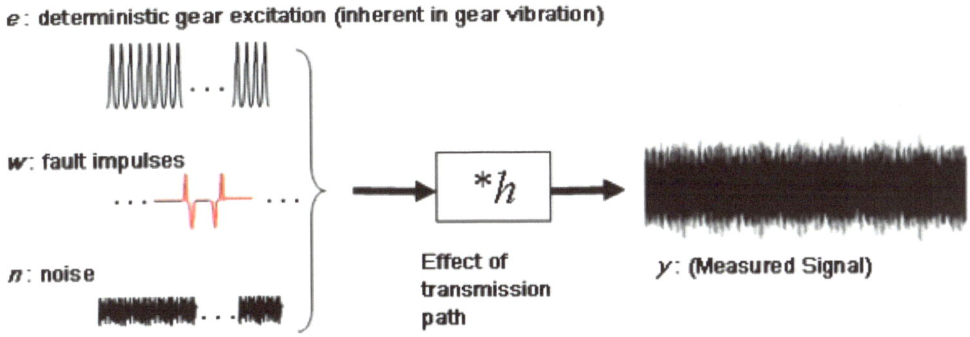

Fig. 3.3.4. Vibration of a Gearbox. [62]

3.4 Modelling a spall in a rolling element bearing

This section discusses a several simple but an effective methods of modelling a spall in the rolling element bearing model introduced previously in section 2.5.1. The ideas behind each modelling approach was discussed by using the example of modelling a spall in the outer race of a bearing. The same method can be easily expanded in to modelling an inner race spall and ball faults. More detailed explanation on this topic is available from the work published by Sawalhi and Randall [29, 30, 31, 34, 35, 66, 67].

The simplest form of a spall model can be implemented to the REB model by assuming instantaneous contact loss between the bearing races and rollig an element(s) as it pass over the spall. So, in reference to the rolling element model described above, the presence of a spall of a depth (Cd) over an angular distance of ($\Delta\Phi d$) can be modelled by using the fault switch βj which defines the contact state of rolling elements over a defined angular position (Φd). In effect, this mothod models the spall as a step function as shown in Figure- and further illustrated by Figure-3.4.2 (a), in which βj is defined as follows:

$$\beta_j = \begin{cases} 1, & if\ \phi_d < \phi_j < \phi_d + \Delta\phi_d \\ 0, & otherwise \end{cases}$$

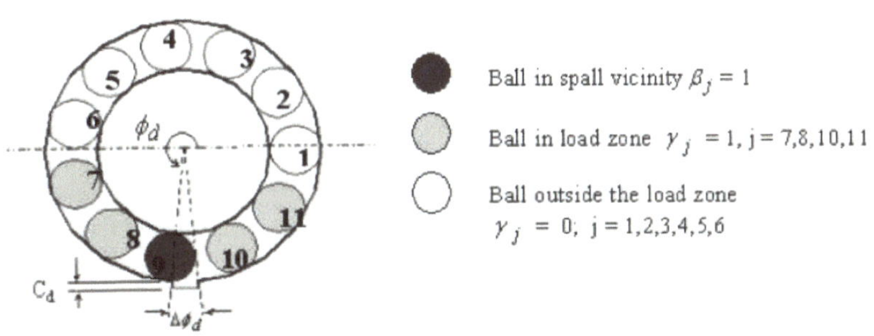

Fig. 3.4.1. Spall definition on the outer race [66]

Fig. 3.4.2. Modified model of a spall based on a more realistic ball trajectory. [66]

The outer race spall is fixed in location between Φd and $\Phi d + \Delta\Phi d$. This normally occurs in the load zone. An inner race spall rotates at the same speed as the rotor, i.e. $\Phi d = \omega c d t + \Phi d o$ (Φdo : initial starting location of the spall).

This model of the spall assumes that the rolling element will lose contact suddenly once it enters the spall region, and will regain contact instantly when exiting from that area (Figure- (a)). The abrupt change in the rolling element positions at the entry and exit of the spall results in very large impulsive forces in the system, which is not quite realistic. An modification on the previous model was introduced in [28] in which the depth of the fault (Cd) was modelled as a function of (Φ_j), Figure-3.4.2 (b). The improvement on the model is to represent more realistic trajectory of the rolling element movement based on the relative size of the rolling element and the depth of the spall. Although, the profile of the trajectory appears much less abrupt than the earlier version; and apears to have only one position that may result in impulse, it still resulted in two impulses which does not agree with the experimental observation.

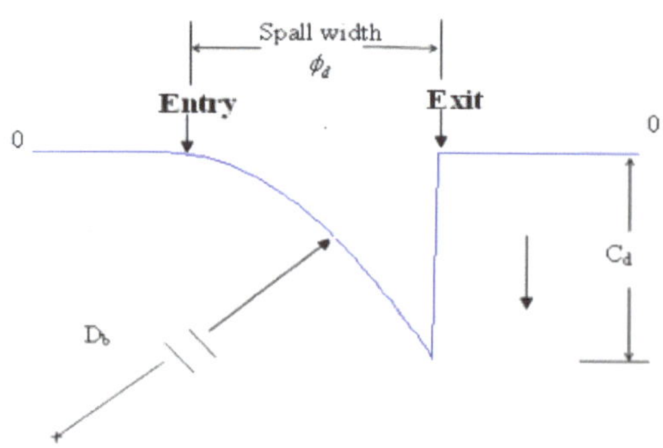

Fig. 3.4.3. A correlated model of a spall based on experimental data.

Careful observation of the interaction between the rolling element and spall leads to the trajectory is shown in Figure-3.4.3. The entry path of the rolling element has been represented as having a fixed radius of curvature (equal to that of the rolling element); entry of the rolling element in to the spall is therefore somewhat smoother. The smoother change in curvature at the entry would then represent a step in acceleration. On exiting the spall, the centre of the rolling element would have to change the direction suddenly, this representing a step change in velocity or an impulse in acceleration. This has been modelled as a sudden change (i.e. similar to the original model [28]). The resulting acceleration signal from this model appears to agree with the experimental observation, however author recommends further validation on this modelling approach and more updates are expected based on the findings.

For more detailed explanations on the modelling of REB fault refer to the works published by Sawalhi & Randall [29, 30, 31, 34, 35, 66, 67].

4. CONCISE SUMMARY

This chapter introduced and addressed the approaches for modeling the effects of gearbox defects, including tooth fillet cracks, tooth face spalls, and bearing spalls. The primary objective of damage modeling is to replicate the impact of faults on the dynamics of a geared transmission system, hence enhancing the comprehension of the diagnostic information evident in the vibration signal collected from a gearbox. Vibration signal analysis techniques for defect detection and diagnosis are an optimal non-destructive method for machine health monitoring, applicable in a minimally invasive manner, for as by affixing an accelerometer to a gearbox case. The interplay among the machine elements of a gearbox is frequently intricate, and the vibration signals recorded from the gearbox are challenging to interpret. Diagnostic information pertinent to an incipient failure in a gear or bearing is generally obscured within the predominant signal components generated by the transmission system mechanisms themselves: For instance, gear meshing signals. Historically, researchers engaged in the development of a signal processing technique for gearbox diagnostics have initiated their efforts by formulating informed hypotheses regarding the characteristics of the diagnostic information pertaining to defects. These assumptions are frequently founded on meticulous examination of observed vibration signals. Nonetheless, the efficacy of this strategy is frequently constrained by the inherent difficulty in discerning the critical features of the defect signals within the signal amalgamation. This chapter highlighted how simulation models may be effectively utilized to examine the properties of fault signals in greater depth. A technique for isolating fault signals from the simulated gearbox signal mixture was detailed in section 3.3. The residual signals derived from this method illustrated the manifestation of flaws in the resultant vibration signals in the "cleaned" condition. The analysis of the simulated residual signals has enhanced the comprehension of the properties of impulses generated by faults and the distorting influence of the transmission line from the fault signal origin to the measurement site. The enhanced comprehension of the fault signal derived from the simulation studies resulted in the creation of more efficient signal processing algorithms [34, 35, 62, 66].

The gearbox defect models discussed in this study necessitate additional refining. Future development areas discussed in the chapter include enhancing the understanding of the effects of plastic deformation in gear TFC, spall forms, and non-linear dynamic interactions between gears and bearings. Enhancing the connection between simulated and measured signals effectively illustrates the comprehension of fault impacts in a geared transmission system. This Knowledge enhances the design and development of vibration signal analysis-based machine condition monitoring solutions. In the imminent future, precisely simulated signals of a defective gearbox can enhance the machine learning process of fault diagnosis algorithms utilizing neural networks. Conducting this work in experiments is a time-consuming and costly endeavor; a simulation model-based method seems far more advantageous.

5. NOMENCLATURE

Unless otherwise stated the following tables defines the symbols and the acronyms used in this chapter.

A_1A_2	A vector connecting points A_1 and A_2
Ø	Pressure Angle
LoA	Line of Action
PoA	Plane of Action
LoC	Line of Centres
R_b,, r_b	Gear Base Radius and Pinion Base Radius
R_{gear}, r_{pinion}	Gear Pitch Radius and Pinion Pitch Radius
R_o and r_o	Gear Outer Radius Pinion Outer Radius
P_b	Base pitch
CR	Contact Ratio
N	Number of teeth on a gear
TE	Transmission Error (Measured experimentally)
GTE	Geometrical Transmission Error
STE	Static Transmission Error
LTC	Long Term Composite of TE
STC	Short Term Composite of TE
ME	Motion Error (Numerically calculated TE)
x_i, y_i, z_i	Translation at i_{th} Degrees of Freedom
θ_{xi}, θ_{yi}, θ zi	Rotation about a translational axis at i_{th} Degrees of Freedom
K, Kmb	Linear stiffness elements. The subscript '*mb*' refers to stiffness at the gearmesh
C, Cmb	Damping matrices. The subscript '*mb*' refers to damping at the gearmesh.

H	An 'on/off' switch governing the contact state of the meshing gear teeth.
et	A vector representing the combined effect of tooth topography deviations and misalignment of the gear pair.
T	Torque
x, *x* , *x*	Vectors of displacement, velocity and acceleration (translation)
Σ	Stress
ε	Strain

REFERENCES

1. J. D. Smith, and D. B. Welbourn, 2000, "Gearing Research in Cambridge", Cambridge Gearing Publications
2. R. E. Smith, 1983, "Solving Gear Noise Problems with Single Flank Composite Inspection", The Gleason Works, Rochester, NY, USA
3. R. G. Munro, D. Houser, 2004, "Transmission Error Concepts", Notes from Gear Noise Intensive Workshop, 19th ~ 21st July 2004, Melbourne, Australia
4. R. W. Gregory, S. L. Harris and R. G. Munro, 1963a, "Dynamic Behaviour of Spur Gears", Proc IMechE, Vol. 178(1), pp 207-218
5. R. W. Gregory, S. L. Harris and R. G. Munro, 1963b, "Torsional Motions of a Pair of Spur Gears", Proc IMechE Applied Mechanics Convention, Vol. 178 (3J), pp 166-173
6. H. N. Ozguven and D. R. Houser, 1988a, "Mathematical Models Used in Gear Dynamics – A Review", Journal of Sound and Vibration, 121(3), pp 383-411
7. C. C. Wang, 1985, "On Analytical Evaluation of Gear Dynamic Factors Based on Rigid Body Dynamics", Journal of Mechanism, Transmission and Automation in Design,Vol. 107, pp 301-311
8. W. Bartelmus, 2001, "Mathematical Modelling and Computer Simulations as an Aid to Gearbox Diagnostics", Mechanical Systems and Signal Processing, 15 (5), pp 855- 871
9. J. Lin and R. G. Parker, 2002, "Mesh Stiffness Variation Instabilities in Two-Stage Gear Systems", Transaction of ASME, Jan 2002, Vol. 124, pp 68-76
10. R. G. Parker and J. Lin, 2001, "Modelling, Modal Properties, and Mesh Stiffness Variation Instabilities of Planetary Gears", NASA, NASA/CR-2001-210939
11. Y. Gao and R. B, Randall, 1999, "The Effect of Bearing and Gear Faults in Rolling Element Bearing Supported Gear Systems", DSTO Aeronautical and Maritime Research Laboratory, July 1999, CEVA-99-02

12. Y. Gao and R. B. Randall, 2000, "Simulation of Geometric, Static and Dynamic Gear Transmission Errors", DSTO Aeronautical and Maritime Research Laboratory, Jan 2000, CEVA-2000-01
13. M. Amabili, and A. Rivola, 1997, "Dynamic Analysis of Spur Gear Pairs: Steady-State Response and Stability of the SDOF Model with Time Varying Meshing Damping", Mechanical System and Signal Processing, 11 (3), pp 375-390
14. I. Howard, S. Jia and J. Wang, 2001, "The Dynamic Modelling of a Spur Gear in Mesh Including Friction and A Crack", Mechanical Systems and Signal Processing, 15 (5), pp 831-853
15. P. Velex and M. Maatar, 1996, "A Mathematical Model for Analysing the Influence of Shape Deviations and Mounting Error on Gear Dynamic Behaviour", Journal of Sound and Vibration, 191(5), pp 629-660
16. S. M. Vijayakar, 1991, "A Combined Surface Integral and Finite Element Solution for a Three-Dimensional Contact Problem", International Journal of Numerical Methods in Engineering, 31, pp 524-546
17. G. W. Blankenship and R. Singh, 1995, "A New Gear Mesh Interface Dynamic Model to Predict Multi-Dimensional Force Coupling and Excitation", Mechanical, Machine Theory, Vol. 30, No.1, pp43-57
18. A.KahramanandG.W.Blankenship,1996,"InteractionsbetweenCommensurate parametric and Forcing Excitations in A System with Clearance", Journal of Sound and Vibration, 194 (3), pp 317-336
19. R. G. Parker, 2000, "Non-Linear Dynamic Response of a Spur Gear Pair: Modelling and Experimental Comparisons", Journal of Sound and Vibration, 237(3), pp 435-455
20. R. G. Parker, V. Agashe and S. M. Vijayakar, 2000, "Dynamic Response of a Planetary Gear System Using a Finite Element/Contact Mechanics Model", Transaction of ASME, Sep 2000, Vol. 122, pp 304-310
21. CALYX, 2002, CALYX USER MANUAL, Advanced Numerical Solutions, Hilliard Ohio, U.S.A
22. HELICAL, 2002, HELICAL 3D USER's MANUAL, Advanced Numerical Solutions, Hilliard Ohio, U.S.A
23. S. Du, 1997, "Dynamic Modelling and Simulation of Gear Transmission Error for Gearbox Vibration Analysis", PhD dissertation, University of New South Wales
24. N. S. Feng, E. J. Hahn, and R. B. Randall, 2002, "Using transient analysis software to simulate vibration signals due to rolling element bearing defects", Proceedings of the 3rd Australian congress on applied Mechanics, 689-694, Sydney.

25. I. K. Epps and H. McCallion, 1994, "An investigation into the characteristics of vibration excited by discrete faults in rolling element bearings" (extract from PhD thesis of I.K. Epps), Annual Conference of the Vibration Association of New Zealand, Christchurch.

26. D. Ho, 1999, "Bearing Diagnostics and Self Adaptive Noise Cancellation", PhD dissertation, UNSW.

27. S. Fukata, E. H. Gad, T. Kondou, T. Ayabe and H. Tamura, 1985, "On the Vibration of Ball Bearings", Bulletin of JSME, 28 (239), pp. 899-904.

28. A. Liew, N.S. Feng and E. J. Hahn, 2002, "Transient rolling element bearing systems", Trans. ASME Turbines and Power, 124(4), pp. 984-991

29. N. Sawalhi and R.B. Randall, 2011, Vibration response of spalled rolling element bearings: observations, simulations and signal processing techniques to track the spall size. Mechanical Systems and Processing, 25 (3), (2011)

30. N. Sawalhi and R.B. Randall, 2010, " Improved simulations for fault size estimation in rolling element bearings", paper presented at the seventh International Conference on Condition Monitoring and Machinery Failure Prevention Technologies, Stratford-upon-Avon, England, 22-24 June (2010).

31. N. Sawalhi and R.B. Randall, 2008, "Localised fault diagnosis in rolling element bearings in gearboxes", paper presented at the Fifth International Conference on Condition Monitoring and Machinery Failure Prevention Technologies, Edinburgh, 15-18 July (2008).

32. J. Sopanen & A Mikkola, 2003, "Dynamic model of a deep-groove ball bearing including localized and distributed defects. Prt 1: Theory", Proc of IMechE, vol. 217, Prt K, J of Multi-body Dynamics, 2003, pp 201-211.

33. T. Tiwari, K. Gupta, O. Prakash, 2000, "Dynamic response of an unbalanced rotor supported on ball bearings", J of Sound & Vibration 238 (2000), pp757-779.

34. N. Sawalhi & R. B. Randall, 2008, "Simulating gear and bearing interaction in the presence of faults: Prt II - Simulation of the vibrations produced by extended bearing faults", MSSP, Vol. 22, pp1952-1996 (2008).

35. N. Sawalhi & R. B. Randall, 2008, "Simulating gear and bearing interaction in the presence of faults: Prt I - The combined gear bearing dynamic model and simulation of localized bearing faults", MSSP, Vol. 22, pp1924-1951 (2008).

36. Z. -Q. Qu, 2004, "Model order reduction technique: with applications in Finite Element Analysis", Springer, 2004.

37. R. Craig & M. Bumpton, 1968, "Coupling of substructures of dynamic analysis", AIAA Journal, 67 (1968), pp1313-1319.

38. B. M. Irons, 1967, "Structural eigenvalue problem-elimination of unwanted variables", AIAA Journal, 3(5), pp961-962.

39. L. Despnade, N.Sawalhi, R.B.Randall, "Improved Gearbox Simulations for Diagnostic and prognostics Purposes Using Finite Element Model Reduction Techniques", paper presented at the 6th Australasian Congress on Applied Mechanics, Perth, Australia, 12-15 December (2010).

40. C. Carmignani, P. Forte & G. Melani, 2009, "Component modal synthesis modelling of a gearbox fro vibration monitoring simulation", The 6th Int'l conf on Condition Monitoring Manhinery Failure Prevention Technology, Dublin, Ireland.

41. N. Sawalhi, L. Deshpande and R. B. Randall, 2011, "Improved simulation of faults in gearboxes for diagnostic and prognostic purposes using a reduced finite element model of casing", AIAC 14 Fourteenth Australian International Aerospace Congress.

42. R. Singh, 1989, "Analysis of Automotive Neutral Gear Rattle", Journal of Sound and Vibration, Vol. 131 (2), pp177-196

43. A. Kahraman and R. Singh, 1991, "Interaction between Time-Varying Mesh Stiffness and Clearance Non-Linearity in Geared System", Journal of Sound and Vibration", 146, pp135-156

44. D. P. Townsend, 1997, "Gear and Transmission Research at NASA", NASA-TM-107428

45. A. K. Wong, 2001, "Vibration-Based Helicopter Health Monitoring – An Overview of the Research Program in DSTO", DSTO-HUMS2001

46. L. E. Alban, 1985, "Systematic Analysis of Gear Failure", American Society of Metals, Metals Park OH, ISBN-0871702002

47. G. DeLange, 1999, "Analyzing Gear Failures", Hydrocarbon Processing, Features: 4/00, www.hydrocarbonprocessing.com

48. DANA, 2002, Raodranger, "Understanding Spur Gear Life", EATON and DANA Corporation, TRSM-0913, (www.roadranger.com)

49. ANSI/AGMA, 1995, "Appearance of Gear Teeth – Terminology of Wear and Failure", American Gear Manufacturers Association, ANSI/AGMA 1010-E95

50. HyGears, HyGears © V2.0 Gear Design and Analysis Software, Involute Simulation Software Inc

51. D. G. Lewicki, 2001, "Gear Crack Propagation Path Studies – Guidelines for Ultra-Safe Design", NASA-TM- 2001 211073

52. T. J. Curtin, R. A. Adey, J. M. W. Bayman and P. Marais, 1998, Computational Mechanics Inc, Billerica, Massachusetts, 01821, USA

53. D. G. Lewicki, 1998, "Three-Dimensional Gear Crack Propagation Studies", NASA-TM- 1998-208827

54. S. Pehan, T. K. Hellen, J. Flasker and S. Glodes, 1997, "Numerical Methods for Determining Stress Intensity Factors VS Crack Depth in Gear Tooth Roots", Int J of Fatigue, Vol. 19, No. 10, pp677-685

55. M. Guagliano and L. Vergani, 2001, "Effect of Crack Closure on Gear Crack Propagation", Int J or Fatigue, Vol. 23, pp67-73

56. H. Endo, C. Gosselin, R. B. Randall, (Sept 2004), "The effects of localized gear tooth damage on the gear dynamics – A comparison of the effect of a gear tooth root crack and a spall on the gear transmission error", IMechE, 8th International Conference on Vibrations in Rotating Machinery, University of Wales, Swansea

57. W. D. Mark, C. P. Reagor and D. R. McPherson, 2007, "Assessing the role of plastic deformation in gear-health monitoring by precision measurement of failed gears", MSSP, 21: pp177-192 (2007)

58. T. E. Tallian, 1992, "Failure Atlas fir Hertz Contact Machine Elements", ASME Press, ISBN 0-7918-0008-3

59. Y. Ding and N. F. Rieger, 2003, "Spalling formation Mechanism for Gears", Wear, 254 2003), pp1307-1317

60. W. D. Mark and C. P. Reagor, 2007, "Static transmission error vibratory excitation contribution from plastically deformed gear teeth caused by tooth bending fatigue damage", MSSP, 21: pp885-905 (2007)

61. H. Endo, R. B. Randall and C. Gosselin, 2004, "Differential Diagnosis of Spall vs. Cracks in the Gear Tooth Fillet Region", Journal of Failure Analysis and Prevention, Vol4,Issue 5, Oct 2004

62. H. Endo, R. B. Randall and C. Gosselin, 2008, "Differential Diagnosis of Spall vs. Cracks in the Gear Tooth Fillet region: Experimental Validation", Journal of Mechanical System and Signal Processing, due to be published on 19/ Jan/2009, Vol23 Issue3.

63. M. EL Badaoui, J. Antoni, F. Guillet and J. Daniere, 2001a, "Use of the Moving Cepstrum Integral to Detect and Localise Tooth Spalls in Gears", Mechanical System and Signal Processing, 2001, 15 (5) , 873-885

64. M. EL Badaoui, V. Cahouet, F. Guillet, J. Daniere and P. Velex, 2001b, "Modelling and Detection of Localized Tooth Defects in Geared Systems", Transaction of ASME, Sep, 2001, Vol. 123, pp422-430

65. J. Mahfoudh, C. Bard, M. Alattass and D. Play, 1995, "Simulation of Gearbox Dynamic Behaviour with Gear Faults", I Mech E, C492/045/95

66. N.Sawalhi, R.B.Randall and H.Endo, "Gear and bearing fault simulation applied to diagnostics and prognostics", paper presented at The 19th International Congress and Exhibition on Condition Monitoring and Diagnostic Engineering Management, Luleå, Sweden, June (2006).

67. N. Sawalhi, and R.B. Randall, Vibration response of spalled rolling element bearings: observations, simulations and signal processing techniques to track the spall size. Mechanical Systems and Processing, 25 (3), (2011).

3 Chapter

SPLIT TORQUE GEARBOXES: SPECIFICATIONS, EFFICIENCY, AND UTILIZATIONS

1. INTRODUCTION: AN OVERVIEW

While the most basic gear systems consist of a single engagement area between two gears, variants exist for situations requiring the transmission of exceptionally high torque inside a confined space. One method to enhance power density is the utilization of split torque systems primarily developed for the aviation sector. These gear systems are founded on a fundamental concept: the distribution of force transmission across multiple contact points, hence enhancing the contact ratio. This, however, presents the issue of interlocking four gears. (Fig. 1).

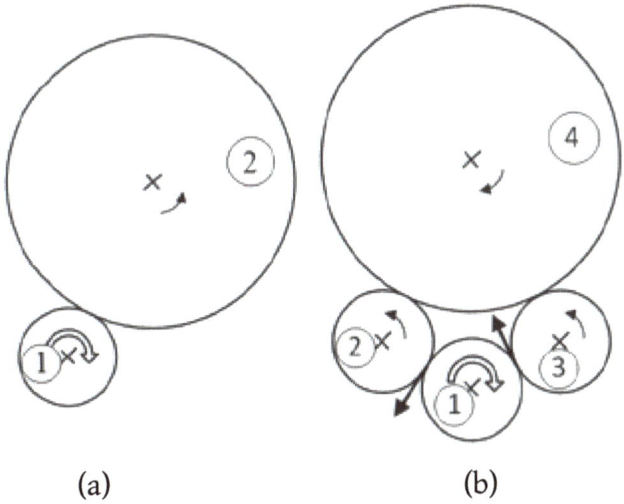

Fig. 1. (a) Standard gearbox assembly; (b) Split torque gearbox assembly

Split torque gearboxes are arrangements in which a driving pinion (1) engages with two intermediate idler pinions (2, 3), which concurrently interact with an additional gear (4). This assembly will thereafter be referred to as four-gear meshing. The torque distribution is from gear (1) to gears (2) and (3), which engage with gear (4). This gear assembly decreases gear speed, hence increasing available torque; consequently, the split torque transmission allows for the utilization of lower gears.

The more gears that engage the same crown, the lesser the torque exerted by each pinion. Gear assemblies may comprise up to 14 gears engaging a single crown, as observed in tunnel boring machines, for instance.

This chapter examines four-gear meshing within a gear assembly that guarantees a 50% torque distribution for each meshing region. Split torque gears are examined from two viewpoints: firstly, the prevalent applications of split torque gearboxes in the aerospace industry, and secondly, the two most limiting factors regarding their utilization, specifically:

- The geometric limitation of the four-gear assembly that requires simultaneous engagement for all four gears. Note that the four gears do not mesh correctly in just any position, although they may seem to do so initially. We will describe the conditions for simultaneous meshing of the four gears in general terms below.
- Torque split between the two gearbox paths must be as balanced as possible to ensure that neither of the paths is overloaded. The technology available to ensure proper torque split between two paths will be discussed below.

2. APPLICATIONS

The gear transmission specifications for airplanes are stringent, typically necessitating a gear ratio of 60:1 between the engine and rotor (Krantz, 1996). Furthermore, the gear transmission system must be secure, dependable, lightweight, and devoid of vibrations. Weight is a significant limiting factor, influenced by three key transmission characteristics.:

1. The number of transmission stages. The greater the number of stages used to achieve the final gear ratio, the heavier the transmission, given that more common elements such as shafts and bearings are necessary.
2. The number of transmission paths, the basis for split torque gearboxes. Torque is divided between several transmission paths, resulting in a contact force in the smaller gear that means that smaller, and consequently lighter, gears can be used.
3. The final stage transmission ratio. Using a greater transmission ratio in the final stage enables weight to be reduced. This is because torque in previous steps is lower, making it possible to use smaller gears.

Helicopters commonly employ planetary gear systems for the final transmission stage, featuring between 3 and 18 gears, with transmission ratios ranging from 5:1 to 7:1 (Krantz, 1996; White, 1989).

The implementation of split-path configurations with fixed shafts in the final transmission stage is a recent innovation that provides various advantages over traditional systems, particularly in terms of total weight reduction of the transmission.:

- It allows torque to be transmitted through various paths. This is a major advantage because when torque is split, the contact force between teeth is less and, hence, smaller, lighter gears can be used, therefore reducing the overall weight. Split torque however has the disadvantage that the torque must be shared equally between the paths. The problems associated with split torque are discussed in Section 4.
- It allows transmission path redundancy. Thus, if one transmission path fails during flight, operation can always be assured through another path. In many cases, consequently, gear transmissions are sized so that a single path can handle 100% of engine power.
- It achieves final-stage transmission ratios of around 10:1 to 14:1 (Krantz, 1996; White, 1989). This improvement over the 5:1 to 7:1 ratios for planetary gearboxes (Krantz, 1996; White, 1989) is reflected in a corresponding reduction in the weight of the transmission system.

Several patents for transmission systems that apply split torque have been filed by Sikorky Aircraft Corporation and McDonnell Douglas Helicopters (Gmirya & Kish, 2003; Gmirya, 2005; Craig et al., 1998) that refer either to complete or improved power transmission systems from the rotorcraft or aircraft engine to the rotor or propeller. Other studies that describe various aspects of split torque transmission systems, particularly their use in helicopter gearboxes (White, 1974, 1983, 1989, 1993, 1998), conclude that such gears have a number of advantages over traditional gear systems.

Below we describe two helicopter transmission systems that use multiple path gearboxes. The first is a helicopter gearbox used for laboratory tests of torque divided into two stages, and the second is a commercial helicopter three-stage gearbox that combines bevel, spur and helical gears.

2.1 Helicopter gearbox for laboratory testing

The gear transmission described below was used to perform numerous tests on the operation of split-torque transmissions (Krantz et al., 1992; Krantz, 1994, 1996; Krantz & Delgado, 1996), which can be considered a standard for aeronautical applications. The full assembly is depicted in Fig. 2.

The transmission is sized for input of 373 kW at a speed of 8780 rpm. As can be observed in Fig. 2, the transmission has two stages:

- First reduction stage. The first stage is a helical gear with a input pinion with 32 teeth and two output gears with 124 teeth each. The gear ratio is 3.875:1, resulting in an output speed of 2256.806 rpm. This is the stage where torque is split between the input pinion and the two output gears.
- High torque reduction stage. The output shaft is driven by a gear which is driven simultaneously by two spur pinions, each coaxial to the gear in the first reduction stage. The ratio between the gear teeth is 27/176, so the transmission ratio is 6.518:1, resulting in an output shaft speed of 347.6 rpm.

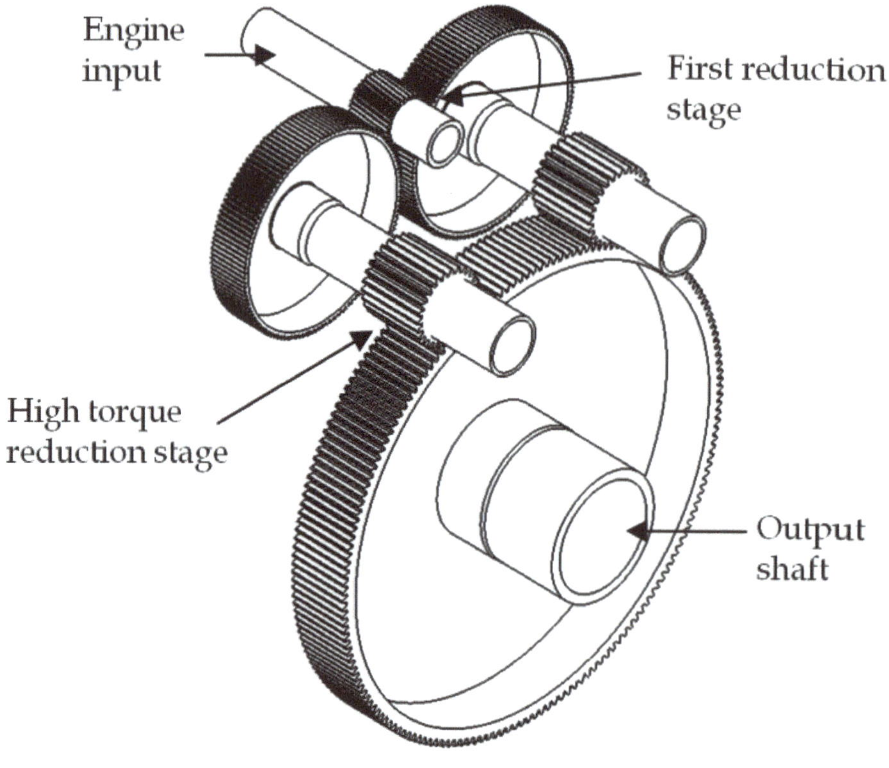

Fig. 2. Helicopter transmission for laboratory testing

This configuration results in torque of 9017.56 Nm. being transmitted through two paths.

2.2 Commercial helicopter transmission

This gear transmission, studied in depth by White (1998), is sized for two engines, each with a continuous rating of 1200 kW turbine at a nominal speed of 22976 rpm. The main rotor speed is 350 rpm for an overall speed reduction ratio of nearly 66:1. Fig. 3 depicts a plan view of the gear transmission and Fig. 4 is a three-dimensional view showing the gears.

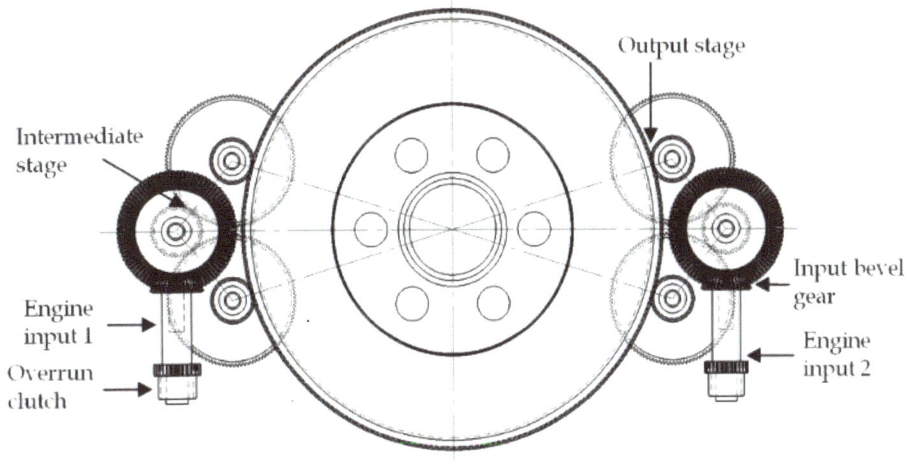

Fig. 3. Arrangement of gear trains between engines.

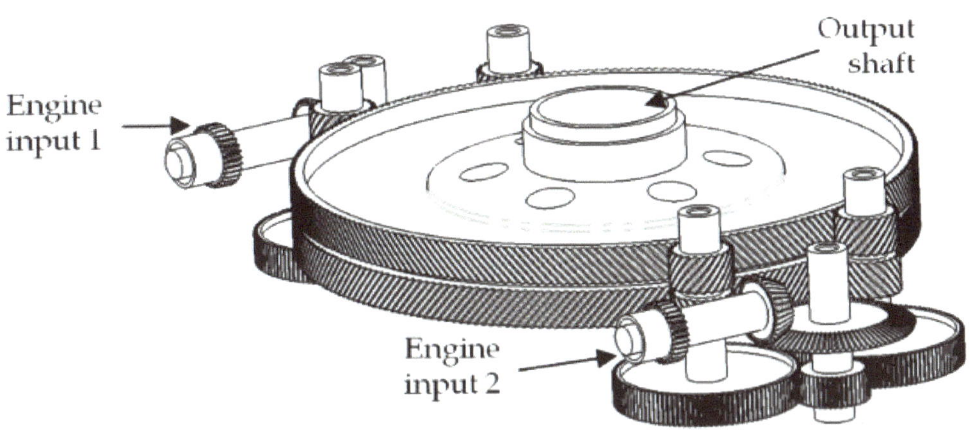

Fig. 4. Three-dimensional view of the gear train arrangement

Total transmission reduction is achieved by three gearing stages, clearly depicted in Fig. 3 and Fig. 4:

- Engine input. Engine torque is accepted by an overrun clutch, mounted with a bevel pinion. This bevel gear, with a between-teeth ratio of 34/84, produces a transmission ratio of 2.470:1. In this stage, the output velocity is 9299 rpm.
- Intermediate stage. Dual offset spur gears are driven by a single pinion. The between- teeth ratio of 41/108 produces a transmission ratio of 2.634:1, resulting in an output shaft speed of 3530 rpm. This meshing results in the first split in torque between the two intermediate gears.
- High-torque output stage. A double-helical gear is driven by a pinion coaxial with each intermediate stage gear. In this stage, the torque is split

again between the two helical pinions, with the result that the output shaft simultaneously receives torque from four pinions for each bevel gear. In this transmission it is very convenient to combine torque split with reduction, as greater torque is transmitted in each stage.

The between-teeth gear ratio is 23/232, so the transmission ratio is 10.087:1, resulting in an output shaft speed of 350 rpm.

This configuration uses double-helical gearing at the output stage to drive the output shaft. The helical pinions have opposing angles, which ensures equilibrium between the axial forces. When a double gear operates on the output shaft, the area of support is twice that of a simple gear. This causes a reduction in contact force, which in turn results in a reduction ratio that is twice that of the simple case, with the corresponding reduction in weight and mechanical load.

Overall, this constitutes a transmission ratio of 65.64:1, with the total torque in the output shaft exercised by each engine of 28818Nm, split between the four pinions that engage the output shaft crown. This calculation is based on estimating overall losses, with each input engine operating independently, of 12%.

One of the main problems in split torque transmission is ensuring equal torque split between the paths. To ensure correct torque split, a long, torsionally flexible shaft is used between the intermediate-stage spur gear and the output-stage helical pinions..

3. FEASIBLE GEOMETRIC CONFIGURATIONS

To ensure simultaneous meshing of four gears (Fig. 1), configuration must comply with certain geometric constraints. A number of studies describe the complexity of simultaneous gearing in split torque gearboxes (Kish, 1993a) and in planetary gear systems (Henriot, 1979, Parker & Lin, 2004); other studies approach the problem generically (Vilán-Vilán et al., 2010), describing possible solutions that ensure the simultaneous meshing of four gears.

For four gears to mesh perfectly, the teeth need to mesh simultaneously at the contact points. The curvilinear quadrilateral and the pitch difference are defined below in order to express the meshing condition. From now on we will use this nomenclature of our own devising -that is, curvilinear quadrilateral - to indicate the zone defined by portions of pitch circles in the meshing area (Fig. 5). The pitch difference is the sum of pitches in the input and output gears minus the sum of pitches in the idler gears at the curvilinear quadrilateral. For perfect engagement between the four gears, the pitch difference must coincide with a whole number of pitches.

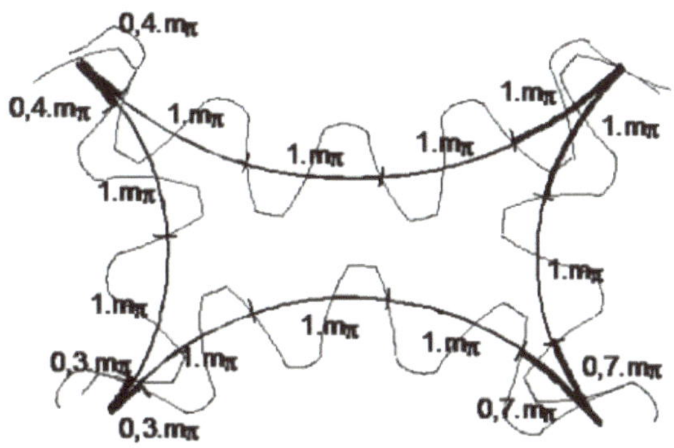

Fig. 5. General conditions for simultaneous meshing of four gears

A relationship is thus established between the position of the gears, as defined by the relative distance between centres, and the number of teeth in each of the gears. Below we explore two possible cases of over-constrained gears:

- CASE 1. Four outside gears.
- CASE 2. Three outside gears and one ring gear.

3.1 Case 1. Four outside gears

For a gearbox with the geometry illustrated in Fig. 6, it is possible to locate the different positions that will produce suitable meshing between gears, in function of the number of teeth in each gear, by defining the value of the angles α, β, δ and Γ.

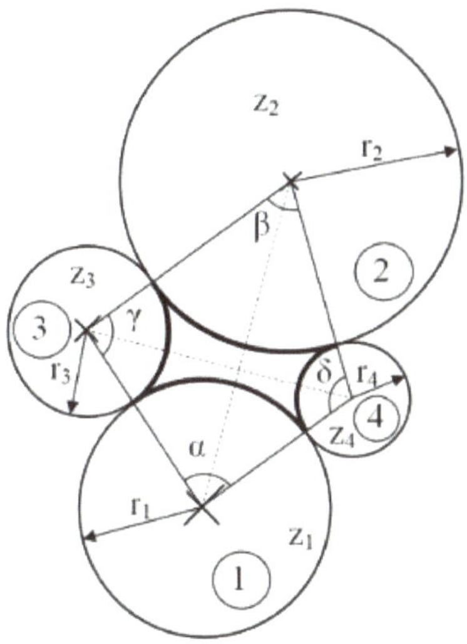

Fig. 6. Nomenclature for the four-gear case

The condition described in the previous section can be mathematically expressed as follows (see Nomenclature):

$$r_1 \cdot \alpha + r_2 \cdot \beta - r_3 \cdot \Gamma - r_4 \cdot \delta = n \cdot (m\varpi)\ n \in Z \quad (1)$$

where n is the pitch difference in the curvilinear quadrilateral. As previously mentioned, n must be a whole number to ensure suitable meshing between gears.

We thus obtain an equation with four unknowns (α, β, Γ, δ). The three remaining relationships can be obtained from the quadrilateral that joins the centres of the pitch circles (this quadrilateral will be denoted the rectilinear quadrilateral). Finally, we come to a transcendental equation (2) from which α can be obtained according to the number of teeth in the gears.

$$e_1 - f \cdot \cos\left[A_1 \cdot \alpha + B_1 \cdot \arccos\left(\frac{c_1 - a_1 + b_1 \cdot \cos\alpha}{d_1}\right) + C_1\right] = g_1 - \\ -h_1 \cdot \cos\left[A'_1 \cdot \alpha + B'_1 \cdot \arccos\left(\frac{c_1 - a_1 + b_1 . \cos\alpha}{d_1}\right) + C'_1\right] \quad (2)$$

Once the angle α has been determined, we can calculate:

$$\beta = \arccos\left(\frac{c_1 - a_1 + b_1 \cdot \cos\alpha}{d_1}\right) \quad (3)$$

$$\gamma = A_1 \cdot \alpha + B_1 \cdot \beta + C_1 \quad (4)$$

$$\delta = A'_1 \cdot \alpha + B'_1 \cdot \beta + C'_1 \quad (5)$$

a_1, b_1, c_1, d_1, e_1, f_1, g_1, h_1, A_1, B_1, C_1, A_1', B_1' and C_1' are numerical relationships among the teeth number from each wheel that must mesh simultaneously. The value of each coefficient is listed in the Appendix.

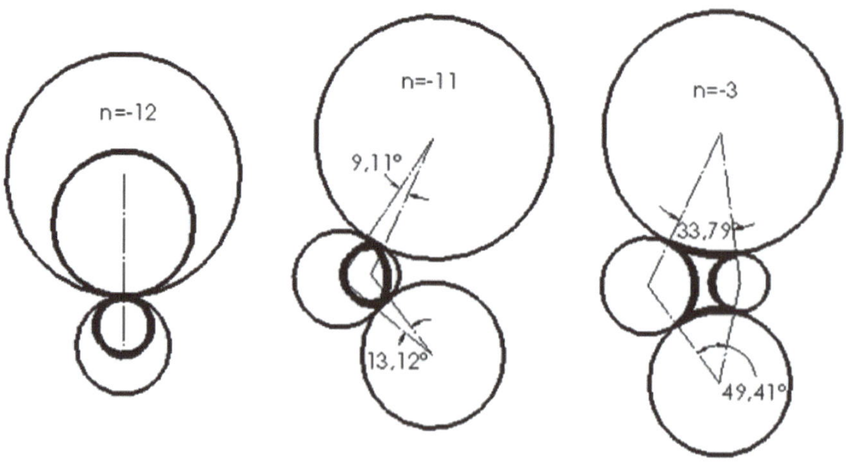

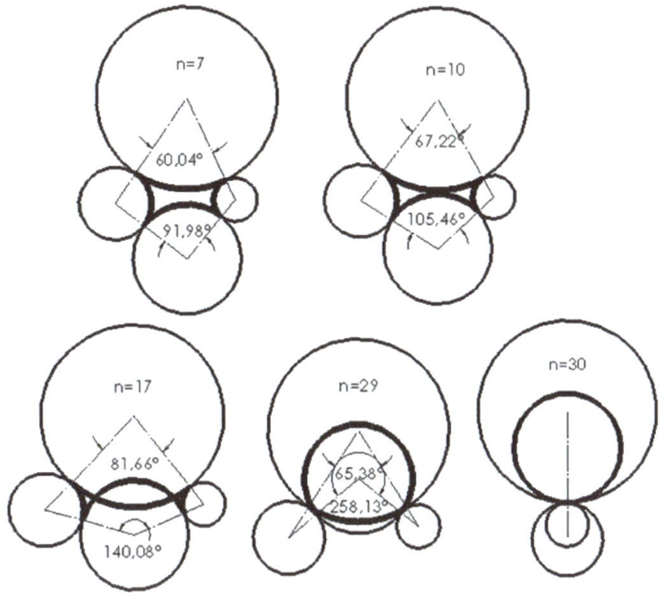

Fig. 7. Feasible solutions for given numbers of teeth

The transcendental equation for obtaining α has several solutions, all representing possible assemblies for the starting gears. For example, for four-gear meshing with the next teeth numbers: z_1=30, z_2=50, z_3=20 and z_4=12 (see Nomenclature), all the possible solutions for the gear can be encountered. In this case solutions are n = -12, -11, -3, -2, -1, 0, 1, 2, 3, 4, 7, 29 and 30, where n is the pitch difference between the two sides of the curvilinear quadrilateral (a whole number that ensures suitable meshing). Fig. 7 shows some of the possible meshing solutions.

3.2 Case 2. Three outside gears and one ring gear

In this case torque is transmitted from a driving pinion (1) to a ring gear (2) through two idler pinions (3) and (4). Two solutions are available depending on the geometry of the rectilinear quadrilateral that joins the centres of the pitch circles, either crossed (Fig. 8 (a)) or non-crossed (Fig. 8 (b)). The starting equation is different for each of these cases.

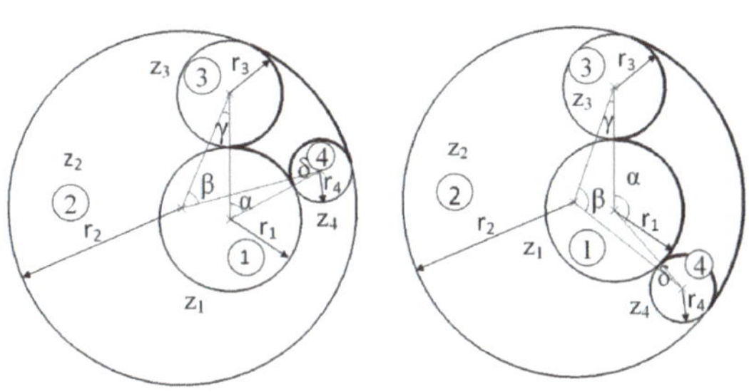

(a) (b)

Fig. 8. Solutions for three outside gears and one ring gear: (a) crossed quadrilateral (b) non- crossed quadrilateral

For the crossed quadrilateral configuration, the starting equation is (see Nomenclature):

$$z_1 \cdot \alpha + z_2 \cdot \beta + z_3 \cdot \Gamma - z_4 \cdot \delta = \varpi \cdot (2 \cdot n + z_3 + z_4) \quad (6)$$

Finally, we come to the same transcendental equation (2), where the coefficients are a_2, b_2, c_2, d_2, e_2, f_2, g_2, h_2, A_2, B_2, C_2, A_2', B_2' and C_2', whose values are listed in the Appendix.

For the non-crossed quadrilateral configuration, the starting equation is:

$$z_1 \cdot \alpha - z_2 \cdot \beta - z_3 \cdot \Gamma - z_4 \cdot \delta = \varpi \cdot (2 \cdot n - 2 \cdot z_2 + z_3 + z_4) \quad (7)$$

Finally, we come to the same transcendental equation (2), where the coefficients become a_2, b_2, c_2, d_2, e_2, f_2, g_2, h_2, A_3, B_3, C_3, A_3', B_3' and C_3', whose values are listed in the Appendix.

3.3 A particular case: Outside meshing with equal intermediate pinions

A common split torque gear assembly is one with two equally sized idler pinions (Fig. 9).

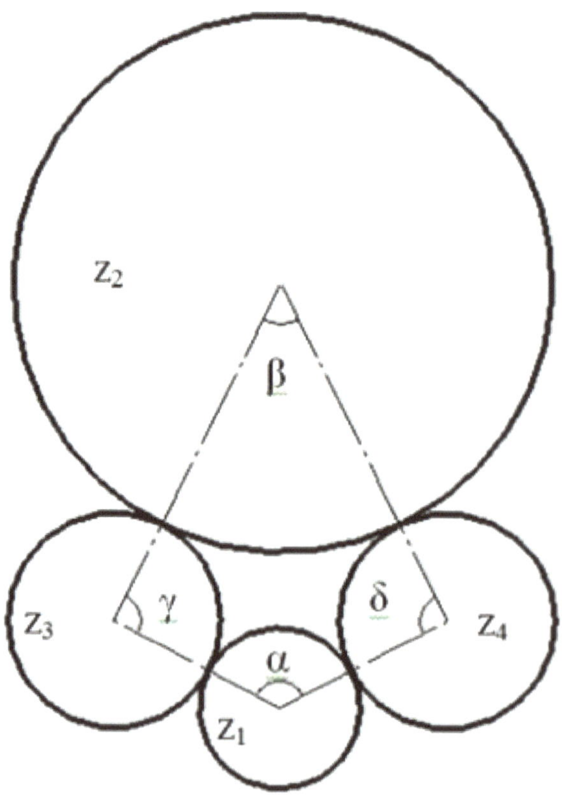

Fig. 9. Idler pinions in an outside gear

The solution is obtained by particularizing the general solution for four outside wheels and imposing the condition $z_3 = z_4$, or $\Gamma = \delta$. The following equations are defined for the curvilinear quadrilateral:

$$z1 \cdot \alpha + z2 \cdot \beta - 2 \cdot z3 \cdot \Gamma = n \cdot 2\varpi \quad (8)$$

$$\alpha + \beta + 2 \cdot \Gamma = 2\varpi \quad (9)$$

$$(z_1 + z_3) \cdot \sin\left(\frac{\alpha}{2}\right) = (z_2 + z_3) \cdot \sin\left(\frac{\beta}{2}\right) \quad (10)$$

Resolving the system, the following transcendental function in α is obtained:

$$\frac{z_1 + z_3}{z_2 + z_3} \cdot \sin\left(\frac{\alpha}{2}\right) = \sin\left[\frac{z_3 + n}{z_2 + z_3} \cdot \pi - \frac{z_1 + z_3}{z_2 + z_3} \cdot \left(\frac{\alpha}{2}\right)\right] \quad (11)$$

The solutions for the other angles can now be obtained:

$$\beta = 2 \cdot \arcsin\left[\frac{z_1 + z_3}{z_2 + z_3} \cdot \sin\left(\frac{\alpha}{2}\right)\right] \quad (12)$$

$$\gamma = \pi - \frac{\alpha}{2} - \frac{\beta}{2} \quad (13)$$

4. LOAD SHARING

The main problem in the design of split torque gearboxes is to ensure that torque is equally split between different paths. Small deviations in machining can result in one of the paths with 100% of torque and the other path operating entirely freely (Kish & Webb, 1992). This situation causes excessive wear in one of the paths and renders the torque split system ineffective.

Below we describe approaches to ensuring equal torque split between different paths in split torque gear arrangements. The main types are:

1. Geared differential. This differential mechanism, frequently used in the automotive sector, delivers equal torques to the drive gears of a vehicle.
2. Pivoted systems. These use a semi-floating pinion constrained both to pivot normal to the line of action and to seek a position where tooth loads are equal.
3. Quill shafts. A torsion divider with a separate gear and pinion, each supported on its own bearings, are connected through the quill shaft, which allows torsional flexibility.

The use of any of these systems to ensure correct torque split makes the gearbox heavier and assembly and maintenance more complex, which is why a number of authors do not support the use of systems that ensure torque split. Described below are the main systems that ensure correct torque split and discussed also are the proposals of authors who advocate for not using special systems.

4.1 Geared differential

One way to ensure correct torque split between two branches is to use a differential system. The great disadvantage of this system, however, is that resistive torque lost in one branch leads to loss of the full engine torque. Different differential mechanisms can be used, with assemblies very similar to those in vehicles or to the system depicted in Fig. 10. Assembled at the entry point to the gearbox is an input planetary system that acts as a differential that ensures load sharing. This transmission accepts power from three input engines, each of which has a differential system that ensures balanced torque splitting. Power is input from each engine to the sun gear of the differential planetary system. The carrier is the output to a bevel pinion that drives one torque splitting branch while the ring gear drives the other torque splitting branch. As the carrier and the ring gear rotate in opposite directions, the bevel pinions are arranged on opposite sides to ensure correct rotation direction. Each output bevel gear drives one pinion which then combines power into the output gear.

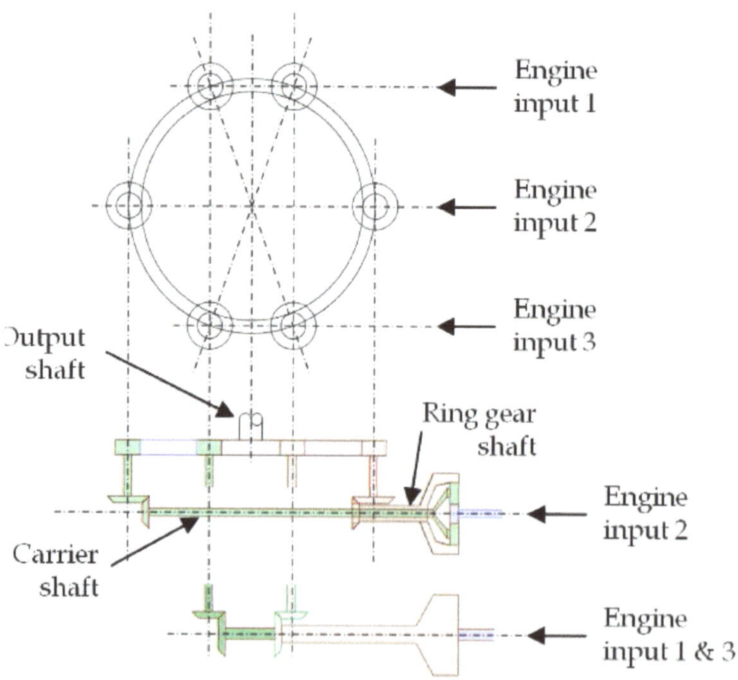

Fig. 10. Schematic view of split torque main transmission

4.2 Pivoted systems

One type of pivoted systems is described in detail in a patent (Gmirya, 2005) for split torque reduction applied to an aerial vehicle propulsion system (Fig. 11). "*The input pinion (64) engages with gears (66) and (68). The input pinion is defined along the gear shaft A_G, the first gear* (66) defines a first gear rotation shaft A_1 and the second gear (68) defines a second gear rotation shaft A_2.

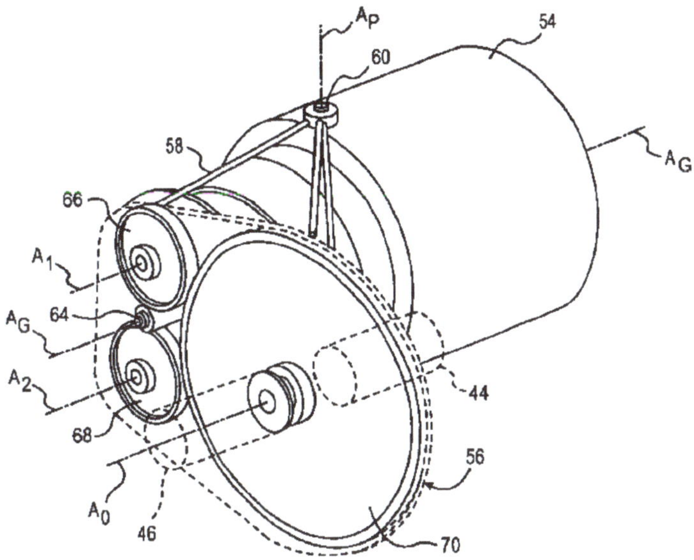

Fig. 11. Perspective view of the split torque gearbox with pivoted engine support (Gmirya, 2005)

The axes A_G, A_1 and A_2 are preferably located transversally to the pivot axis A_p. The first gear (66) and the second gear (68) engage an output gear (70). The output gear (70) defines an output rotation shaft A_0 and is rotationally connected to the translational driveshaft (44) and the rotor driveshaft (46) to power, respectively, the translational propulsion system and the rotor system".

The assembly transmits torque from the pinion (64), which operates at very high revolutions, to the output shaft (44 -46) via two paths. The pivot system works as follows: since the input pinion (64) meshes with two gears (66) and (68), the pivoted engine arrangement permits the input pinion (64) to float until gear loads between the input gear (64), the first gear (66) and the second gear (68) are balanced. Irrespective of gear teeth errors or gearbox shaft misalignments, the input pinion will float and split torque between the two gears.

4.3 Quill shafts

Below we describe assemblies used in systems that allow some torsion in the split torque shafts (Smirnov, 1990; Cocking, 1986) in order to minimize the difference in torque split between paths. These systems achieve their goal in several ways:

- Conventional systems (Kish, 1993a) assemble intermediate shafts with some torsional flexibility so that angular deviation produced between the input and output pinions adjusts the torque transmitted via the two paths.
- Other systems are based on elastomeric elements in the shaft (Isabelle et al., 1992, Kish & Webb, 1992) or materials with a lower elastic modulus (Southcott, 1999), such as an idler pinion constructed of nylon or a similar material (Southcott, 1999). This solution is not explored here because the torque transmitted is reduced.
- Yet other systems operate on the basis of spring elements (Gmirya & Vinayak, 2004).

The use of such elements in the design adds weight and makes both initial assembly and maintenance more complex, thereby losing to some degree the advantages of split torque gearboxes. Described below are the most representative types of quill shaft.

4.3.1 Conventional quill shafts

Conventional quill shaft design involves assembly on three different shafts (Fig. 12). The input shaft (1) is assembled with two separate bearings (2) and the input gear (3).The output shaft (4) is assembled with two separate bearings (5) and, in this case, two output pinions (6). The quill shaft is a third shaft (7) that connects the other two shafts. Due to a lower polar moment of inertia, it admits torsional flexibility, resulting in a small angular deviation between the input and output shafts. The value of the angular deviation is proportional to the transmitted torque; thus, if one path transmits more torque than the other, the angular deviation is greater, allowing the shaft that transmits less torque to increase its load.

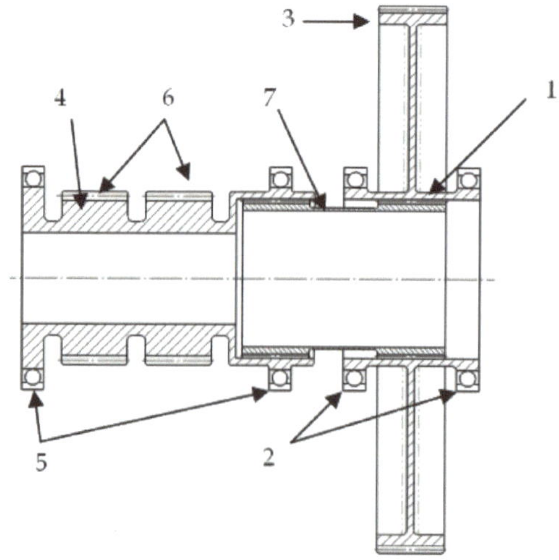

Fig. 12. Conventional assembly of a quill shaft

4.3.2 Quill shafts based on elastomeric elements

Elastomeric elements are frequently used in quill shafts given their low elastic modulus. For example, one system (Isabelle et al., 1992), based on using elastomers (Fig. 13), consists of *"an annular cylindrical elastomeric bearing (14) and several rectangular elastomeric bearing pads (16). The elastomeric bearing (14) and bearing pads (16) have one or more layers (60); each layer (60) has an elastomer (62) with a metal backing strip (64) secured by conventional means such as vulcanization, bonding or lamination".*

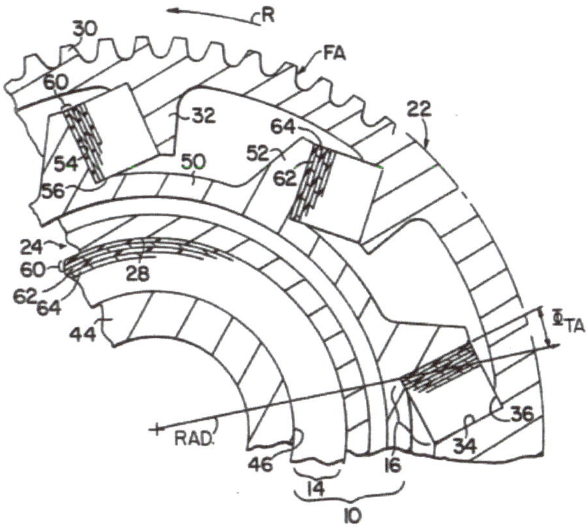

Fig. 13. Elastomeric load sharing device (Isabelle et al., 1991)

The annular cylindrical elastomeric bearing (14) absorbs possible misalignments between shafts resulting from defects in assembly. The rectangular elastomeric bearing pads (16) are responsible for providing torsional flexibility to the shafts of the possible gear paths in order to ensure equal torque transmission.

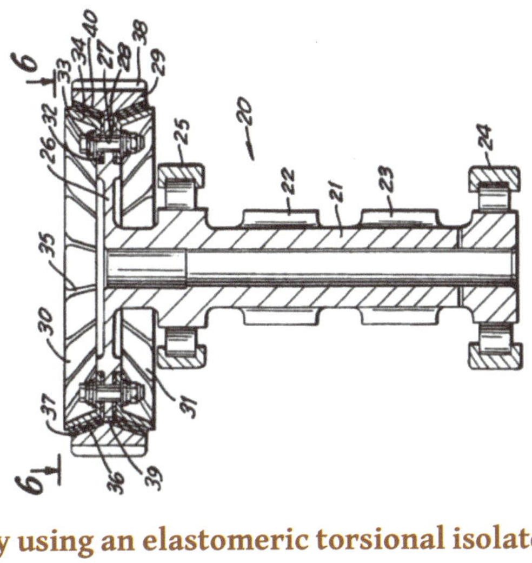

Fig. 14. Gear assembly using an elastomeric torsional isolator (Kish & Webb, 1992)

Another elastomer-based system (Kish & Webb, 1992) (Fig. 14) consists of an assembly with *"a central shaft (21) and a pair of bull pinions (22) and (23). The shaft (21) is supported by the bearings (24) and (25); a gear flange (26) at the end of the shaft has bolt holes (27) and teeth (28) on the outer circumference. A spur gear (29) is held to the flange (26) using upper and lower rims (30) and (31), consisting of flat circular disks (32) with bolt holes (33) and an angled outer wall (34). Gussets (35) between the wall and the disk increase rim stiffness to minimize deflection. One or more elastomer layers (36), bonded to the outer surface (37) of the wall (34), act as an elastomeric torsional isolator"*.

Helicopters often utilize planetary gear systems for the final transmission stage, including 3 to 18 gears, with transmission ratios between 5:1 and 7:1 (Krantz, 1996; White, 1989).

The adoption of split-path designs with fixed shafts in the final transmission stage is a new advancement that offers numerous benefits over conventional systems, especially regarding overall weight reduction of the transmission.

4.3.3 Quill shafts based on spring elements

Some authors propose the use of metallic elements to achieve the same effect as the quill shaft. One such system (Gmirya & Vinayak, 2004) (Fig. 15) is based on achieving this effect by using *"at least one spring element (30) placed between and structurally connecting the gear shaft (32) and the outer ring of gear teeth (34). The gear shaft (32) has flange elements (36) that project radially outboard of the shaft. The ring of gear teeth (34), similarly, has a flange element (38) that projects radially inward towards the gear shaft"*. In this case, a pair of spring elements (30) is arranged on each side of the gear teeth flange element (38)".

This assembly is designed in such a way that the spring elements absorb torsional deflection between the gears, thereby ensuring proportional torque split between paths.

4.4 No use of special systems

Split torque gearboxes are used in order to reduce the weight of the gear system, so the simplest option is assembly without special systems for regulating torque split. Several authors support this option, for example, Kish (1993a, 1993b), who concluded from tests that acceptable values can be achieved without using any special torque split system, simply by ensuring manufacturing according to strict tolerances and correct assembly. Krantz (1996) proposed the use of the clocking angle as a design parameter to achieve adequate torque split between paths. This author has studied the effects of gearshaft twisting and bending, and also tooth bending, Hertzian deformations within bearings and the impact of bearing support movement on load sharing.

Krantz (1996) defined the clocking angle as β and described the assembly prepared for measurement (Fig. 16): *"The output gear is fixed from rotating and a nominal counter-clockwise torque is applied to the input pinion so that the gear teeth come into contact. When all the gear teeth for both power paths come into contact, then the clocking angle β is, by definition, equal to zero. If the teeth of one power path are not in contact, then the clocking angle β is equal to the angle that the first- stage gear would have to be rotated relative to the second-stage pinion to bring all the teeth into contact".*

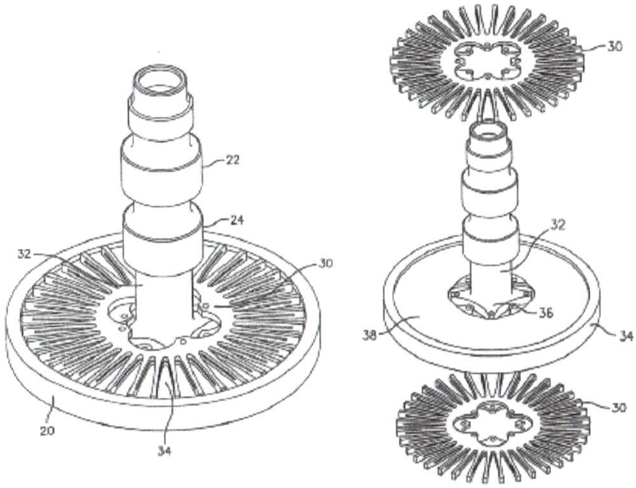

Fig. 15. Load sharing gear in combination with a double helical pinion (Gmirya & Vinayak, 2004)

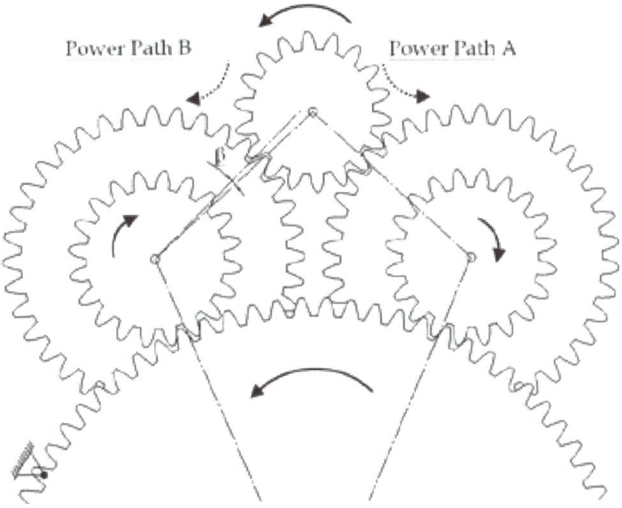

Fig. 16. Assembly for measurement of the clocking angle

The experiments indicate that appropriate load sharing (47 percent/53 percent) may be attained solely by considering the clocking angle and guaranteeing precise machining and assembly. This investigation of the clocking angle has been further examined by other writers (Parker & Lin, 2004), who analyzed the sequencing of contact among several planetary gears.

5. CONCISE SUMMARY

Selecting the appropriate assembly for aircraft power transmission is crucial for achieving weight reduction. While advancements in mechanical components can facilitate weight reduction in gear systems, their impact is far inferior to that of selecting the appropriate gear assembly.

Planetary gear or split torque systems are commonly employed in helicopter gear gearboxes. The primary benefit of split torque systems is the reduction in weight achieved through uniform torque transmission and gear ratios. This advantage is mostly founded on the following reasoning.:

- In the final transmission stage where the greatest torque is achieved, the use of several paths for the transmission ratio means that, given equal torque and stress levels in the teeth, the ratio between output torque/weight will be better in torque split gear systems than in planetary gear systems.
- In the final transmission stage, transmission ratios of around 5:1 or 7:1 are achieved by planetary gearboxes used with a single stage, compared to 10:1 or 14:1 for split torque gears used in the final stage.
- The possibility of achieving higher transmission ratios in split torque gearboxes makes it possible to use a smaller number of gear stages, resulting in lighter gear systems.
- Split torque gearboxes need fewer gears and bearings that planetary gearboxes, which means lower transmission losses.
- A key factor for aircraft use is that split torque gearboxes improve reliability by using multiple power paths; thus, if one path fails, operation is always assured through another path.
- The main disadvantage of the split torque gearboxes is when torque split between the possible paths is uneven; however, several solutions are available to ensure correct torque split.

These arguments would indicate the advisability of using this type of transmission in aircraft gear systems.

6. NOMENCLATURE

m	gear module
r_i	radius of the pitch circle of wheel i
z_i	number of teeth in wheel i

α	angle formed by the lines between centres, between wheels 3 4 1
β	angle formed by the lines between centres, between wheels 3 2 4
Γ	angle formed by the lines between centres, between wheels 2 3 1
δ	angle formed by the lines between centres, between wheels 1 4 2
n	pitch difference between the two sides of the curvilinear quadrilateral

7. APPENDIX

The numerical relationships among the teeth number used in the text are listed below. C_1, C_1', C_2, C_2', C_3 and C_3' are functionS of n, a whole number which represents the pitch difference in the curvilinear quadrilateral.

$$a_1 = (z_1 + z_3)^2 + (z_1 + z_4)^2 \tag{14}$$

$$b_1 = 2 \cdot (z_1 + z_3) \cdot (z_1 + z_4) \tag{15}$$

$$c_1 = (z_2 + z_3)^2 + (z_2 + z_4)^2 \tag{16}$$

$$d_1 = 2 \cdot (z_2 + z_3) \cdot (z_2 + z_4) \tag{17}$$

$$e_1 = (z_1 + z_3)^2 + (z_2 + z_3)^2 \tag{18}$$

$$f_1 = 2 \cdot (z_1 + z_3) \cdot (z_2 + z_3) \tag{19}$$

$$g_1 = (z_1 + z_4)^2 + (z_2 + z_4)^2 \tag{20}$$

$$h_1 = 2 \cdot (z_1 + z_4) \cdot (z_2 + z_4) \tag{21}$$

$$A_1 = \frac{z_1 + z_4}{z_3 - z_4} \tag{22}$$

$$B_1 = \frac{z_2 + z_4}{z_3 - z_4} \tag{23}$$

$$C_1 = 2\pi \cdot \frac{z_4 + n}{z_4 - z_3} \tag{24}$$

$$A_1' = \frac{z_1 + z_3}{z_4 - z_3} \tag{25}$$

$$B_1' = \frac{z_2 + z_3}{z_4 - z_3} \tag{26}$$

$$C_1' = 2\pi \cdot \frac{z_3 + n}{z_3 - z_4} \tag{27}$$

$$a_2 = (z_1 + z_3)^2 + (z_1 + z_4)^2 \tag{28}$$

$$b_2 = 2 \cdot (z_1 + z_3) \cdot (z_1 + z_4) \tag{29}$$

$$c_2 = (z_2 - z_3)^2 + (z_2 - z_4)^2 \tag{30}$$

$$d_2 = 2 \cdot (z_2 - z_3) \cdot (z_2 - z_4) \tag{31}$$

$$e_2 = (z_1 + z_3)^2 + (z_2 - z_3)^2 \tag{32}$$

$$f_2 = 2 \cdot (z_1 + z_3) \cdot (z_2 - z_3) \tag{33}$$

$$g_2 = (z_1 + z_4)^2 + (z_2 - z_4)^2 \tag{34}$$

$$h_2 = 2 \cdot (z_1 + z_4) \cdot (z_2 - z_4) \tag{35}$$

$$A_2 = \frac{z_1 + z_4}{z_4 - z_3} \tag{36}$$

$$B_2 = \frac{z_2 - z_4}{z_4 - z_3} \tag{37}$$

$$C_2 = \pi \cdot \frac{2 \cdot n + z_3 + z_4}{z_3 - z_4} \tag{38}$$

$$A_2' = \frac{z_1 + z_3}{z_4 - z_3} \tag{39}$$

$$B_2' = \frac{z_2 - z_3}{z_4 - z_3} \tag{40}$$

$$C_2' = \pi \cdot \frac{2 \cdot n + z_3 + z_4}{z_3 - z_4} \tag{41}$$

$$A_3 = \frac{z_1 + z_4}{z_3 - z_4} \tag{42}$$

$$B_3 = \frac{z_4 - z_2}{z_3 - z_4} \tag{43}$$

$$C_3 = \pi \cdot \frac{2 \cdot n - 2 \cdot z_2 + z_3 + 3 \cdot z_4}{z_4 - z_3} \tag{44}$$

$$A_3' = \frac{z_1 + z_3}{z_4 - z_3} \tag{45}$$

$$B_3' = \frac{z_3 - z_2}{z_4 - z_3} \tag{46}$$

$$C_3' = \pi \cdot \frac{2 \cdot n - 2 \cdot z_2 + 3 \cdot z_3 + z_4}{z_3 - z_4} \tag{47}$$

REFERENCES

1. Cocking, H. (1986). The Design of an Advanced Engineering Gearbox. *Vertica.* Vol 10, No. 2, Westland Helicopters and Hovercraft PLC, Yeovil, England, pp. 213-215
2. Craig, G. A.; Heath, G. F. & Sheth, V. J. (1998). Split Torque Proprotor Transmission, McDonnell Douglas Helicopter Co., *U.S. Patent* Number 5,823,470
3. Gmirya, Y.; Kish, J.G. (2003). Split-Torque Face Gear Transmission, Sikorsky Aircraft Corporation, *U.S. Patent* Number 6,612,195
4. Gmirya, Y. & Vinayak, H. (2004). Load Sharing Gear for High Torque, Split-Path Transmissions. Sikorsky Aircraft Corporation. *International Patent PCT.* WO 2004/094093
5. Gmirya, Y. (2005). Split Torque Gearbox With Pivoted Engine Support, Sikorsky Aircraft Corp., *U.S. Patent Number* 6,883,750
6. Henriot G. (1979). *Traité théorique et pratique des engrenages (I)*, Ed. Dunod, 6th Edition, ISBN 2-04-015607-0, Paris, France, pp. 587-662
7. Isabelle, C.J; Kish & J.G, Stone, R.A. (1992). Elastomeric Load Sharing Device. United Technologies Corporation. *U.S. Patent* Number 5,113,713
8. Kish, J.G. & Webb, L.G. (1992). Elastomeric Torsional Isolator. United Technologies Corporation. *U.S. Patent* Number 5,117,704
9. Kish, J.G. (1993a). *Sikorsky Aircraft Advanced Rotorcraft Transmission (ART) Program* – Final Report. NASA CR-191079, NASA Lewis Research Center, Cleveland, OH
10. Kish, J.G. (1993b). Comanche Drive System. *Proceedings of the Rotary Wing Propulsion Specialists Meeting*, American Helicopter Society, Williamsburg, VA, pp. 7
11. Krantz, T.L.; Rashidi, M. & Kish, J.G. (1992). *Split Torque Transmission Load Sharing*, in: Technical Memorandum 105,884, NASA Lewis Research Center, Cleveland, Ohio. Army Research Laboratory Technical Report 92-C-030
12. Krantz, T.L. (1994). *Dynamics of a Split Torque Helicopter Transmission*, in: Technical Memorandum 106,410, NASA Lewis Research Center, Cleveland, Ohio. Army Research Laboratory Technical Report ARL-TR-291

13. Krantz, T.L. (1996). *A Method to Analyze and Optimize the Load Sharing of Split Path Transmissions*, in: Technical Memorandum 107,201, NASA Lewis Research Center, Cleveland, Ohio. Army Research Laboratory Technical Report ARL-TR-1066

14. Krantz, T.L. & Delgado, I.R. (1996). *Experimental Study of Split-Path Transmission Load Sharing*, in: Technical Memorandum 107,202, NASA Lewis Research Center, Cleveland, Ohio. Army Research Laboratory Technical Report ARL-TR-1067

15. Parker, R.G., Lin, J. (2004). Mesh phasing relationships in planetary and epicyclic gears, *Journal of Mechanical Design*. Vol. 126, Issue 2, pp. 365-370, ISSN 1050-0472, eISSN 1528-9001

16. Smirnov, G. (1990). *Multiple-Power-Path Nonplanetary Main Gearbox of the Mi-26 Heavy-Lift Transport Helicopter*, in: Vertiflite, Mil Design Bureau, Moscow, Vol. 36, pp. 20-23, ISSN 0042-4455

17. Southcott, B. (1999). Gear Arrangement. Adelaide Gear Pty. Ltd. *U.S. Patent* Number 5,896,775

18. Vilán-Vilán, J.A.; Segade-Robleda, A.; López-Lago, M & Casarejos-Ruiz, E. (2010), Feasible Geometrical Configurations for Split Torque Gearboxes With Idler Pinions, *Journal of Mechanical Design*, Vol. 132, Issue 12, pp. 121,011, 8 pages, ISSN 1050-0472, eISSN 1528-9001

19. White, G. (1974). New Family of High-Ratio Reduction Gear With Multiple Drive Paths, *Proceedings of the Institution of Mechanical Engineers.*, Vol. 188, pp. 281-288, ISSN 020- 3483

20. White, G. (1983). A 2400 kW Lightweight Helicopter Transmission Whit Split Torque Gear Trains, ASME Paper 84 Det-91

21. White, G. (1989). Split-Torque Helicopter Transmission With Widely Separated Engines. *Proceedings of the Institution of Mechanical Engineers.* Vol 203, Number G1, pp. 53-65, ISSN 0954-4100

22. White, G. (1993). *3600 HP Split-Torque Helicopter Transmission*, in: Mechanical Systems Technology Branch Research Summary, 1985-1992, NASA Technical Memorandum 106,329

23. White, G. (1998). Design study of a split-torque helicopter transmission, in: *Proceedings of the Institution of Mechanical Engineers, Part G: Journal of Aerospace Engineering*. Vol. 212, No. 2, pp. 117-123, ISSN 0954-4100, eISSN 2041-3025.

4 Chapter

ON THE MODELLING OF SPUR AND HELICAL GEAR DYNAMIC BEHAVIOUR

1. INTRODUCTION: AN OVERVIEW

This chapter aims to introduce the foundations of spur and helical gear dynamics. The main equations of motion for single-stage spur or helical gears are described using three-dimensional lumped models and a thin-slice technique for mesh elasticity. Specific instances, including the quintessential one degree-of-freedom model, are analyzed to elucidate and exemplify the fundamental phenomena. The concept of transmission errors is examined, and several practical considerations are derived. Deliberate emphasis is focused on analytical conclusions that, while approximate, offer a clearer comprehension of gear dynamics than comprehensive numerical simulations. Several expansions to continuous models are offered.

2. NOMENCLATURE

b : face width

Cm ,Cr : pinion, gear torque

$e(M)$, $E_{MAX}(t)$: composite normal deviation at M , maximum of $e(M)$ at time t .

E,E^* : actual and normalized depth of modification at tooth tips

$\mathbf{F_e}(t) = \int_{L(t,\mathbf{q})} k(M)\delta e(M)\mathbf{V}(\mathbf{M})dM$: time-varying, possibly non-linear forcing term associated with tooth shape modifications and errors

$\mathbf{G} = \mathbf{V}\,\mathbf{V}^T$

$H(x)$: unit Heaviside step function ($H(x) = 1$ *if* $x > 1$; $H(x) = 0$ *otherwise*)

k_m , $k(t,\mathbf{q})$: average and time-varying, non-linear mesh stiffness

$k(t) = km(1 + \alpha\Phi(t))$, linear time-varying mesh stiffness

k_0 : mesh stiffness per unit of contact length

$k(M)$, mesh stiffness per unit of contact length at M $k\Phi p$: modal stiffness associated with (ωp , $\mathbf{\Phi p}$)

$\mathbf{KG}(t) = \int_{L(t,\mathbf{q})} k(M)\mathbf{V}(\mathbf{M})\,\mathbf{V}(\mathbf{M})T\,dM$: time-varying, possibly non-linear gear mesh stiffness matrix

$L\,(t,\mathbf{q})$: time-varying, possibly non-linear, contact length

$Lm = \varepsilon_\alpha \dfrac{b}{\cos \beta b}$: average contact length

$\tilde{m} = \dfrac{I_{02} I_{01}}{Rb_1^2 I_{02} + Rb_2^2 I_{01}}$: : equivalent mass

$m\Phi p$: modal mass associated with (ωp , $\mathbf{\Phi p}$)

$\mathbf{n}_1$: outward unit normal vector with respect to pinion flanks

NLTE : no-load transmission error

01 ,*02* : pinion, gear centre

Pba : apparent base pitch

*Rb*1 , *Rb*2 : base radius of pinion, of gear

(**s, t, z**) : coordinate system attached to the pinion-gear centre line, see Figs. 1&2

Tm : mesh period.

TE , *TES* : transmission error, quasi-static transmission error under load

V (M) , $\mathbf{V}_0$, structural vector, averaged structural vector

W : projection vector for the expression of transmission error, see (44-1)

(**X, Y, z**) : coordinate system associated with the base plane, see Fig. 2

$\mathbf{X} = \mathbf{K}^{-1}\,\mathbf{F}$: static solution with averaged mesh stiffness (constant)

XS , **XD** **X** : quasi-static, dynamic and total (elastic) displacement vector (time-dependent)

*Z*1 , *Z*2 : tooth number on pinion, on gear

α : small parameter representative of mesh stiffness variations, see (30)

αp : apparent pressure angle

β_b : base helix angle

$\delta_{ms} = \dfrac{F_S}{k_{...}} = \mathbf{V}^T\mathbf{X}_0$: static mesh deflection with average mesh stiffness

$\delta e\,(M) = E_{MAX}\,(t) - e\,(M)$: instantaneous initial equivalent normal gap at *M*

$\Delta(\,M\,)$: mesh deflection at point *M*

ε_α : theoretical profile contact ratio

ε_β : overlap contact ratio

$\Lambda = \dfrac{Cm}{Rb_1 b\, k_0}$, deflection of reference

Φp : p^{th} eigenvector of the system with constant averaged stiffness matrix

Σ_p : damping factor associated with the pth eigenfrequency

Γ : dimensionless extent of profile modification (measured on base plane)

$\tau = \frac{t}{T_m}$, , dimensionless time

ω_p : p^{th} eigenfrequency of the system with constant averaged stiffness matrix

$\varpi_{pn} = \frac{\omega_p}{n\Omega_1}$, dimensionless eigenfrequency

Ω1 ,Ω2 : pinion, gear angular velocity

A : vector **A** completed by zeros to the total system dimension

$(\bullet)^* = \frac{(\bullet)}{\delta_m}$,normalized displacement with respect to the average static mesh deflection

$(\tilde{\bullet}) = \frac{(\bullet)}{k_m}$, normalized stiffness with respect to the average mesh stiffness

3.THREE-DIMENSIONALLUMPEDPARAMETERMODELSFORSPUR AND HELICAL GEARS

3.1 Rigid-body rotations – State of reference

The speed ratio for a pinion-gear combination with ideal involute spur or helical teeth remains constant, provided that deflections are disregarded. Nonetheless, shape inaccuracies are inherent in all gears due to machining mistakes, thermal deformations post-heat treatment, and similar factors. Consequently, certain alterations to the ideal tooth flanks, such as profile and/or lead adjustments and topping, are frequently required to mitigate elastic or thermal distortions, deflections, misalignments, and positioning inaccuracies. From a simulation perspective, rigid-body rotations will serve as the reference points around which minor elastic displacements can be placed. It is essential to characterize rigid-body motion transfer between a pinion and a gear with tooth inaccuracies and/or geometric alterations. In the subsequent discussion, e(M) denotes the equivalent normal deviation at the prospective contact point M (the aggregate of deviations on the pinion and the gear) and is typically positive when there is an excess of material, while it is negative when material is deficient relative to the ideal geometry. Under rigid-body conditions, or in the absence of load, contacts will occur at the points on the contact lines where e(M) is maximized, resulting in a modification of the velocity transfer from the pinion to the gear compared to perfect gears, such that:

$$(Rb_1\Omega_1 + Rb_2\Omega_2)\cos\beta_b + \frac{dE_{MAX}(t)}{dt} = 0 \qquad (1)$$

where $E_{MAX}(t) = \max_M(e(M))$ with $\max M()$, maximum over all the potential point of contact at time t

The difference with respect to ideal motion transfer is often related to the notion of no-load transmission error $NLTE$ via:

$$\frac{d}{dt}(NLTE) = Rb_1\Omega_1 + Rb_2\Omega_2 = -\frac{1}{\cos\beta_b}\frac{dE_{MAX}(t)}{dt} \tag{2}$$

Using the Kinetic Energy Theorem, the rigid-body dynamic behaviour for frictionless gears is controlled by:

$$J1\ \Omega 1\ \Omega 1 + J2\ \Omega 2\ \Omega 2 = Cm\ \Omega 1 + Cr\ \Omega 2 \tag{3}$$

with J_1, J_2: the polar moments of inertia of the pinion shaft line and the gear shaft line respectively. Cm, Cr: pinion and gear torques.

The system with 4 unknowns ($\Omega 1$, $\Omega 2$, Cm, Cr) is characterised by equations (2) - (3) only, and 2 parameters have to be imposed.

3.2 Deformed state – Principles

Modular models based on the definition of gear elements (pinion-gear pairs), shaft elements and lumped parameter elements (mass, inertia, stiffness) have proved to be effective in the simulation of complex gear units (Küçükay, 1987), (Baud & Velex, 2002). In this section, the theoretical foundations upon which classic gear elements are based are presented and the corresponding elemental stiffness and mass matrices along with the possible elemental forcing term vectors are derived and explicitly given.

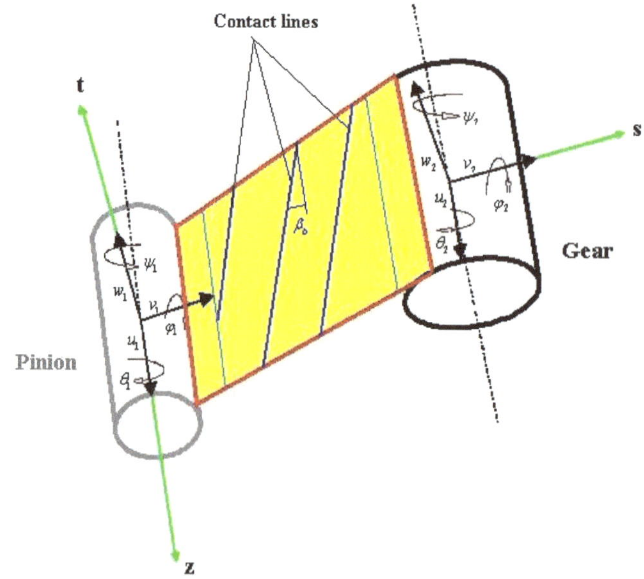

Fig. 1. A 3D lumped parameter model of pinion-gear pair.

The simplest and most frequently used 3D representation corresponds to the pinion-gear model shown in Figure 1. Assuming that the geometry is not affected by deflections (small displacements hypothesis) and provided that mesh elasticity (and to a certain extent, gear body elasticity) can be transferred onto the base plane, a rigid-body approach can be employed. The pinion and the gear can therefore be assimilated to two rigid cylinders with 6 degrees of freedom each, which are connected by a stiffness element or a distribution of stiffness elements (the discussion of the issues associated with damping and energy dissipation will be dealt with in section 4.3). From a physical point of view, the 12 degrees of freedom of a pair represent the generalised= displacements of i) traction: u_1 ,u_2 (axial displacements), ii) bending:$v1$, $w1$, $v2$, $w2$(translations in two perpendicular directions of the pinion/gear centre),$\Phi 1$,ψ 1 ,$\Phi 2$,ψ 2(bending rotations which can be assimilated to misalignment angles) and finally, iii) torsion:

θ_1 ,θ_2 which are small angles associated with deflections superimposed on rigid-body rotations t $\Theta_1 = \int \Omega_1 (\Sigma) d\Sigma$ (pinion) and 0 t $\Theta_2 = \int \Omega_2 (\Sigma) d\Sigma$ (gear). Following Velex and Maatar 0 (1996), screws of infinitesimal displacements are introduced whose co-ordinates for solid k (*conventionally k=1 for the pinion, k=2 for the gear*) can be expressed in two privileged coordinate systems: i) (**s, t, z**) such that **z** is in the shaft axis direction (from the motor to the load machine), **s** is in the centre-line direction from the pinion centre to the gear centre and **t** = **z** × **s** (Fig. 1) or, ii) (**X, Y, z**) attached to the base plane (Fig. 1):

$$\{S_k\}\begin{Bmatrix} \mathbf{u_k}(\mathbf{O_k}) = v_k\,\mathbf{s} + w_k\mathbf{t} + u_k\mathbf{z} \\ \boldsymbol{\omega}_\mathbf{k} == \varphi_k\,\mathbf{s} + \psi_k\,\mathbf{t} + \theta_k\,\mathbf{z} \end{Bmatrix} or \begin{Bmatrix} \mathbf{u_k}(\mathbf{O_k}) = V_k\,\mathbf{X} + W_k\mathbf{Y} + u_k\mathbf{z} \\ \boldsymbol{\omega}_\mathbf{k} == \Phi_k\,\mathbf{X} + \Psi_k\,\mathbf{Y} + \theta_k\,\mathbf{z} \end{Bmatrix} \quad k=1,2 \tag{4}$$

where O_1 ,O_2 are the pinion and gear centres respectively

3.3 Deflection at a point of contact – Structural vectors for external gears

Depending on the direction of rotation, the direction of the base plane changes as illustrated in Figure 2 where the thicker line corresponds to a positive rotation of the pinion and the finer line to a negative pinion rotation about axis $(O_1, \mathbf{z})$.

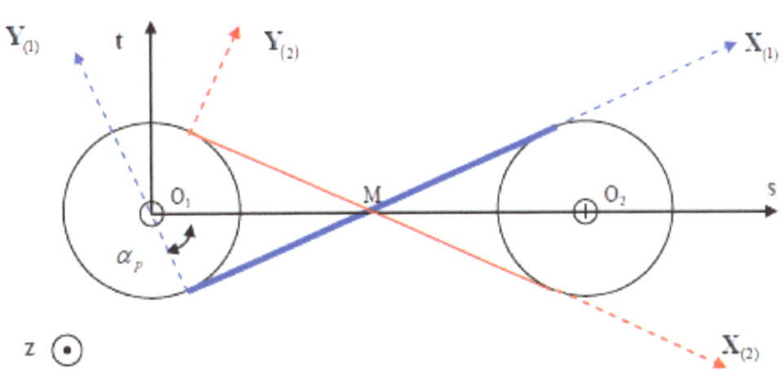

Fig. 2. Directions of rotation and planes (lines) of action. (*the thicker line corresponds to a positive rotation of pinion*)

For a given helical gear, the sign of the helix angle on the base plane depends also on the direction of rotation and, here again; two configurations are possible as shown in Figure 3.

Since a rigid-body mechanics approach is considered, contact deflections correspond to the interpenetrations of the parts which are deduced from the contributions of the degrees-of- freedom and the initial separations both measured in the normal direction with respect to the tooth flanks. Assuming that all the contacts occur in the theoretical base plane (or plane of action), the normal deflection $\Delta(M)$ at any point M, potential point of contact, is therefore expressed as:

$$\Delta(M) = \mathbf{u}_1(\mathbf{M}).\mathbf{n}_1 - \mathbf{u}_2(\mathbf{M}).\mathbf{n}_1 - \delta e(M) \tag{5}$$

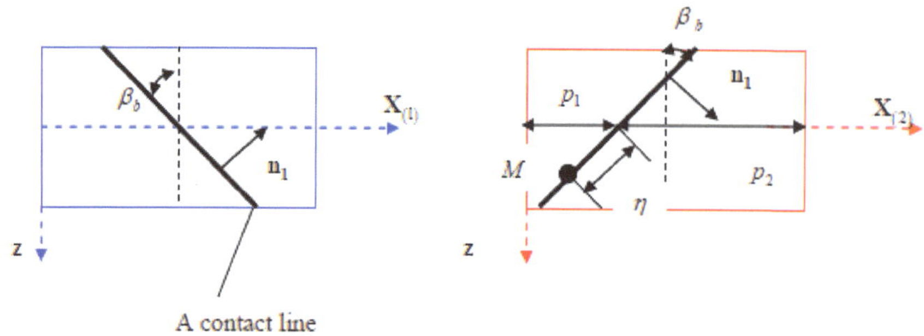

Fig. 3. Helix angles on the base plane.

where $\delta e(M) = \max_M(e(M)) - e(M)$ is the equivalent initial normal gap at M caused by tooth modifications and/or errors for example, $\mathbf{n}_1$ is the outward unit normal vector to pinion tooth flanks (Fig.3)

Using the shifting property of screws, one obtains the expression of $\Delta(M)$ in terms of the screw co-ordinates as:

$$\Delta(M) = \mathbf{u}_1(\mathbf{O}_1).\mathbf{n}_1 + (\omega_1 \times \mathbf{O}_1\mathbf{M}).\mathbf{n}_1 - \mathbf{u}_2(\mathbf{O}_2).\mathbf{n}_1 - (\omega_2 \times \mathbf{O}_2\mathbf{M}).\mathbf{n}_1 - \delta e(M) \tag{6}$$

which is finally expressed as:

$$\Delta(M) = \begin{bmatrix} \mathbf{n}_1 \\ \mathrm{O_1M} \times \mathbf{n}_1 \\ -\mathbf{n}_1 \\ -\mathrm{O_2M} \times \mathbf{n}_1 \end{bmatrix}^T . \begin{bmatrix} \mathbf{u}_1(\mathrm{O}_1) \\ \omega_1 \\ \mathbf{u}_2(\mathrm{O}_2) \\ \omega_2 \end{bmatrix} - \delta e(M) \tag{7}$$

or, in a matrix form:

$$\Delta(M) = \mathbf{V}(\mathbf{M})T\,\mathbf{q} - \delta e(M) \tag{8}$$

where **V** **(M)** is a structural vector which accounts for gear geometry (Küçükay, 1987) and **q** is the vector of the pinion-gear pair degrees of freedom (*superscript T refers to the transpose of vectors and matrices*)

The simplest expression is that derived in the (**X, Y, z**) coordinate system associated with the base plane leading to:

$$\begin{aligned} \mathrm{V}(\mathrm{M})^T = \langle & \cos\beta_b, \ 0, \ \varepsilon\sin\beta_b, \ -\zeta\,\varepsilon Rb_1\sin\beta_b, \ \eta-\varepsilon p_1\sin\beta_b, \ \zeta Rb_1\cos\beta_b, \\ & -\cos\beta_b, \ 0, \ -\varepsilon\sin\beta_b, \ -\zeta\,\varepsilon Rb_2\sin\beta_b, \ -[\eta+\varepsilon p_2\sin\beta_b], \ \zeta Rb_2\cos\beta_b \ \rangle \end{aligned} \tag{9}$$

$$\mathbf{q}^T = \langle V_1 \ \ W_1 \ \ u_1 \ \ \Phi_1 \ \ \Psi_1 \ \ \theta_1 \ \ V_2 \ \ W_2 \ \ u_2 \ \ \Phi_2 \ \ \Psi_2 \ \ \theta_2 \rangle$$

where Rb_1 , Rb_2 are the pinion, gear base radii; β_b is the base helix angle (*always considered as positive in this context*); p_1 , p_2 ,η are defined in Figure 3; $\varepsilon = \pm 1$ depending on the sign of the helix angle; $\zeta = +1$ for a positive rotation of the pinion and $\zeta = -1$ for a negative rotation of the pinion.

An alternative form of interest is obtained when projecting in the (**s, t, z**) frame attached to the pinion-gear centre line:

$$\begin{aligned} \mathrm{V}(\mathrm{M})^T = \langle & \cos\beta_b\sin\alpha_p, \ \zeta\cos\beta_b\cos\alpha_p, \ \varepsilon\sin\beta_b, \ -\zeta\,\varepsilon Rb_1\sin\beta_b\sin\alpha_p - \zeta(\eta-\varepsilon p_1\sin\beta_b)\cos\alpha_p \\ & \left[-\varepsilon Rb_1\sin\beta_b\cos\alpha_p + (\eta-\varepsilon p_1\sin\beta_b)\sin\alpha_p\right], \ \zeta Rb_1\cos\beta_b, \ -\cos\beta_b\sin\alpha_p, \ -\zeta\cos\beta_b\cos\alpha_p, \\ & -\varepsilon\sin\beta_b, \ -\zeta\,\varepsilon Rb_2\sin\beta_b\sin\alpha_p + \zeta(\eta+\varepsilon p_2\sin\beta_b)\cos\alpha_p \\ & -\left[\varepsilon Rb_2\sin\beta_b\cos\alpha_p + (\eta+\varepsilon p_2\sin\beta_b)\sin\alpha_p\right], \ \zeta Rb_2\cos\beta_b \ \rangle \end{aligned}$$

$$\mathbf{q}^T = \langle v_1 \ \ w_1 \ \ u_1 \ \ \varphi_1 \ \ \psi_1 \ \ \theta_1 \ \ v_2 \ \ w_2 \ \ u_2 \ \ \varphi_2 \ \ \psi_2 \ \ \theta_2 \rangle \tag{10}$$

3.4 Stiffness matrix of the mesh and external gear forcing parameters

For a given direction of rotation, the usual contact conditions in gears correspond to single- sided contacts between the mating flanks which do not account for momentary tooth separations which may appear if dynamic displacements are large (of the same order of magnitude as static displacements). A review of the mesh stiffness models is beyond the scope of this chapter but one usually separates the simulations accounting for elastic convection (i.e., the deflection at one point *M* depends on the entire load distribution on the tooth or all the mating teeth (Seager, 1967)) from the simpler (and classic) thin-slice approach (the deflection at point *M* depends on the load at the same point only). A discussion of the limits of this theory can be found in Haddad (1991), Ajmi & Velex (2005) but it seems that, for solid gears, it is sufficiently accurate as far as dynamic phenomena such as critical speeds are considered as opposed to exact load or stress distributions in the teeth which are more dependent on local conditions. Neglecting contact damping and friction forces compared with the normal elastic components on tooth flanks, the elemental force transmitted from the pinion onto the gear at one point of contact *M* reads:

$$\mathbf{dF}_{1/2}(\mathbf{M}) = k(M)\Delta(M)\,dM\,\mathbf{n}_1 \tag{11}$$

with $k(M)$: mesh stiffness at point *M* per unit of contact length

The resulting total mesh force and moment at the gear centre O_2 are deduced by integrating over the time-varying and possibly deflection-dependent contact length $L\,(t,\mathbf{q})$ as:

$$\{F_{1/2}\}\begin{cases} \mathbf{F}_{1/2} = \int\limits_{L(t,\mathbf{q})} k(M)\Delta(M)dM\,\mathbf{n}_1 \\ \mathbf{M}_{1/2}(\mathrm{O}_2) = \int\limits_{L(t,\mathbf{q})} k(M)\Delta(M)\mathrm{O}_2\mathbf{M}\times\mathbf{n}_1\,dM \end{cases} \tag{12-1}$$

Conversely the mesh force wrench at the pinion centre O_1 is:

$$\{F_{2/1}\}\begin{cases} \mathbf{F}_{2/1} = -\int\limits_{L(t,\mathbf{q})} k(M)\Delta(M)dM\,\mathbf{n}_1 \\ \mathbf{M}_{2/1}(\mathrm{O}_1) = -\int\limits_{L(t,\mathbf{q})} k(M)\Delta(M)\mathrm{O}_1\mathbf{M}\times\mathbf{n}_1\,dM \end{cases} \tag{12-2}$$

The mesh inter-force wrench can be deduced in a compact form as:

$$\{F_M\}\begin{cases}\{F_{2/1}\}\\ \{F_{1/2}\}\end{cases} = -\int\limits_{L(t,\mathbf{q})} k(M)\Delta(M)\mathbf{V}(\mathbf{M})dM \tag{13}$$

and introducing the contact normal deflection $\Delta(M) = \mathbf{V}(\mathbf{M})^T\,\mathbf{q} - \delta e(M)$ finally leads to:

$$\{F_M\} = -\left[\mathbf{K}_G(t)\right]\mathbf{q} + \mathbf{F}_e(t) \tag{14}$$

where $\left[\mathbf{K}_G(t)\right] = \int\limits_{L(t,\mathbf{q})} k(M)\mathbf{V}(\mathbf{M})\,\mathbf{V}(\mathbf{M})^T dM$ is the time-varying gear mesh stiffness matrix

$\mathbf{F}_e(t) = \int\limits_{L(t,\mathbf{q})} k(M)\delta e(M)\mathbf{V}(\mathbf{M})dM$ is the excitation vector associated with tooth shape

modifications and errors

3.5 Mass matrix of external gear elements–Additional forcing (inertial) terms

For solid k (pinion or gear), the dynamic sum with respect to the inertial frame can be expressed as:

$$\Sigma_k^0 = m_k\left[\left(\ddot{v}_k - e_k\dot{\Omega}_k\sin\Theta_k - e_k\Omega_k^2\cos\Theta_k\right)\mathbf{s} + \left(\ddot{w}_k + e_k\dot{\Omega}_k\cos\Theta_k - e_k\Omega_k^2\sin\Theta_k\right)\mathbf{t} + \ddot{u}_k\mathbf{z}\right] \tag{15}$$

where mk and ek are respectively the mass and the eccentricity of solid k

A simple expression of the dynamic moment at point Ok can be obtained by assuming that Ok is the centre of inertia of solid k and neglecting gyroscopic components (complementary information can be found in specialised textbooks on rotor dynamics (see for instance (Lalanne & Ferraris, 1998)):

$$\delta_{\mathbf{k}}^{0}(O_k) \cong I_k \ddot{\phi}_k \mathbf{s} + I_k \ddot{\psi}_k \mathbf{t} + I_{0k}\left(\dot{\Omega}_k + \ddot{\theta}_k\right)\mathbf{z} \tag{16}$$

where Ik is the cross section moment of inertia and I_0k is the polar moment of solid k

Using the same DOF arrangement as for the stiffness matrices, a mass matrix for the pinion- gear system can be deduced as (note that the same mass matrix is obtained in the (**X, Y, z**) coordinate system):

$$[\mathbf{M}_G] = \mathbf{diag}\,(m_1, m_1, m_1, I_1, I_1, I_{01}, m_2, m_2, m_2, I_2, I_2, I_{02}) \tag{17-1}$$

along with a forcing term associated with inertial forces (whose expression in (**X, Y, z**) the same form on the condition that angles Θ1,2 are measured from **X** and **Y**):

$$\begin{aligned}\mathbf{F}_G(t) = \langle & m_1 e_1\left(\dot{\Omega}_1 \sin\Theta_1 + \Omega_1^2 \cos\Theta_1\right) \quad -m_1 e_1\left(\dot{\Omega}_1 \cos\Theta_1 - \Omega_1^2 \sin\Theta_1\right) \quad 0 \quad 0 \quad 0 \quad -I_{01}\dot{\Omega}_1 \\ & m_2 e_2\left(\dot{\Omega}_2 \sin\Theta_2 + \Omega_2^2 \cos\Theta_2\right) \quad -m_2 e_2\left(\dot{\Omega}_2 \cos\Theta_2 - \Omega_2^2 \sin\Theta_2\right) \quad 0 \quad 0 \quad 0 \quad -I_{02}\dot{\Omega}_2 \rangle\end{aligned} \tag{17-2}$$

3.6 Usual simplifications

Examining the components of the structural vectors in (9) and (10), it can be noticed that most of them are independent of the position of the point of contact M with the exception of those related to bending slopes Ψ 1,2 or Φ1,2 ,ψ 1,2 . Their influence is usually discarded especially for narrow-faced gears so that the mesh stiffness matrix can be simplified as:

$$\left[\mathbf{K}_G(t)\right] \cong \int_{L(t,\mathbf{q})} k(M)\,dM\; \mathbf{V}_0 \mathbf{V}_0^T = k(t,\mathbf{q})\mathbf{G} \tag{18}$$

where $\mathbf{V}_0$ represents an average structural vector and $k\,(t,\mathbf{q})$ is the time-varying, possibly non-linear, mesh stiffness function (scalar) which plays a fundamental role in gear dynamics.

3.6.1 Classic one-DOF torsional model

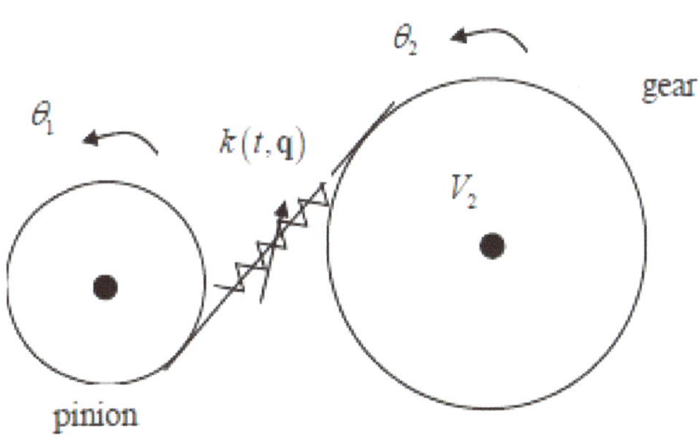

Fig. 4. Basic torsional model.

Considering the torsional degrees-of-freedom only (Figure 4), the structural vector reads (keeping solely the non-zero components):

$$\mathbf{V}(\mathbf{M}) = \mathbf{V}_0 = \begin{bmatrix} \zeta Rb_1 \\ \zeta Rb_2 \end{bmatrix} \cos\beta_b \tag{19}$$

and the following differential system is derived ($\zeta^2 = 1$) :

$$\begin{bmatrix} I_{01} & 0 \\ 0 & I_{02} \end{bmatrix} \begin{bmatrix} \ddot{\theta}_1 \\ \ddot{\theta}_2 \end{bmatrix} + k(t,\theta_1,\theta_2)\cos^2\beta_b \begin{bmatrix} Rb_1^2 & Rb_1Rb_2 \\ Rb_1Rb_2 & Rb_2^2 \end{bmatrix} \begin{bmatrix} \theta_1 \\ \theta_2 \end{bmatrix} = \\ = \begin{bmatrix} Cm \\ Cr \end{bmatrix} + \int_{L(t,q)} k(M)\delta e(M)dM \begin{bmatrix} \zeta Rb_1 \\ \zeta Rb_2 \end{bmatrix} \cos\beta_b - \begin{bmatrix} I_{01}\dot{\Omega}_1 \\ I_{02}\dot{\Omega}_2 \end{bmatrix} \tag{20}$$

Note that the determinant of the stiffness matrix is zero which indicates a rigid-body mode (the mass matrix being diagonal). After multiplying the first line in (20) by Rb_1I_{02} , the second line by Rb_2I_{01}, , adding the two equations and dividing all the terms by $(I_{10}Rb_2^2 + I_{20}Rb_1^2)$, the semi-definite system (20) is transformed into the differential equation:

$$\tilde{m}\ddot{x} + k(t,x)x = F_t + \zeta\cos\beta_b \int_{L(t,x)} k(M)\delta e(M)dM - \kappa\frac{d^2}{dt^2}(NLTE) \tag{21}$$

With $x = Rb1\theta_1 + Rb2\theta_2$, relative apparent displacement

$\tilde{m} = \dfrac{I_{02}I_{01}}{Rb_1^2I_{02} + Rb_2^2I_{01}}$, equivalent mass

$\kappa = \Omega_1^2 \dfrac{I_{02}}{Rb_2^2}$ when the pinion speed Ω and the output torque C are supposed to be constant.

3.6.2 A simple torsional-flexural model for spur gears

The simplest model which accounts for torsion and bending in spur gears is shown in Figure 5.

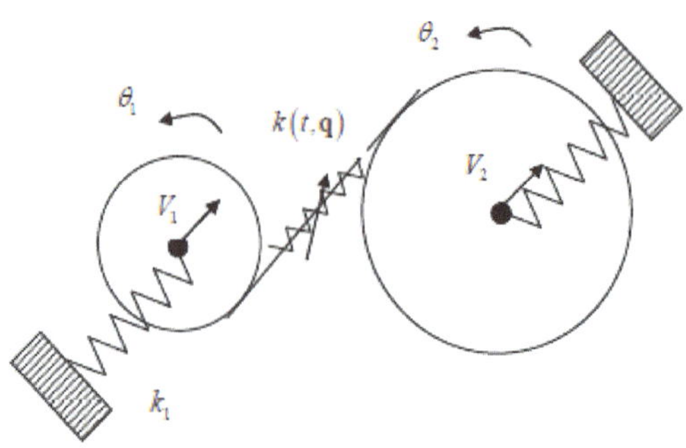

Fig. 5. Simplified torsional-flexural spur gear model.

It comprises 4 degrees of freedom, namely: 2 translations in the direction of the line of action V_1, V_2 (at pinion and gear centres respectively) and 2 rotations about the pinion and gear axes of rotation θ_1, θ_2. Because of the introduction of bending DOFs, some supports (bearing/shaft equivalent stiffness elements for instance) must be added.

The general expression of the structural vector **V** (**M**) (9) reduces to:

$$\mathbf{V}_0^T = \langle 1 \quad \zeta Rb_1 \quad -1 \quad \zeta Rb_2 \rangle \tag{22}$$

Re-writing the degree of freedom vector as $\mathbf{q}^{*T} = v1\ Rb1\theta_1\ v2\ Rb2\theta_2$, the following parametrically excited differential system is obtained for linear free vibrations:

$$\mathbf{M}\mathbf{q}^* + \mathbf{K}(t)\mathbf{q}^* = \mathbf{0} \tag{23-1}$$

$$\mathbf{M} = \begin{bmatrix} m_1 & & & \\ & I_{01}/Rb_1^2 & & \\ & & m_2 & \\ & & & I_{02}/Rb_2^2 \end{bmatrix}; \quad \mathbf{K}(t) = \begin{bmatrix} k(t)+k_1 & \zeta k(t) & -k(t) & \zeta k(t) \\ & k(t) & -\zeta k(t) & k(t) \\ & & k(t)+k_2 & -\zeta k(t) \\ & & & k(t) \end{bmatrix} \tag{23-2}$$

Remark: The system is ill-conditioned since rigid-body rotations are still possible (no unique static solution). In the context of 3D models with many degrees of freedom, it is not interesting to solve for the normal approach $Rb1\theta_1 + Rb2\theta_1$ as is done for single DOF models.

The problem can be resolved by introducing additional torsional stiffness element(s) which can represent shafts; couplings etc. thus eliminating rigid-body rotations.

4. MESH STIFFNESS MODELS – PARAMETRIC EXCITATIONS

4.1 Classic thin-slice approaches

From the results in section 2-5, it can be observed that, in the context of gear dynamic simulations, the mesh stiffness function defined as $k(t,\mathbf{q}) = \int L(t,\mathbf{q})\, k(M)\, dM$ plays a key role.

This function stems from a 'thin-slice' approach whereby the contact lines between the mating teeth are divided in a number of independent stiffness elements (with the limiting case presented here of an infinite set of non-linear time-varying elemental stiffness elements) as schematically represented in Figure 6.

Since the positions of the teeth (and consequently the contact lines) evolve with time (or angular positions), the profiles slide with respect to each other and the stiffness varies because of the contact length and the individual tooth stiffness evolutions.

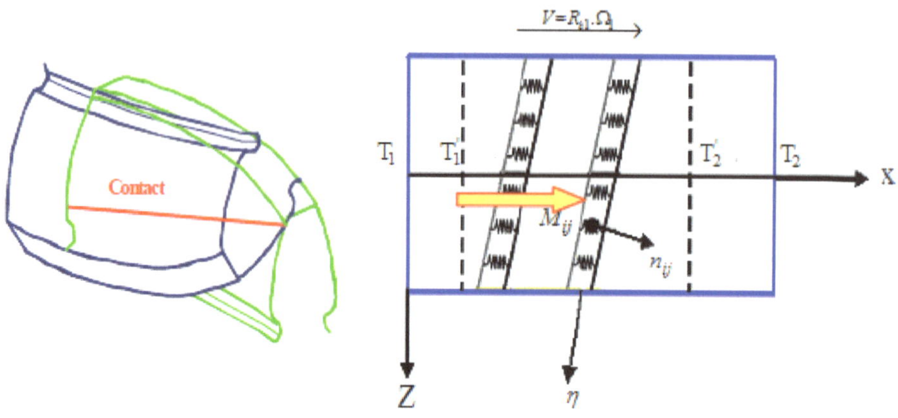

Fig. 6. 'Thin-slice' model for time-varying mesh stiffness.

The definition of mesh stiffness has generated considerable interest but mostly with the objective of calculating accurate static tooth load distributions and stress distributions. It has been shown by Ajmi and Velex (2005) that a classic 'thin-slice' model is sufficient for dynamic calculations as long as local disturbances (especially near the tooth edges) can be ignored. In this context, Weber and Banascheck (1953) proposed a analytical method of calculating tooth deflections of spur gears by superimposing displacements which arise from i) the contact between the teeth, ii) the tooth itself considered as a beam and, iii) the gear body (or foundation) influence. An analytical expression of the contact compliance was obtained using the 2D Hertzian theory for cylinders in contact which is singular as far as the normal approach between the parts (contact deflection) is concerned. The other widely-used formulae for tooth contact deflection comprise the analytical formula of Lundberg (1939), the approximate Hertzian approach originally used at Hamilton Standard (Cornell, 1981) and the semi-empirical formula developed by Palmgren (1959) for rollers. The tooth bending radial and tangential displacements were derived by equating the work produced by one individual force acting on the tooth profile and the strain energy of the tooth assimilated to a cantilever of variable thickness. Extensions and variants of the methodology were introduced by Attia (1964), Cornell (1981) and O'Donnell (1960, 1963) with regard to the foundation effects. Gear body contributions were initially evaluated by approximating them as part of an elastic semi-infinite plane loaded by the reactions at the junction with the tooth. A more accurate expression for this base deflection has been proposed by Sainsot *et al.* (2004) where the gear body is simulated by an elastic annulus instead of a half-plane. Figure 7 shows two examples of mesh stiffness functions (no contact loss) calculated by combining Weber's and Lundberg's results for a spur and a helical gear example. It can be observed that the stiffness fluctuations are stronger in the case of conventional spur gears compared with helical gears for which the contact variations between the teeth are smoother.

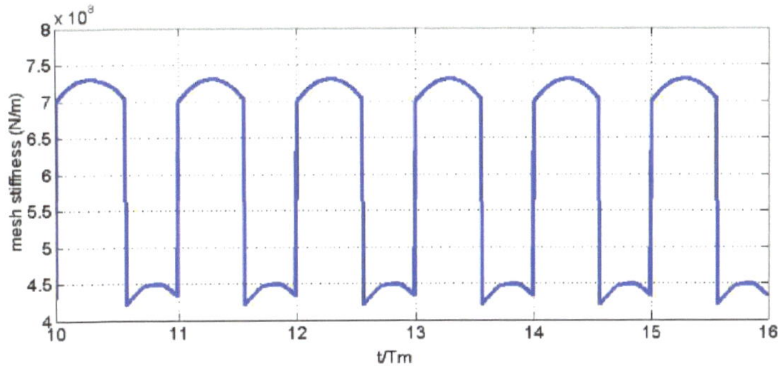

a - Spur gear

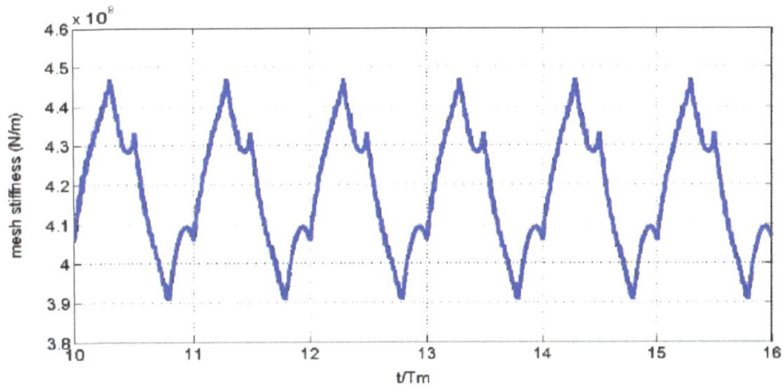

b - Helical gear $(\beta = 30°)$

Fig. 7. Examples of mesh stiffness functions for errorless gears.

Although the results above are based on simplified bi-dimensional approaches, they are still widely used in gear design. For example, the mesh stiffness formulae in the ISO standard 6336 stem from Weber's analytical formulae which were modified to bring the values in closer agreement with the experimental results. Another important simplification brought by the ISO formulae is that the mesh stiffness per unit of contact length k_0 is considered as approximately constant so that the following approximation can be introduced:

$$\int_{L(t,\mathbf{q})} k(M)\,dM \cong k_0 \int_{L(t,\mathbf{q})} dM = k_0 L(t,\mathbf{q}) \tag{24}$$

where $L(t,\mathbf{q})$ is the time-varying (possibly non-linear) contact length.

4.2 Contact length variations for external spur and helical gears

Considering involute profiles, the contact lines in the base plane are inclined by the base helix angle β_b (Figure 8) which is nil for spur gears. All contact lines are spaced by integer multiples of the apparent base pitch *Pba* and, when the pinion and the gear rotate, they all undergo a translation in the **X** direction at a speed equal to $Rb_1\,\Omega_1$.

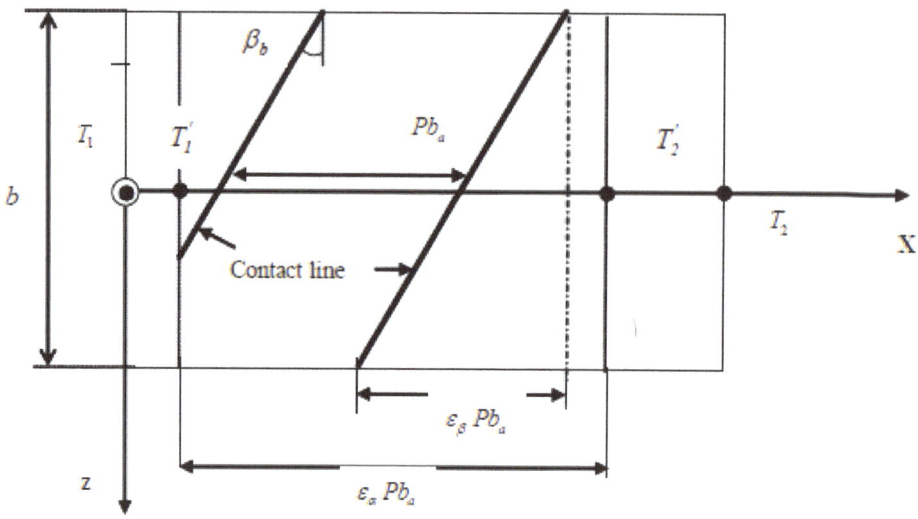

Fig. 8. Base plane and contact lines (*b* : face width; z : axial direction (direction of the axes of rotation); *T* ,*T* : points of tangency on pinion and gear base circles and *T* ',*T* ' : limits of 1 2 1 2 the contact area on base plane).

It transpires from this geometrical representation that the total length of contact between the pinion and the gear is likely to vary with time and, based on the simple stiffness equation (24), that mesh stiffness is time-varying and, consequently, contributes to the system excitation via parametric excitations.

The extent of action on the base plane is an important property measured by the contact ratio ε_α which, in simple terms, represents the 'average number' of tooth pairs in contact (possibly non integer) and is defined by:

$$\varepsilon_\alpha = \frac{T_1'T_2'}{Pb_a} = \frac{\sqrt{Ra_1^{\,2} - Rb_1^{\,2}} + \sqrt{Ra_2^{\,2} - Rb_2^{\,2}} - E\sin\alpha_p}{\pi\, m \cos\alpha_p} \tag{25-1}$$

with *Ra*1 , *Ra*2 : external radius of pinion, of gear; *Rb*1 , *Rb*2 : base radius of pinion, of gear; *E* = || O1O2 || : centre distance

In the case of helical gears, the overlap due to the helix is taken into account by introducing the overlap ratio εβ defined as:

$$\varepsilon_\beta = \frac{b\tan\beta_b}{Pb_a} = \frac{1}{\pi}\frac{b}{m}\frac{\tan\beta_b}{\cos\alpha_p} \tag{25-2}$$

and the sum ε = εα + εβ is defined as the total contact ratio.

Introducing the dimensionless time $\tau = \frac{t}{T_m}$ where $\frac{Pb_a}{Rb_1\Omega_1}$ is the mesh period i.e. the time needed for a contact line to move by a base pitch on the base plane, a closed form expression of the contact length $L(\tau)$ for ideal gears is obtained under the form (Maatar & Velex, 1996), (Velex et al., 2011):

$$\frac{L(\tau)}{L_m} = 1 + 2\sum_{k=1}^{\infty} Sinc(k\varepsilon_\alpha)Sinc(k\varepsilon_\beta)\cos\left(\pi k\left(\varepsilon_\alpha + \varepsilon_\beta - 2\tau\right)\right) \tag{26}$$

with: $L_m = \varepsilon_\alpha \dfrac{b}{\cos\beta_b}$, average contact length

$Sinc\,(x) = \dfrac{\sin(\pi x)}{\pi x}$ is the classic sine cardinal function which is represented in Figure 9.

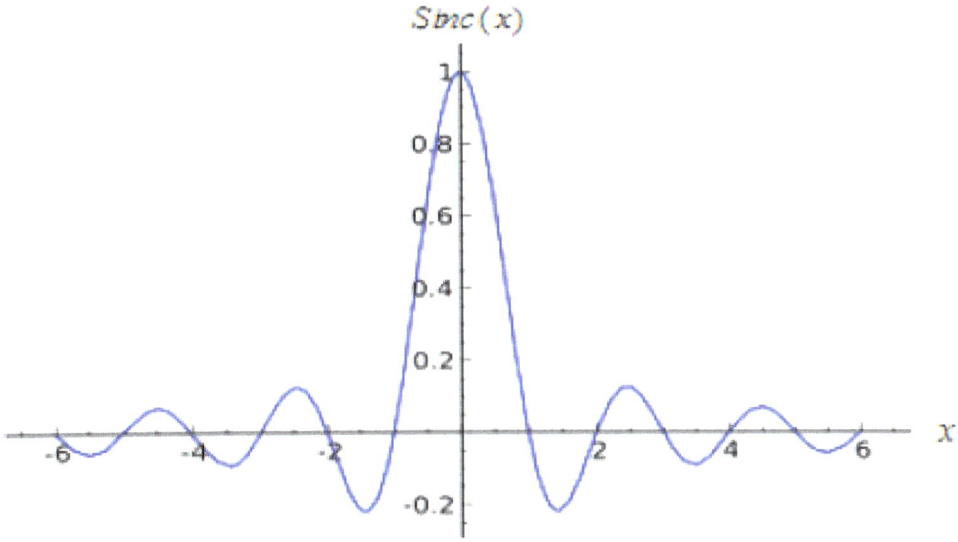

Fig. 9. Evolutions of $Sinc\,(x) = \dfrac{\sin(\pi x)}{\pi x}$.

The following conclusions can be drawn:

a. for spur gears, $\varepsilon\beta = 0$ and $Sinc\,(k\varepsilon\beta) = 1$
b. it can observed that the time-varying part of the contact length disappears when either ε_α or $\varepsilon\beta$ is an integer
c. harmonic analysis is possible by setting $k = 1, 2,...$ in (27) and it is possible to represent the contact length variations for all possible values of profile and overlap contact ratios on a unique diagram. Figure 10 represents the RMS of contact length variations for a realistic range of contact and overlap ratios. It shows that:
   - contact length variations are significant when ε_α is below 2 and ε_β below 1
   - contact length is constant when $\varepsilon_\alpha = 2$ ($\varepsilon_\alpha = 1$ has to avoided for a continuous motion transfer) and /or $\varepsilon_\beta = 1$
   - for overlap ratios ε_β above 1, contact length variations are very limited.

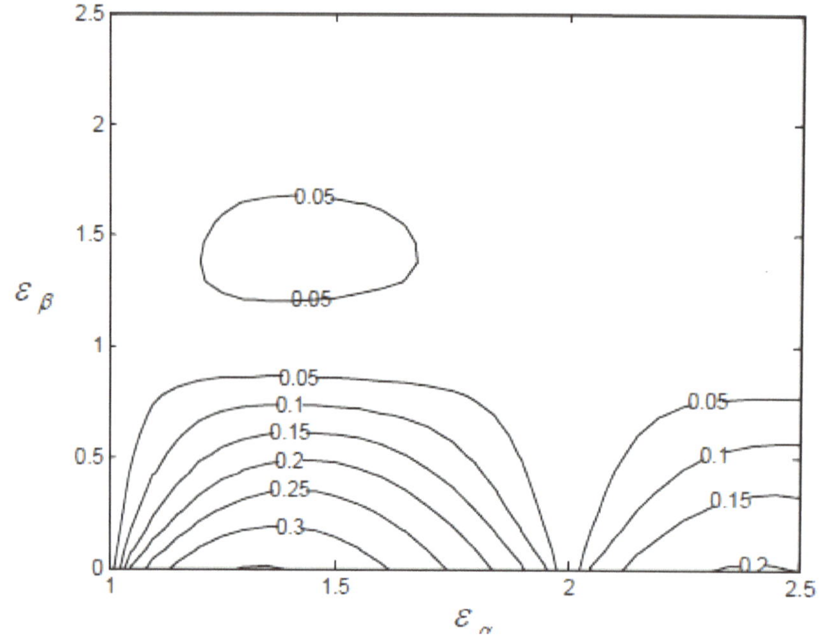

Fig. 10. Contour plot of the R.M.S. of $L(\tau)/L_m$ for a range of profile and transverse contact ratios.

4.3 Approximate expressions – Orders of magnitude

Mesh stiffness can be determined using the Finite Elements Method but it is interesting to have orders of magnitude or approximate values at the design stage. For solid gears made of steel, an order of magnitude of the mesh stiffness per unit of contact length k_0 is ≅ 1.3 1010 N/m². More accurate expressions can be derived from the ISO 6336 standard which, for solid gears, gives:

$$k_0 \cong \cos\beta \frac{0.8}{q} \tag{27}$$

with

β : helix angle (on pitch cylinder)

$$q = C_1 + \frac{C_2}{Zn_1} + \frac{C_3}{Zn_2} + C_4 x_1 + C_5 \frac{x_1}{Zn_1} + C_6 x_2 + C_7 \frac{x_2}{Zn_2} + C_8 x_1^2 + C_9 x_2^2$$

coefficients C1 ,...,C9 have been tabulated and are listed in Table 1 below

$Zn_i = \frac{Z_i}{\cos^3\beta}$, $i = 1,2$ are the number of teeth of the equivalent virtual spur pinion ($i = 1$)

and gear ($i = 2$).

, $i = 1, 2$, are the profile shift coefficients on pinion($i = 1$) and gear($i = 2$)

Table 1. Tabulated coefficients for mesh stiffness calculations according to ISO 6336.

C1	C2	C3	C4	C5	C6	C7	C8	C9
0,04723	0,15551	0,25791	-0,00635	-0,11654	-0,00193	-0,24188	0,00529	0,00182

5. EQUATIONS OF MOTION – DYNAMIC BEHAVIOUR

5.1 Differential system

The equations of motion for undamped systems are derived by assembling all the elemental matrices and forcing term vectors associated with the gears but also the supporting members (shafts, bearings, casing, etc.) leading to a parametrically excited non-linear differential system of the form:

$$[\mathbf{M}]\ddot{\mathbf{X}}+\left[\mathbf{K}(t,\mathbf{X})\right]\mathbf{X}=\mathbf{F}_0+\mathbf{F}_1\left(t,\mathbf{X},\delta e(M)\right)+\mathbf{F}_2\left(t,\dot{\Omega}_{1,2}\right) \qquad (28)$$

where $\mathbf{X}$ is the total DOF vector, [**M**] and [**K** (t, **X**)] are the global mass and stiffness matrices. Note that, because of the contact conditions between the teeth, the stiffness matrix can be non-linear (partial or total contact losses may occur depending on shape deviations= and speed regimes). $\mathbf{F}_0$ comprises the constant nominal torques; $\mathbf{F}_1$ (t, **X**,δ e (M)) includes the contributions of shape deviations (errors, shape modifications, etc.); **F2** (t,Ω 1,2) represents the inertial effects due to unsteady rotational speeds

5.2 Linear behaviour - Modal analysis

Considering linear (or quasi-linear) behaviour, the differential system can be re-written as:

$$[\mathbf{M}]\ddot{\mathbf{X}}+\left[\mathbf{K}(t)\right]\mathbf{X}=\mathbf{F}_0(t)+\mathbf{F}_1\left(t,\delta e(M)\right)+\mathbf{F}_2\left(t,\dot{\Omega}_{1,2}\right) \qquad (29)$$

The time variations in the stiffness matrix $[\mathbf{K}(t)]$ are caused by the meshing and, using the formulation based on structural vectors, the constant and time-varying components can be separated as:

$$\left[\mathbf{K}(\mathrm{t})\right]=[\mathbf{K}_0]+\int_{L(t)} k(M)\bar{\mathbf{V}}(\mathbf{M})\bar{\mathbf{V}}(\mathbf{M})^T\, dM \qquad (30)$$

where **V** (**M**) is the extended structural vector: structural vector completed by zeros to the total number of DOF of the model

Using an averaged structural vector as in (18):

$$\bar{\mathbf{V}}_0=\frac{1}{T_m}\int_0^{T_m}\bar{\mathbf{V}}(M)\,dt \qquad (31)$$

(30) can be simplified as:

$$[K(t)] = [K_0] + \int_{L(t)} k(M)dM\,\overline{V}_0\,\overline{V}_0^T = [K_0] + k_m(t)\overline{V}_0\,\overline{V}_0^T \tag{32}$$

The separation of the average and time-varying contributions in the mesh stiffness function as $k(t) = k_m(1 + \alpha\Phi(t))$ leads to the following state equations:

$$[M]\ddot{X} + \left[[K_0] + k_m(1+\alpha\varphi(t))\overline{V}_0\,\overline{V}_0^T\right]X = F_0 + F_1(t,\delta e(M)) + F_2(t,\dot{\Omega}_{1,2}) \tag{33}$$

For most gears, α is usually a small parameter ($\alpha \ll 1$) and an asymptotic expansion of the solution can be sought as a straightforward expansion of the form:

$$\mathbf{X} = \mathbf{X} + \alpha\,\mathbf{X} + \alpha^2\mathbf{X} + \dots \tag{34}$$

which, when re-injected into (33) and after identifying like order terms leads to the following series of constant coefficient differential systems:

Main order:

$$[M]\ddot{X}_0 + \left[[K_0] + k_m\overline{V}_0\,\overline{V}_0^T\right]X_0 = F_0 + F_1(t,\delta e(M)) + F_2(t,\dot{\Omega}_{1,2}) \tag{35-1}$$

1th order:

$$[M]\ddot{X}_\ell + \left[[K_0] + k_m\overline{V}_0\,\overline{V}_0^T\right]X_\ell = -k_m\varphi(t)\overline{V}_0\,\overline{V}_0^T X_{\ell-1} \tag{35-2}$$

Interestingly, the left-hand sides of all the differential systems are identical and the analysis of the eigenvalues and corresponding eigenvectors of the homogeneous systems will provide useful information on the dynamic behaviour of the geared systems under consideration (critical speeds, modeshapes).

The following system is considered (the influence of damping on critical speeds being ignored):

$$[M]\ddot{X}_\ell + \left[[K_0] + k_m\overline{V}_0\,\overline{V}_0^T\right]X_\ell = 0 \tag{36}$$

from which the eigenvalues and eigenvectors are determined. The technical problems associated with the solution of (36) are not examined here and the reader may refer to specialised textbooks. It is further assumed that a set of real eigenvalues ω_p and real orthogonal eigenvectors $\mathbf{\Phi P}$ have been determined which, to a great extent, control the gear set dynamic behaviour.

Focusing on dynamic tooth loads, it is interesting to introduce the percentage of modal strain energy stored in the gear mesh which, for a given pair (ω_p , $\mathbf{\Phi P}$), is defined as:

$$\rho_p = k_m \frac{\Phi_P^T \bar{V}_0 \bar{V}_0^T \Phi_P}{\Phi_P^T \left[[K_0] + k_m \bar{V}_0 \bar{V}_0^T \right] \Phi_P} = v_{\Phi p}^2 \frac{k_m}{k_{\Phi p}} \tag{37}$$

with $v_{\Phi p} = \Phi_P^T \bar{V} = \bar{V}^T \Phi_P$

km , *kΦp* : average mesh stiffness and modal stiffness associated with (ω_p , **Φp**)

It as been shown (Velex & Berthe., 1989) that ρ_p is a reliable indicator of the severity of one frequency with regard to the pinion-gear mesh and it can be used to identify the potentially critical speeds ω_p for tooth loading which are those with the largest percentages of modal strain energy in the tooth mesh. If the only excitations are those generated by the meshing (the mesh frequency is $Z_1\Omega_1$), the tooth critical speeds can be expressed in terms of pinion speed as:

$$\Omega 1 = \omega_p / kZ1 \quad k = 1, 2,... \tag{38}$$

Based on the contact length variations and on the transmission error spectrum, the relative severity of the excitations can be anticipated.

Remark: The critical frequencies are supposed to be constant over the speed range (gyroscopic effects are neglected). Note that some variations can appear with the evolution of the torque versus speed (a change in the torque or load can modify the average mesh stiffness especially for modified teeth).

For the one DOF tosional model in Figure 4, there is a single critical frequency ω = whose expression can be developed for solid gears of identical face width leading to:

$$\Omega_1 \cong \frac{\Lambda}{k} \frac{\cos\alpha_p}{MZ_1^2} \sqrt{\frac{b}{B}} \sqrt{\cos\beta_b} \sqrt{\varepsilon_\alpha} \sqrt{1+u^2} \tag{39}$$

where $k = 1, 2,...$ represents the harmonic order; $\Lambda = \sqrt{\frac{8k_0}{\pi\rho}}$ (ρ is the density), for steel gears

$\Lambda \cong 2\ 10^3\ ms^{-1}$; *M* is the module (in meter); *B* is the pinion or gear thickness (supposed identical); *b* is the effective contact width (which can be shorter than *B* because of chamfers for example); $u = \frac{Z_1}{Z_2}$, speed ratio.

5.3 Dynamic response

5.3.1 The problem of damping

Energy dissipation is present in all geared systems and the amount of damping largely controls the amplification at critical speeds. Unfortunately, the prediction of damping is still a challenge and, most of the time; it is adjusted in order to fit with experimental evidence. Two classical procedures are frequently employed:

a. the assumption of proportional damping (Rayleigh's damping) which, in this case, leads to:

$$[C] = a[M] + b\left[[K_0] + k_m \bar{V}_0 \bar{V}_0^T\right] \tag{40}$$

with: a,b , two constants to be adjusted from experimental results

b. the use of (a limited number of) modal damping factors Σp :

The damping matrix is supposed to be orthogonal with respect to the mode-shapes of the undamped system with the averaged stiffness matrix such that:

$$\Phi_p^T [C] \Phi_p = 2\varsigma_p \sqrt{k_{\Phi p}\, m_{\Phi p}} \tag{41-1}$$

$$\Phi_p^T [C] \Phi_q = 0 \tag{41-2}$$

with: Σp : modal damping factor associated with mode p

$k\Phi p$,$m\Phi p$: modal stiffness and mass associated with mode p

or introducing the modal damping matrix $[C_\Phi]$:

$$[C_\Phi] = diag\left(2\varsigma_p \sqrt{k_{\Phi p}\, m_{\Phi p}}\right),\ p = 1, N\mathrm{mod} \tag{41-3}$$

Following Graig (1981), the damping matrix can be deduced by a truncated summation on a limited number of modes *Nr* leading to the formula:

$$[C] = \sum_{p=1}^{Nr} \frac{2\varsigma_p \omega_p}{m_{\Phi p}} \left([M]\Phi_p\right)\left([M]\Phi_p\right)^T \tag{42}$$

with: $\omega p = \sqrt{\dfrac{k_{\Phi p}}{m_{\Phi p}}}$

Regardless of the technique employed, it should be stressed that both (41) and (42) depend on estimated or measured modal damping factors Σ_p for which the data in the literature is rather sparse. It seems that $0.02 \le \Sigma_p \le 0.1$ corresponds to the range of variation for modes with significant percentages of strain energy in the meshing teeth. The methods also rely on the assumption of orthogonal mode shapes which is realistic when the modal density (number of modes per frequency range) is moderate so that inter-modal couplings can be neglected.

5.3.2 Linear response

Based on the previous developments, the linear response of gears to mesh parametric excitations can be qualitatively assessed. Response peaks are to be expected at all tooth critical speeds and every sub-harmonic of these critical speeds because mesh stiffness time variations may exhibit several harmonics with significant amplitudes. Figure 11, taken from Cai and Hayashi (1994), is a clear example of such typical dynamic response curves when the gear dynamic behaviour is dominated by one major tooth frequency ω_n (and can be simulated by using the classic one DOF model). The amplifications associated with each peak depends on i) the excitation amplitude (Eq. (27) can provide some information on the amplitude associated with each mesh frequency harmonic) and ii) the level of damping for this frequency. For more complex gear sets, interactions between several frequencies can happen but, as far as the author is aware, the number of frequencies exhibiting a significant percentage of modal strain energy in the tooth mesh seems very limited (frequently less than 5) thus making it possible to anticipate the potential dangerous frequency coincidences for tooth durability.

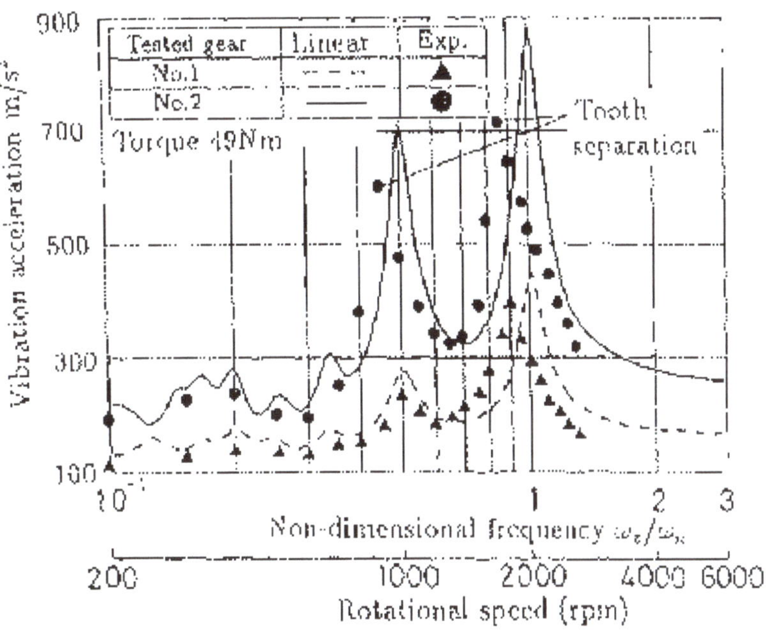

Fig. 11. Examples of dynamic response curves (Cay & Hayashi, 1994).

5.3.3 Contact condition – Contact losses and shocks

Only compressive contact forces can exist on tooth flanks and using (11), this imposes the following unilateral condition in case of contact at point *M*:

$$\mathbf{dF}_{1/2}(\mathbf{M}) \cdot \mathbf{n}_1 = k(M)\Delta(M)dM > 0 \qquad (43)$$

or, more simply, a positive mesh deflection $\Delta(M)$.

If $\Delta(M) \leq 0$, the contact at M is lost (permanently or temporarily) and the associated contact force is nil. These constraints can be incorporated in the contact force expression by introducing the unit Heaviside step function $H(x)$ such that $H(x) = 1$ if $x > 0$ and $H(x) = 0$ otherwise. Finally, one obtains:

$$\mathbf{dF}_{1/2}(\mathbf{M}) = k(M)\Delta(M)\,H(\Delta(M))dM\,\mathbf{n}_1 \qquad (44)$$

It can observed that contact losses are related to the sign of $\Delta(M) \cong \mathbf{V}^T\mathbf{X} - \delta e(M)$, from which, it can be deduced that two cases have to be considered:

α. $\delta e(M)$ is larger than the normal approach $\mathbf{V}^T\mathbf{X}$ which, typically, corresponds to large amplitudes of tooth modifications reducing the actual contact patterns, to spalls on the flanks (pitting) where contact can be lost, etc.

b. the amplitude of the dynamic displacement $\mathbf{X}$ is sufficiently large so that the teeth can separate ($\mathbf{X}$ is periodic and can become negative in some part of the cycle).

Momentary contact losses can therefore occur when vibration amplitudes are sufficiently large; they are followed by a sequence of free flight within the tooth clearance until the teeth collide either on the driving flanks or on the back of the teeth (back strike). Such shocks are particularly noisy (rattle noise) and should be avoided whenever possible. Analytical investigations are possible using harmonic balance methods and approximations of $H(x)$ (Singh et al., 1989), (Comparin & Singh, 1989), (Kahraman & Singh, 1990), (Kahraman & Singh, 1991), and numerical integrations can be performed by time-step schemes (Runge- Kutta, Newmark, etc.). The most important conclusions are:

a. contact losses move the tooth critical frequencies towards the lower speeds (softening effect) which means that predictions based on a purely linear approach might be irrelevant. The phenomenon can be observed in Fig. 11 where the experimental peaks are at lower speed than those predicted by the linear theory.

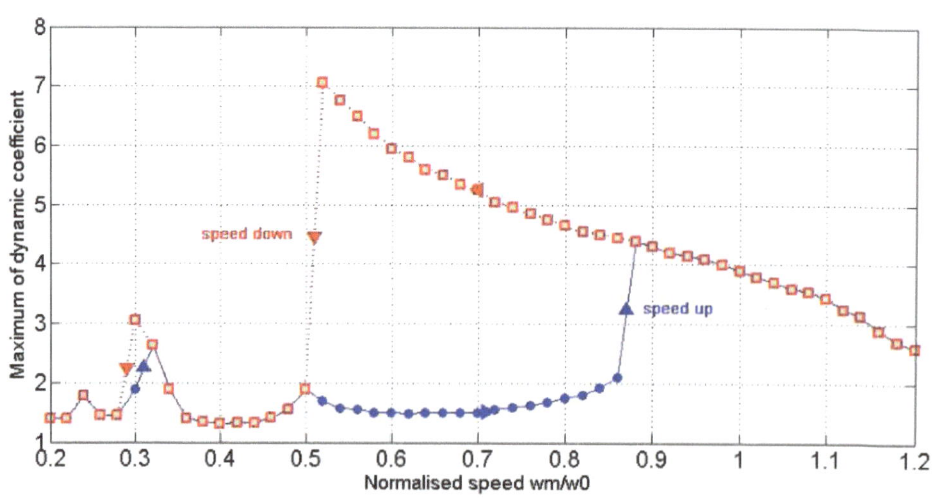

Fig. 12. Dynamic response curves by numerical simulations – Amplitude jumps – Influence of the initial conditions (speed up vs speed down), *2% of the critical damping, Spur gears* $Z_1 = 30$, $Z_2 = 45$, $M = 2mm$, *standard tooth proportions.*

b. When contact losses occur, response curves exhibit amplitude jumps (sudden amplitude variations for a small speed variation),

c. Because of a possibly strong sensitivity to initial conditions, several solutions may exist depending on the kinematic conditions i.e., speed is either increased or decreased

d. damping reduces the importance of the frequency shift and the magnification at critical tooth frequency.

These phenomena are illustrated in the response curves in Figure 12.

6. TRANSMISSION ERRORS

6.1 Definitions

The concept of transmission error (*TE*) was first introduced by Harris (1958) in relation to the study of gear dynamic tooth forces. He realised that, for high speed applications, the problem was one of continuous vibrations rather than a series of impacts as had been thought before. Harris showed that the measure of departure from perfect motion transfer between two gears (which is the definition of *TE*) was strongly correlated with excitations and dynamic responses. *TE* is classically defined as the deviation in the position of the driven gear (for any given position of the driving gear), relative to the position that the driven gear would occupy if both gears were geometrically perfect and rigid.

NB: *The concept embodies both rigid-body and elastic displacements which can sometimes be confusing.*

Figure 13 illustrates the concept of transmission error which (either at no-load or under load) can be expressed as angular deviations usually measured (calculated) on the driven member (gear) or as distances on the base plane.

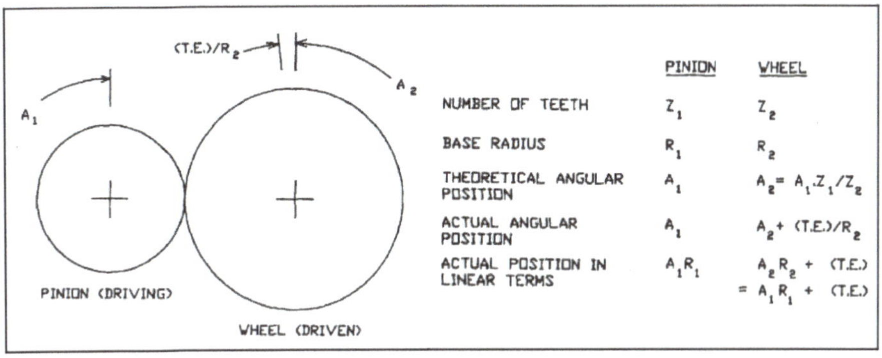

Fig. 13. Concept of transmission error and possible expressions (after Munro, (1989)).

Figures 14 and 15 show typical quasi-static *T.E.* traces for spur and helical gears respectively. The dominant features are a cyclic variation at tooth frequency (mesh frequency) and higher harmonics combined with a longer term error repeating over one revolution of one or both gears.

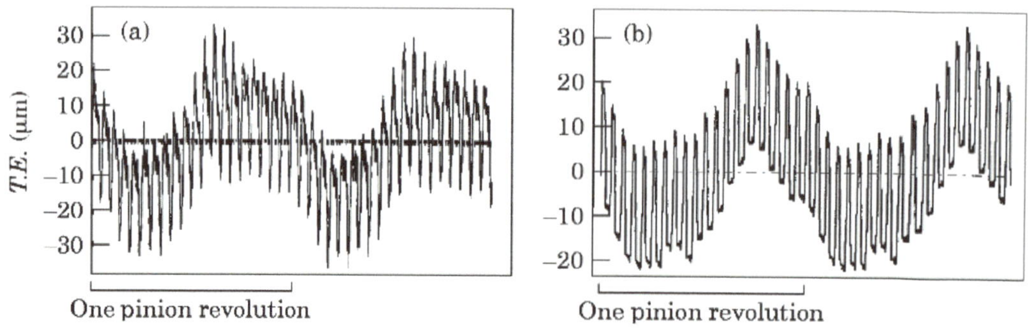

Fig. 14. Examples of quasi-static T.E. measurements and simulations – Spur gear (Velex and Maatar, 1996).

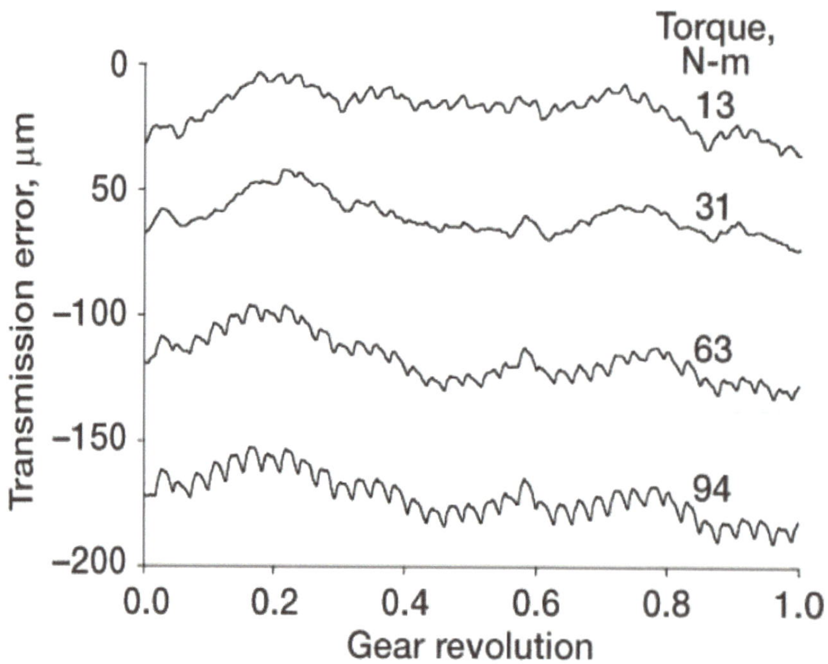

Fig. 15. T.E. measurements at various loads – Helical gear example. NASA measurements from www.grc.nasa.gov/WWW/RT2001/5000/5950oswald1.html.

6.2 No-load transmission error (NLTE)

No-load *T.E.* (*NLTE*) has already been introduced in (2); it can be linked to the results of gear testing equipment (single flank gear tester) and is representative of geometrical deviations. From a mathematical point of view, *NLTE* is derived by integrating (2) and is expressed as:

$$NLTE = -\frac{E_{MAX}(t)}{\cos \beta_b} \tag{45}$$

6.3 Transmission errors under load

The concept of transmission error under load (TE) is clear when using the classic single degree of freedom torsional model (as Harris did) since it directly relies on the angles of torsion of the pinion and the gear. For other models (even purely torsional ones), the definition of *TE* is ambiguous or at least not intrinsic because it depends on the chosen cross-sections (or nodes) of reference for measuring or calculating deviations between actual and perfect rotation transfers from the pinion to the gear. Following Velex and Ajmi (2006), transmission error can be defined by extrapolating the usual experimental practice based on encoders or accelerometers, i.e., from the actual total angles of rotation, either measured or calculated at one section of reference on the pinion shaft (subscript *I*) and on the gear shaft (subscript *II*). *TE* as a displacement on the base plane reads therefore:

$$TE = Rb_1\left[\int_0^t \Omega_1\, d\xi + \theta_I\right] + Rb_2\left[\int_0^t \Omega_2\, d\xi + \theta_{II}\right] = Rb_1\,\theta_I + Rb_2\,\theta_{II} + NLTE \qquad (46)$$

with ξ, a dummy integration variable and θ_I, θ_{II}, the torsional perturbations with respect to rigid-body rotations (degrees of freedom) at node I on the pinion shaft and at node II on the gear shaft.

Introducing a projection vector **W** of components Rb_1 and $Rb2$ at the positions corresponding to the torsional degrees of freedom at nodes I and II and with zeros elsewhere, transmission error under load can finally be expressed as:

$$TE = \mathbf{W}^T\,\mathbf{X} + NLTE \qquad (47\text{-}1)$$

which, for the one DOF model, reduces to:

$$TE = x + NLTE \qquad (47\text{-}2)$$

6.4 Equations of motion in terms of transmission errors

For the sake of clarity the developments are conducted on the one-DOF torsional model. Assuming that the dynamic contact conditions are the same as those at very low speed, one obtains from (21) the following equation for quasi-static conditions (i.e., when Ω_1 zero):

$$k(t,x)x_S = F_t + \zeta\cos\beta_b \int_{L(t,x)} k(M)\delta e(M)\,dM \qquad (48)$$

which, re-injected in the dynamic equation (21), gives:

$$\bar{m}\ddot{x} + k(t,x)x = k(t,x)x_S - \kappa\frac{d^2}{dt^2}(NLTE) \qquad (49)$$

From (47-2), quasi-static transmission error under load can be introduced such that *xS* = *TES* - *NLTE* and the equation of motion is transformed into:

$$\tilde{m}\ddot{x}+k(t,x)x=k(t,x)[TE_S-NLTE]-\kappa\frac{d^2}{dt^2}(NLTE) \quad (50)$$

An alternative form of interest can be derived by introducing the dynamic displacement *xD* defined by *x* = *xS* + *xD* as:

$$\tilde{m}\ddot{x}_D+k(t,x)x_D=-\tilde{m}\frac{d^2}{dt^2}(TE_S)+(\tilde{m}-\kappa)\frac{d^2}{dt^2}(NLTE) \quad (51)$$

The theory for 3D models is more complicated mainly because there is no one to one correspondence between transmission error and the degree of freedom vector. It can be demonstrated (Velex and Ajmi, 2006) that, under the same conditions as for the one DOF model, the corresponding differential system is:

$$[\mathbf{M}]\ddot{\mathbf{X}}_D+[\mathbf{K}(t,\mathbf{X})]\mathbf{X}_D\cong-[\mathbf{M}]\hat{\mathbf{D}}\frac{d^2}{dt^2}(TE_S)+\left[\frac{1}{Rb_2}\mathbf{I_P}+[\mathbf{M}]\hat{\mathbf{D}}\right]\frac{d^2}{dt^2}(NLTE) \quad (52)$$

where $\mathbf{D}\hat{}$ = $km\cos\beta_b\,[\mathbf{K}]^{-1}\,\mathbf{V}$, $\mathbf{X}_D=\mathbf{X}-\mathbf{X}_S$, dynamic displacement vector

6.5 Practical consequences

From (51) and (52), it appears that the excitations in geared systems are mainly controlled by the fluctuations of the quasi-static transmission error and those of the no-load transmission error as long as the contact conditions on the teeth are close to the quasi-static conditions (this hypothesis is not verified in the presence of amplitude jumps and shocks). The typical frequency contents of *NLTE* mostly comprise low-frequency component associated with run-out, eccentricities whose contributions to the second-order time-derivative of NLTE can be neglected. It can therefore be postulated that the mesh excitations are dominated by $\frac{d^2}{dt^2}(TE_S)$.

This point has a considerable practical importance as it shows that reducing the dynamic response amplitudes is, to a certain extent, equivalent to reducing the fluctuations of *TES* . Profile and lead modifications are one way to reach this objective. Equation (50) stresses the fact that, when total displacements have to be determined, the forcing terms are proportional to the product of the mesh stiffness and the difference between *TES* and *NLTE* (and not *TES* !). It has been demonstrated by Velex et al. (2011) that a unique dimensionless equation for quasi-static transmission error independent of the number of degrees of freedom can be derived under the form:

$$\cos\beta_b\,\hat{k}(t,X_S)\,TE_S^*(t) = 1 - \int_{L(t,X_S)} \hat{k}(M)e^*(M)dM \tag{53}$$

with, $\hat{A} = \frac{A}{k_m}$, $A^* = \frac{A}{\delta_m}$, for any generic variable A (normalization with respect to the average mesh stiffness and the average static deflection).

Assuming that the mesh stiffness per unit of contact length is approximately constant (see section 2-5), analytical expressions for symmetric profile modifications (identical on pinion and gear tooth tips as defined in Fig. 16) rendering $TE_S(t)$ constant (hence cancelling most of the excitations in the gear system) valid for spur and helical gears with $\varepsilon_\alpha \le 2$ can be found under the form:

$$E = \frac{\Gamma\Lambda}{2\Gamma - 1 + \frac{1}{\varepsilon_\alpha}} \tag{54}$$

submitted to the condition $\Gamma \ge \frac{\varepsilon_\alpha - 1}{2\varepsilon_\alpha}$ with E : tip relief amplitude; Γ : dimensionless extent of modification (such that the length of modification on the base plane is $\Gamma\varepsilon_\alpha Pb$) and $\Lambda = \frac{Cm}{Rb_1 b k_0}$: deflection of reference.

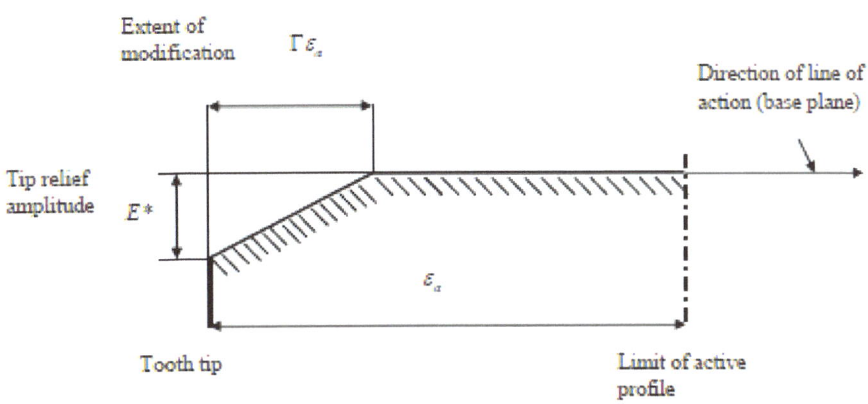

Fig. 16. Definition of profile relief parameters

Based on these theoretical results, it can be shown that quasi-static transmission error fluctuations for ideal gears with profile relief depend on a very limited number of parameters:

i) the profile and lead contact ratios which account for gear geometry and ii) the normalised depth and extent of modification. These findings, even though approximate, suggest that rather general performance diagrams can be constructed which all exhibit a zone of minimum TE variations defined by (54) as illustrated in Figure 17 (Velex et al., 2011).It is to be noticed that similar results have been obtained by a number of authors using very different models (Velex & Maatar, 1996), (Sundaresan et al., 1991), (Komori et al., 2003), etc.

The dynamic factor defined as the maximum dynamic tooth load to the maximum static tooth load ratio is another important factor in terms of stress and reliability. Here again, an approximate expression can be derived from (51-52) by using the same asymptotic expansion as in (34) and keeping first-order terms only (Velex & Ajmi, 2007). Assuming that TE_S and $NLTE$ are periodic functions of a period equal to one pinion revolution; all forcing terms can be decomposed into a Fourier series of the form:

$$_S)+\left[\frac{1}{Rb_2}\mathbf{I_P}+[\mathbf{M}]\hat{\mathbf{D}}\right]\frac{d^2}{dt^2}(NLTE)=-\Omega_1^2\sum_{n\geq 1}n^2\left[A^*{}_n\sin n\Omega_1 t+B^*{}_n\cos n\Omega_1 t\right] \tag{55}$$

and an approximate expression of the dimensionless dynamic tooth load can be derived under the form:

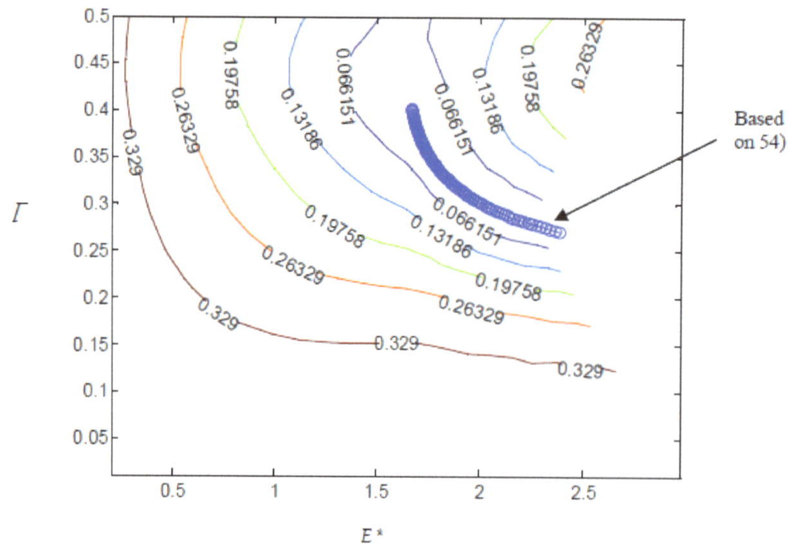

Fig. 17. Example of performance diagram: contour lines of the RMS of quasi-static transmission error under load - Spur gear ε_α = 1.67.

with

$$Y_{pn}(t)=\sum_{n\geq 1}\frac{\bar{A}^*{}_n\left[(\varpi_{pn})^2-1\right]+2\bar{B}^*{}_n\varsigma_p(\varpi_{pn})}{\left[(\varpi_{pn})^2-1\right]^2+4\varsigma_p^2(\varpi_{pn})^2}\sin n\Omega_1 t+\frac{\bar{B}^*{}_n\left[(\varpi_{pn})^2-1\right]-2\bar{A}^*{}_n\varsigma_p(\varpi_{pn})}{\left[(\varpi_{pn})^2-1\right]^2+4\varsigma_p^2(\varpi_{pn})^2}\cos n\Omega_1 t$$

$$\varpi_{pn}=\frac{\omega_p}{n\Omega_1}$$

Equation (56) makes it possible to estimate dynamic tooth loads with minimum computational effort provided that the modal properties of the system with averaged stiffness matrix and the spectrum of *TES* (predominantly) are known. One can notice that the individual contribution of a given mode is directly related to its percentage of strain energy in the meshing teeth and to the ratio of its modal stiffness to the average mesh stiffness. These properties can be used for identifying the usually limited number of critical mode shapes and frequencies with respect to tooth contact loads. They may also serve to test the structural modifications aimed at avoiding critical loading conditions over a range of speeds. It is worth noting that, since α is supposed to be a small parameter, the proposed methodology is more suited for helical gears.

7. TOWARDS CONTINUOUS MODELS

7.1 Pinion, gear distortions

In the case of wide-faced gears, gear body deflections (especially those of the pinion) cannot be neglected and the torsion/bending distortions must be modelled since they can strongly affect the contact conditions between the teeth. For solid gears, one of the simplest approaches consists in modelling gear bodies by two node shaft finite elements in bending, torsion and traction as described in Ajmi and Velex (2005) which are connected to the same mesh interface model as that described in section 3 and Fig. 6. Assuming that any transverse section of the pinion or gear body originally plane remains plane after deformation (a fundamental hypothesis in Strength of Materials), gear bodies can then be sliced into elemental discs and infinitesimal gear elements using the same principles as those presented in section 2. The degrees of freedom of every infinitesimal gear element are expressed by using the shape functions of the two-node, six DOFs per node shaft element. By so doing, all the auxiliary DOFs attributed at every infinitesimal pinion and gear are condensed in terms of the degrees of freedom of the shaft nodes leading to a (global) gear element with 24 DOFs.

7.2 Thin-rimmed applications

The approach in 6.1 is valid for solid gears but is irrelevant for deformable structures such as thin-rimmed gears in aeronautical applications for example where the displacement field cannot be approximated by simple polynomial functions as is the case for shafts. Most of the attempts rely on the Finite Element Method applied to 2D cases (Parker et al., 2000), (Kahraman et al., 2003) but actual 3D dynamic calculations are still challenging and do not lend themselves to extensive parameter analyses often required at the design stage. An alternative to these time-consuming methods is to use hybrid FE/lumped models as described by Bettaieb et al, (2007). Figure 18 shows an example of such a model which combines i) shaft elements for the pinion shaft and pinion body, ii) lumped parameter elements for the bearings and finally iii) a FE model of the gear + shaft assembly which is sub-structured and connected to the pinion by a time-varying, non-linear Pasternak foundation model for the mesh stiffness. The computing time has decreased; however, the modeling challenges at the interfaces among the several sub-models remain complex.

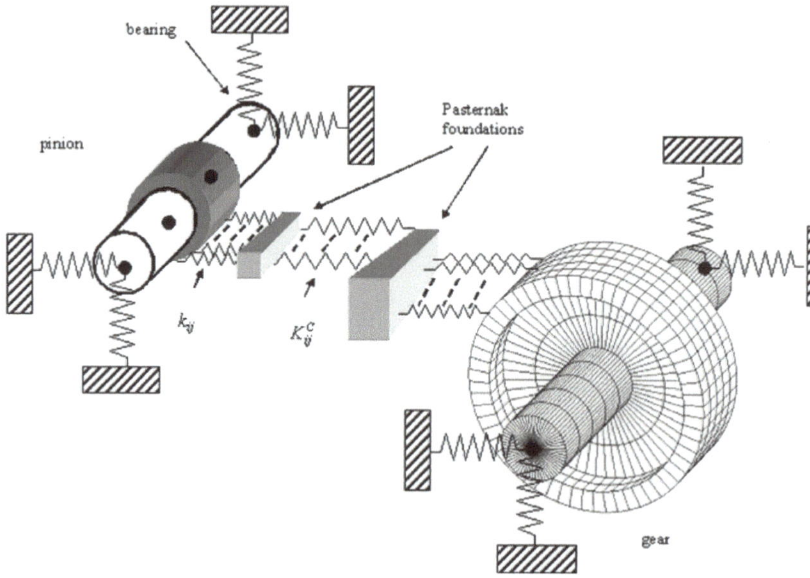

Fig. 18. Example of hybrid model used in gear dynamics (Bettaieb et al., 2007).

8. CONCISE SUMMARY

A systematic approach has been introduced that defines gear elements with all six rigid-body degrees of freedom and time-dependent, potentially non-linear, mesh stiffness functions. Through certain simplifications, some innovative analytical conclusions have been obtained that demonstrate the fundamental processes observed in gear dynamics. These results offer approximate quantitative data on tooth critical frequencies and mesh excitations deemed beneficial during the design phase.

Gear vibration analysis commenced in the late 1950s and encompasses a wide array of study issues and applications that cannot all be addressed in this chapter, including multi-mesh gears, power losses, friction, and bearing-shaft-gear interactions, among others. Gearing is a component of classical mechanics, and a notable disadvantage of its longstanding existence is the pervasive sense of déjà vu, leading to the inclination to assume that gear behavior is fully comprehended and no longer merits investigation (Velex & Singh, 2010). There is a consensus that, despite their existence for centuries, gears will persist well into the 21st century across various machinery and vehicles.

The characterization of damping in geared sets is a priority for the future of gear dynamics, as it significantly influences dynamic load and stress amplitudes. The pressing necessity for enhanced comprehension and modeling of damping in gears was the ultimate conclusion of the seminal paper by Gregory et al. (1963-64). Nearly fifty years later, advancements in this field are scarce, with the notable exception of the findings by Li & Kahraman (2011), which continue to be relevant. A multitude of dynamic models exists in the literature, frequently based on disparate theories. Conversely, experimental findings are notably limited, highlighting an immediate necessity for validated models that extend beyond the foundational results of Munro (1962), Gregory et al. (1963), Kubo (1978), Küçükay (1984 & 1987), Choy et al. (1989), Cai & Hayashi (1994), Kahraman & Blankenship (1997), Baud & Velex (2002), Kubur et al. (2004), among others, particularly for intricate multi-mesh systems. The examination of gear dynamics and noise necessitates multi-scale, interdisciplinary methodologies that incorporate non-linear vibrations, tribology, fluid dynamics, and other relevant fields. The ramifications are evident; significantly increased flexibility will be required, thereby dismantling the conventional boundaries between mechanical engineering, materials science, and chemistry.

REFERENCES

1. Ajmi, M. & Velex, P. (2005). A model for simulating the dynamic behaviour of solid wide- faced spur and helical gear, *Mechanisms and Machine Theory*, vol. 40, n°2, (February 2005), pp. 173-190. ISSN: 0094 -114X.
2. Attia, A. Y. (1964). Deflection of Spur Gear Teeth Cut in Thin Rims, *Journal of Engineering for Industry*, vol. 86, (November 1964), pp. 333-342. ISSN: 0022-0817.
3. Baud, S. & Velex, P. (2002). Static and dynamic tooth loading in spur and helical geared systems—experiments and model validation, *Journal of Mechanical Design*, vol.124, n° 2, (June 2002), pp. 334–346. ISSN: 1050-0472
4. Bettaieb, N.M.; Velex, P. & Ajmi, M. (2007). A static and dynamic model of geared transmissions by combining substructures and elastic foundations - Applications to thin-rimmed gears, *Journal of Mechanical Design*, vol.119, n°2, (February 2007), pp. 184-195, ISSN: 1050-0472
5. Cai, Y. & Hayashi, T. (1994). The linear approximated equation of vibration of a pair of spur gears (theory and experiment), *Journal of Mechanical Design*, Vol. 116, n°2, (June 1994), pp. 558-565, ISSN: 1050-0472
6. Choy F.K., Townsend, D.P. & Oswald F.B. (1989). Experimental and analytical evaluation of dynamic load vibration of a 2240-kW (3000 Hp) rotorcraft transmission, *Journal of the Franklin Institute*, vol. 326, n°5, pp. 721-735, ISSN: 0016-0032.

7. Comparin, R.J & Singh, R.(1989). Non-linear frequency response characteristics of an impact pair, *Journal of Sound and Vibration,* vol. 134, n°2, (October 1989), pp. 259–290, ISSN: 0022-460X
8. Cornell, R.W. (1981). Compliance and stress sensitivity of spur gear teeth, *Journal of Mechanical Design*, Vol. 103, n°2, (April 1981), pp. 447-460, ISSN: 1050-0472
9. Craig, R. R. (1981). *Fundamentals of structural dynamics*, John Wiley, ISBN: 13: 978-0-471- 43044-5, New-York
10. Gregory, R. W.;S. L. Harris, S. L. & Munro, R. G. (1963). Torsional motion of a pair of spur gears. *ARCHIVE: Proceedings of the Institution of Mechanical Engineers, Conference Proceedings*, Vol. 178, n° 3J, pp. 166-173, ISSN: 0367-8849
11. Haddad, C. D. (1991). *The elastic analysis of load distribution in wide-faced helical gears*, PhD dissertation, University of Newcastle, UK.
12. Harris, S. L. (1958). Dynamic loads on the teeth of spur gears, *ARCHIVE: Proceedings of the Institution of Mechanical Engineers*, Vol. 172, (1958), pp. 87-112. ISSN: 0020-3483.
13. Kahraman, A. & Singh, R. (1990). Non-linear dynamics of spur gears, *Journal of Sound and Vibration*, Vol. 142, n°1, (October 1990), pp. 49-75, ISSN: 0022-460X.
14. Kahraman, A. & Singh, R. (1991). Interactions between time-varying mesh stiffness and clearance non-linearities in a geared system, *Journal of Sound and Vibration*, Vol. 146, n°1, (April 1991), pp. 135-156, ISSN: 0022-460X.
15. Kahraman, A. & Blankenship, G.W. (1997). Experiments on nonlinear dynamic behaviour of an oscillator with clearance and periodically time-varying parameters, *Journal of Applied Mechanics,* vol. 64, n°1, (March 1997), pp. 217-227, ISSN: 0021-8936.
16. Kahraman, A; Kharazi, A. A. & Umrani, M. (2003). A deformable body dynamic analysis of planetary gears with thin rims, *Journal of Sound and Vibration*, Vol. 262, n°3, (Mai 2003), pp. 752-768, ISSN: 0022-460X.
17. Komori, M.; Kubo, A. & Suzuki, Y. (2003). Simultaneous optimization of tooth flank form of involute helical gears in terms of both vibration and load carrying capacity, *JSME International Journal, Series C*, vol. 46, n° 4, pp. 1572-1581.
18. Kubo, A. (1978). Stress condition, vibrational exciting force and contact pattern of helical gears with manufacturing and alignment errors, *Journal of Mechanical Design*, Vol. 100, n°1, (1978), pp. 77-84, ISSN: 1050-0472.
19. Kubur, M., Kahraman, A., Zini, D. M. & Kienzle, K. (2004). Dynamic analysis of multi-shaft helical gear transmission by finite elements: model and

experiment, *Journal of Vibration and Acoustics*, vol. 126, n°3, (July 2004), pp. 398-407, ISSN: 1048-9002.

20. Küçükay, F. (1984). Dynamic behaviour of high-speed gears, *Proceedings of the 3rd International Conference on Vibrations in Rotating Machinery*, York, September 1984, pp. 81-90.
21. Küçükay, F. (1987). *Dynamik der Zahnradgetriebe. Modelle, Verfahren, Verhalten*, Springer Verlag, ISBN 3-540-17111-8, Berlin.
22. Lalanne, M. & Ferraris, G. (1998). *Rotordynamics – Prediction in Engineering (2nd edition)*, John Wiley, New-York.
23. Li, S. & Kahraman, A. (2011). A spur gear mesh interface damping model based on elastohydrodynamic contact behaviour, *International Journal of Powertrains*, vol. 1, n°1, (2011), pp. 4-21, ISSN: 1742-4267.
24. Lundberg, G. (1939).Elastische Berührung zweier Halbraüme, *Forschung auf dem Gebiete des Ingenieurwesen*, vol. 10, n°5, (September-October 1939), pp. 201–211.
25. Maatar, M. &Velex, P. (1996). An analytical expression of the time-varying contact length in perfect cylindrical gears-Some possible applications in gear dynamics, *Journal of Mechanical Design*, Vol. 118, n°4, (December 1996), pp. 586-589, ISSN: 1050-0472.
26. Munro, R.G., (1962). *The dynamic behaviour of spur gears.* PhD dissertation, University of Cambridge, UK.
27. Munro, R.G. (1989). The DC component of gear transmission error, *Proceedings of the 1989 ASME International Power Transmission and Gearing Conference*, Chicago, April 1989, pp. 467-470.
28. O'Donnell, W. J. (1960). The Additional Deflection of a Cantilever Due to the Elasticity of the Support, *Journal of Applied Mechanics*, vol. 27 (September 1960), pp. 461-464. ISSN: 0021-8936.
29. O'Donnell, W. J. (1963). Stresses and Deflections in Built-In Beams, *Journal of Engineering for Industry*, vol. 85 (August 1963), pp. 265-273. ISSN: 0022-0817.
30. Ozgüven, H.N. & Houser, D.R. (1988). Mathematical models used in gear dynamics—A review, *Journal of Sound and Vibration*, Vol. 121, n°3, (March 1988), pp. 383–411, ISSN: 0022-460X.
31. Palmgren, A. (1959). *Ball and roller bearing engineering (3rd edition)*, S.H. Burbank & Co, Philadelphia, USA.
32. Parker, R. G.; Vijayakar, S. M. & Imajo, T. (2000). Non-linear dynamic response of a spur gear pair: modelling and experimental comparisons. *Journal of Sound and Vibration*, Vol. 237, n°3, (October 2000), pp. 435-455, ISSN: 0022-460X.

33. Sainsot, P.; Velex, P. & Duverger, O. (2004). Contribution of gear body to tooth deflections- A new bidimensional analytical formula, *Journal of Mechanical Design*, vol. 126, n°4, (July 2004), pp. 748-752, ISSN: 1050-0472.

34. Seager, D. L., (1967). *Some elastic effects in helical gear teeth*, PhD dissertation, University of Cambridge, UK.

35. Singh, R.; Xie, H. & Comparin R.J. (1989). Analysis of automotive neutral gear rattle, *Journal of Sound and Vibration*, vol. 131, n° 2, (June 1989), pp. 177–196. ISSN: 0022-460X.

36. Sundaresan, S.; Ishii, K. & Houser, D. R. (1991). A Procedure Using Manufacturing Variance to Design Gears With Minimum Transmission Error, *Journal of Mechanical Design*, vol. 113, n°3, (September 1991), pp. 318-325, ISSN: 1050-0472.

37. Velex, P. & Berthe,D. (1989). Dynamic tooth loads on geared train, *Proceedings of the1989 ASME International Power Transmission and Gearing Conference*, Chicago, April 1989, pp. 447–454.

38. Velex, P.& Maatar, M. (1996). A mathematical model for analyzing the influence of shape deviations and mounting errors on gear dynamics. *Journal of Sound and Vibration*,Vol. 191, n°5, (April 1996), pp. 629-660, ISSN: 0022-460X.

39. Velex, P. & Ajmi, M. (2006). On the modelling of excitations in geared systems by transmissions errors, *Journal of Sound and Vibration*, Vol. 290, n° 3-5, (March 2006), pp. 882-909, ISSN 0022-460X.

40. Velex, P. & Ajmi, M. (2007). Dynamic tooth loads and quasi-static transmission errors in helical gears – Approximate dynamic factor formulae, *Mechanism and Machine Theory*, vol. 42, n° 11, (November 2007), pp. 1512-1526, ISSN: 0094-114X.

41. Velex, P. & Singh, A. (2010). Top gear, (Guest Editorial) *Journal of Mechanical Design*, vol. 132, n°6, (June 2010), 2 p. , ISSN: 1050-0472.

42. Velex, P.; Bruyère, J. & Houser, D. R. (2011). Some analytical results on transmission errors in narrow-faced spur and helical gears – Influence of profile modifications, *Journal of Mechanical Design*, Vol. 133, n°3, (March 2011), 11 p., ISSN: 1050-0472.

43. Weber, C. & Banaschek, K. (1953). *Formänderung und Profilrücknahme bei Gerad-und Schrägverzahnten Antriebstechnik*, Heft 11, F. Vieweg und Sohn, Braunschweig, Germany.

5 Chapter THE FUNCTION OF THE GEARBOX IN AN AUTOMATIC MECHANISM

1. INTRODUCTION: AN OVERVIEW

A machine is a system composed of various components with distinct functions, interconnected to accomplish a certain goal. A classification of machines having moving parts can be conducted based on the task. Specifically, one can differentiate:

- Drive machines (*motors*): these machines deliver mechanical power from other forms of energy. If their purpose is simply to make placements or generate forces/torques, they are called *actuators*.
- Working machines (*users*): these machines absorb mechanical power to accomplish a specific task (machine tools, transportation, agricultural machinery, textile machinery, machine packaging, etc.).
- Mechanical transmissions: these machines transmit mechanical power by appropriately changing values of torques and speed. Mechanical transmissions are generally made up of *mechanisms* that have been studied (mainly from the point of kinematic view) to connect motors and users.

The combination of a motor, a transmission and a mechanical user is the simplest form of machine.

In servo-actuated machines, the choice of the electric motor required to handle a dynamic load, is closely related to the choice of the transmission Giberti et al. (2011).

The choice of the transmission plays an important role in ensuring the performance of the machine. It must be carried out to meet the limitations imposed by the motor working range and it is subjected to a large number of constraints depending on the motor, through its rotor inertia *JM* or its mechanical speed and on the speed reducer, through its transmission ratio τ, its mechanical efficiency η and its moment of inertia *JT* .

This chapter critically analyzes the role of the transmission on the performance of an automatic machine and clarifies the strategies to choose this component. In particular, it is treated the general case of coupled dynamic addressing the problem of inertia matching and presenting a methodology based on a graphical approach to the choice of the transmission.

The identification of a suitable coupling between motor and transmission for a given load has been addressed by several authors proposing different methods of selection. The most common used procedure are described in Pasch et al. (1984), Van de Straete et al. (1998), Van de Straete et al. (1999), Roos et al. (2006). In these procedures, the transmission is approximated to an ideal system in which power losses are neglected, as the effects of the transmission inertia. A machine is a system comprised of diverse components with specific functions, integrated to achieve a certain objective. A classification of machines with moving components can be performed according to their function. One can specifically differentiate.

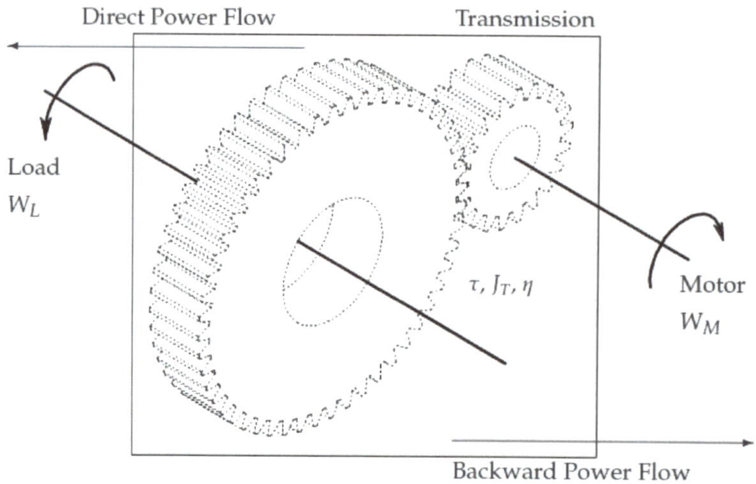

Fig. 1. A scheme of a simple transmission

2. THE TRANSMISSION

To evaluate the effect of the transmission in a motor-load coupling, the transmission can be considered as a black box in which the mechanical power flows (Figures 1, 2, 4). The mechanical power is the product of a torque (extensive factor) for an angular speed (intensive factor).

A mechanical transmission is a mechanism whose aim is:

1. to transmit power
2. to adapt the speed required by the load
3. to adapt the torque.

and it is characterized by a transmission ratio τ and a mechanical efficiency η.

Conventional mechanical transmissions with a constant ratio involve the use of *friction wheels*, *gear*, *belt* or *chains*. The choice of the most suitable transmission for a given application depends on many factors such as dimensions, power, speed, gear ratio, motor and load characteristics, cost, maintenance requirements. Table 1 gives an overview of the most common applications of mechanical transmissions.

Table 1. Typical characteristics of mechanical transmissions.

	Power max (kW)	τ minimum	Dimensions	Cost	Efficiency	Load on bearings
Friction wheels	1/6	20	low	medium	0.90	high
Spur gear	750	1/6	low	high	0.96	low
Helical gears	50000	1/10	low	high	0.98	low
Worm gears	300	1/100	low	high	0.80	medium
Belt	200	1/6	high	low	0.95	high
Trapezoidal belt	350	1/6	medium	low	0.95	high
Toothed belt	100	1/6	medium	low	0.90	low
Linkages	200	1/6	medium	medium	0.90	low

2.1 The transmission ratio

The *transmission ratio* τ is defined as:

$$\tau = \frac{\omega_{out}}{\omega_{in}} \tag{1}$$

where ω*in* and ω*out* are the angular velocity of the input and the output shafts respectively. This value characterizes the transmissions. If τ < 1 the gearbox is a *speed reducer*, while if τ > 1 it isa *speed multiplier.*

The mechanical transmission on the market can be subdivided into three main categories: transmissions with constant ratio τ, transmissions with variable ratio τ and transmissions that change the kind of movement (for example from linear to rotational).

Since it is generally easier to produce mechanical power with small torques at high speeds, the transmission performs the task of changing the distribution of power between its extensive and intensive factors to match the characteristics of the load.

It is possible to define the term μ as the *multiplication factor of force (or torque):*

$$\mu = \frac{T_{out}}{T_{in}} \tag{2}$$

where *Tin*, and *Tout* are respectively the torque upstream and downstream the transmission.

2.2 The mechanical efficiency

If the power losses in the transmission can be considered as negligible, it results:

$$\mu = \frac{1}{\tau}. \tag{3}$$

However, a more realistic model of the transmission has to take into account the inevitable loss of power to evaluate how it affects the correct sizing of the motor-reducer coupling and the resulting performance of the machine.

In general, transmissions are very complex, as the factors responsible for the losses of power. In this chapter they are taken into account just considering the transmission mechanical efficiency η defined as the ratio between the power outgoing from the transmission (*Wout*) and the incoming one (*Win*), or through the extensive factors (*Tin, Tout*) and the intensive ones (ω*in*, ω*out*) of the power itself:

$$\eta = \frac{W_{out}}{W_{in}} = \frac{T_{out}}{T_{in}}\frac{\omega_{out}}{\omega_{in}} = \mu\tau \leq 1 \tag{4}$$

The loss of power within the transmission leads to a reduction of available torque downstream of the transmission. Indeed, if the coefficient of multiplication of forces for an ideal transmission is *μ* = τ−1, when η = 1 it becomes *μ'* < *μ*, thus leading to a consequent reduction of the transmitted torque (*T'o ut* < *Tout*).

Let's define *WM* and *WL* respectively as the power upstream and downstream of the transmission. When the power flows from the motor to the load the machine is said to work with *direct power flow*, otherwise, the functioning is said to be *backward*. Depending on the machine functioning mode, the transmission power losses are described by two different mechanical efficiency values η*d* and η*r* , where:

$$\eta_d = \frac{W_L}{W_M} \text{ (direct power flow)} \qquad \eta_r = \frac{W_M}{W_L} \text{ (backward power flow)}. \tag{5}$$

3. A SINGLE D.O.F. MACHINE

An automatic machine is a system, usually complex, able to fulfill a particular task. Regardless of the type of machine, it may be divided into simpler subsystems, each able to operate only one degree of freedom and summarized by the three key elements: motor, transmission and load (Fig. 2).

3.1 The load and the servo-motor

The power supplied by the motor depends on the external load applied *TL* and on the inertia acting on the system *JL* ω̇*L*. Since different patterns of speed ω*L* and acceleration ω̇*L* generate different loads, the choice of a proper law of motion is the first project parameter that should be taken into account when sizing the motor-reducer unit. For this purpose, specific texts are recommended (Ruggieri et al. (1986), Melchiorri (2000)).

Once the law of motion has been defined, all the characteristics of the load are known.

Electric motor, and among these brushless motors, are the most widespread electrical actuators in the automation field, and their working range can be approximately subdivided into a *continuous* working zone (delimited by the motor rated torque) and a *dynamic* zone (delimited by the maximum motor torque *TM,max*). A typical shape of the working zones is displayed in Fig.3. Usually the motor rated torque decreases slowly with the motor speed ω*M* from *TM,Nc* to *TM,N*. To simplify the rated torque trend and to take a cautious approach, it's usually considered constant and equal to *TM,N* up to the maximum allowed motor speed ω *M, max*, whereas *TM, max* decreases from a certain value of ω*M*.

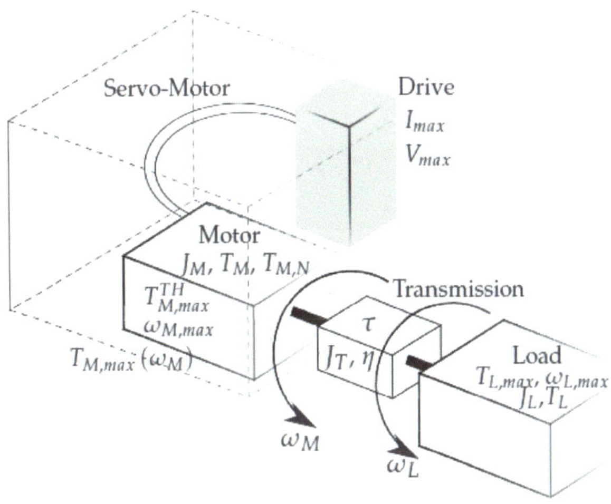

Fig. 2. Scheme of a generic machine actuated by an electric motor

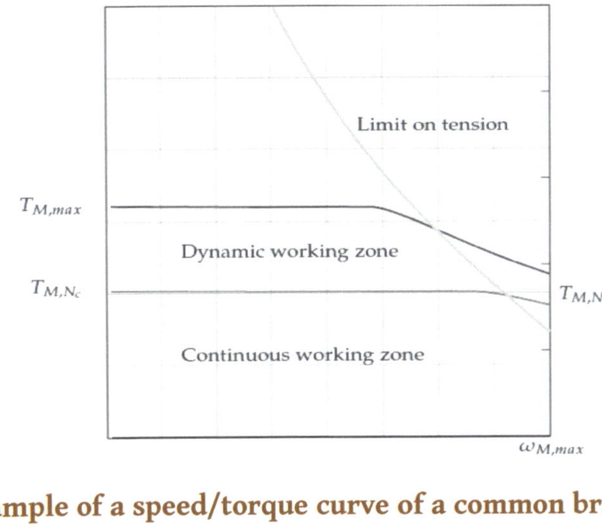

Fig. 3. An example of a speed/torque curve of a common brushless motor

The nominal motor torque $T_{M,N}$, which is specified by the manufacturer in the catalogs, is defined as the torque that can be supplied by the motor for an infinite time, without overheating. Conversely, the trend of the maximum torque $T_{M,max}$ is very complex and depends on many factors. For this reason it is difficult to express it with an equation.

3.2 Conditions to the right coupling between motor and load

Frequently in industrial applications, the machine task is cyclical with period t_a much smaller than the motor thermal time constant. Once the task has been defined, a motor-task

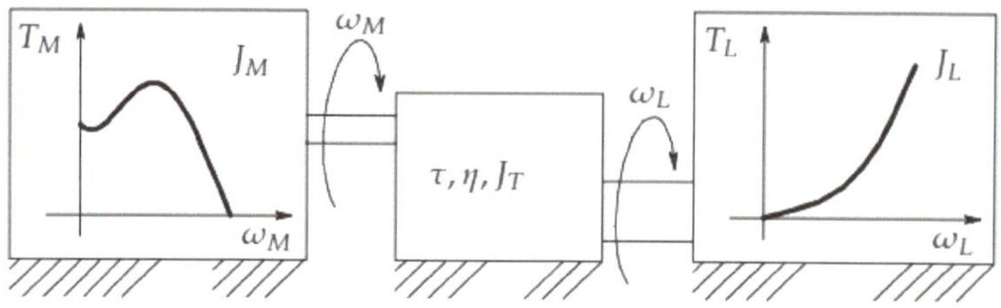

Fig. 4. Scheme of a generic single degree of freedom machine

combination is feasible if there exists a transmission ratio such that:

1. the maximum speed required from the motor is smaller than the maximum speed achievable ($\omega_{M,max}$);

$$\omega_M \leq \omega_{M,max} \tag{6}$$

2. a certain p-norm [1] of the motor torque $\|T_M(t)\|_p$ is smaller than a corresponding motor specific limit T_p. The norms that have most physical meaning for motor torques are Van de Straete et al. (1998):

$$\|T_M(t)\|_2 = T_{rms} = \left(\frac{1}{t_a}\int_0^{t_a} T_M^2(t)dt\right)^{\frac{1}{2}} \tag{7}$$

$$\|T_M(t,\omega)\|_\infty = T_{Max}(\omega) = \max |T_M(t,\omega)| \quad 0 \leq t \leq t_a \tag{8}$$

Since the motor torque is proportional to the current, 2-norm is a measure of mean square current through the windings and it should be limited by the nominal torque $T_{M,N}$ to avoid overheat. The ∞-norm is a measure of peak current and its limit is the maximum torque that can be exerted $T_{M,max}$. These limits translate into:

$$T_{rms} \leq T_{M,N} \tag{9}$$

$$T_{Max}(\omega) \leq T_{M,max}(\omega) \qquad \forall\omega \tag{10}$$

where $T_{M,max}$ depends on ω as shown in Fig. 3.

While the terms on the right of inequalities (6), (9), (10) are characteristics of each motor, ωM, *Trms* and *TMax* depend on the load and on the mechanical features of the motor and the speed reducer.

3.3 Mathematical model of a machine

The machine functioning can be described as an instantaneous balance of power. In the case of direct power flow it can be expressed as:

$$\eta_d\left(\frac{T_M}{\tau}-\frac{J_M\dot{\omega}_L}{\tau^2}\right)=T_L+(J_L+J_T)\,\dot{\omega}_L \qquad (11)$$

while in backward power flow it is:

$$\left(\frac{T_M}{\tau}-\frac{J_M\dot{\omega}_L}{\tau^2}\right)=\eta_r\left[T_L+(J_L+J_T)\,\dot{\omega}_L\right] \qquad (12)$$

Equations (11), (12) can be written as:

$$T_{M,d}=\frac{\tau T_L^*}{\eta_d}+J_M\frac{\dot{\omega}_L}{\tau} \qquad (13)$$

$$T_{M,r}=\tau T_L^*\eta_r+J_M\frac{\dot{\omega}_L}{\tau} \qquad (14)$$

where:

$$T_L^*=T_L+(J_L+J_T)\,\dot{\omega}_L \qquad (15)$$

is the total torque applied to the outgoing transmission shaft.

To unify these different operating conditions, a general mechanical efficiency function is introduced by Legnani et al. (2002). It is defined as:

$$\eta=\begin{cases}\eta_d & \text{(direct power flow)}\\ 1/\eta_r & \text{(backward power flow)}\end{cases} \qquad (16)$$

where η, η*d* and η*r* are constants.

In the case of backward power flow it results η > 1. Note that this does not correspond to a power gain, but it is simply an expedient for unifying the equations of the two working conditions (direct/backward power flow). In fact, in this case, the effective efficiency is η*d* = 1/η < 1.

4. SELECTING THE TRANSMISSION

Conditions (6), (9), (10) can be expressed, for each motor, as constraints on acceptable transmission ratios. However, each transmission is characterized not only by its reduction ratio, but also by its mechanical efficiency and its moment of inertia. This chapter analyzes how these factors affect the conditions (6), (9), (10) and how they reduce the range of suitable transmission ratios for a given motor.

4.1 Limit on the maximum achievable speed

Since each motor has a maximum achievable speed $\omega M,max$, a *limit transmission ratio* τkin can be defined as the minimum transmission ratio below which the system cannot reach the requested speed:

$$\tau_{kin} = \frac{\omega_{L,max}}{\omega_{M,max}} \tag{17}$$

The condition of maximum speed imposed by the system can be rewritten in terms of minimum gear ratio τkin to guarantee the required performance. Eq.(6) becomes:

$$\tau \geq \tau kin \tag{18}$$

4.2 Limit on the root mean square torque

When the direction of power flow during a working cycle is mainly direct or mainly backward, using the notation introduced in eq. (16), the root mean square torque can be expressed as:

$$T_{rms}^2 = \int_0^{t_a} \frac{T_M^2}{t_a} dt = \int_0^{t_a} \frac{1}{t_a} \left(\frac{\tau T_L^*}{\eta} + J_M \frac{\dot{\omega}_L}{\tau} \right)^2 dt \tag{19}$$

Developing the term in brackets and using the properties of the sum of integrals, it is possible to split the previous equation into the following terms:

$$\begin{aligned} \int_0^{t_a} \frac{1}{t_a} \left(\frac{\tau T_L^*}{\eta} \right)^2 dt &= \frac{T_{L,rms}^{*2} \tau^2}{\eta^2} \\ \int_0^{t_a} \frac{1}{t_a} \left(J_M \frac{\dot{\omega}_L}{\tau} \right)^2 dt &= \frac{J_M^2}{\tau^2} \dot{\omega}_{L,rms}^2 \\ \int_0^{t_a} \frac{1}{t_a} \left(\frac{2 T_L^* J_M \dot{\omega}_L}{\eta} \right)^2 dt &= 2 \frac{J_M}{\eta} \left(T_L^* \dot{\omega}_L \right)_{mean} \end{aligned} \tag{20}$$

In cases in which the power flow changes during the cycle it is not possible to choose the proper value of η and eq.(19) is no longer valid. In this circumstance, any individual working cycle must be analyzed to check if one of the two conditions (eq.(11) or eq.(12)) can be reasonably adopted. For example, in the case of purely inertial load (TL = 0) it is2:

$$\left| T_{M,d} \right| \geq \left| T_{M,r} \right| \tag{21}$$

and the equation for direct power flow can be prudentially adopted. Same considerations can be done when the the load is mainly resistant (limited inertia). For all these cases, the condition of eq.(9) becomes:

$$\frac{T_{M,N}^2}{J_M} \geq \frac{\tau^2}{J_M}\frac{T^{*2}_{L,rms}}{\eta^2} + J_M\frac{\dot{\omega}^2_{L,rms}}{\tau^2} + 2\frac{(T_L^*\dot{\omega}_L)_{mean}}{\eta} \tag{22}$$

4.2.1 The accelerating factor and the load factor

Considering eq.(22) two terms can be introduced: the *accelerating factor* Legnani et al. (2002)

$$\alpha = \frac{T_{M,N}^2}{J_M} \tag{23}$$

that characterizes the performance of a motor Giberti et al. (2010b), and the *load factor*:

$$\beta = 2\left[\dot{\omega}_{L,rms}T^*_{L,rms} + (\dot{\omega}_L T_L^*)_{mean}\right] \tag{24}$$

defining the performance required by the task. The unit of measurement of both factors is *W/s*.

Coefficient α is exclusively defined by parameters related to the motor and therefore it does not depend on the machine task. It can be calculated for each motor using the information collected in the manufacturers' catalogs. On the other hand, coefficient β depends only on the working conditions (applied load and law of motion) and it is a term that defines the power rate required by the system.

4.2.2 Range of suitable transmission ratios

Introducing α and β, equation (22) becomes:

$$\alpha \geq \frac{\beta}{\eta} + \left[\frac{T^*_{L,rms}}{\eta}\left(\frac{\tau}{\sqrt{J_M}}\right) - \dot{\omega}_{L,rms}\left(\frac{\sqrt{J_M}}{\tau}\right)\right]^2 \tag{25}$$

Since the term in brackets is always positive, or null, the load factor β represents the minimum value of the right hand side of equation (25). It means that the motor accelerating factor α must be sufficiently greater than the load factor β/η, so that inequality (22) is verified. This check is a first criterion for discarding some motors. If $\alpha > \beta/\eta$, a range of useful transmission ratio exists and can be obtained by solving the biquadratic inequality:

$$\left(\frac{T^{*\,2}_{L,rms}}{\eta^2 J_M}\right)\tau^4 + \left(\frac{\beta}{\eta} - \alpha - 2\frac{T^*_{L,rms}}{\eta}\dot{\omega}_{L,rms}\right)\tau^2 + J_M\dot{\omega}^2_{L,rms} \leq 0 \tag{26}$$

Inequality (26) has 4 different real solutions for τ. As the direction of the rotation is not of interest, only the positive values of τ are considered. A range of suitable transmission ratios is included between a minimum τ_{min} and a maximum gear ratio τ_{max} for which the condition in equation (9) is verified:

$$\left(\frac{T_{L,rms}^{*}{}^{2}}{\eta^2 J_M}\right)\tau^4 + \left(\frac{\beta}{\eta} - \alpha - 2\frac{T_{L,rms}^{*}}{\eta}\dot{\omega}_{L,rms}\right)\tau^2 + J_M\dot{\omega}_{L,rms}^2 \leq 0 \tag{27}$$

$$\tau_{min}, \tau_{max} = \eta\sqrt{J_M}\frac{\sqrt{\alpha - \frac{\beta}{\eta} + \frac{4\dot{\omega}_{L,rms}T_{L,rms}^{*}}{\eta}} \pm \sqrt{\alpha - \frac{\beta}{\eta}}}{2T_{L,rms}^{*}} \tag{27}$$

$$\tau_{min} \leq \tau \leq \tau_{max} \tag{28}$$

>From equation 27 it is evident that a solution exists only if $\alpha \geq \frac{\beta}{\eta}$.

4.2.3 The optimum transmission ratio

The constraint imposed by equation (25) becomes less onerous when a suitable transmission is selected, with a transmission ratio τ that annuls the terms in brackets. This value of τ is called *optimum transmission ratio*. Considering an ideal transmission (η = 1, *JT* = 0 *kgm*²) one gets:

$$\tau = \tau_{opt} = \sqrt{\frac{J_M\dot{\omega}_{L,rms}}{T_{L,rms}^{*}}} \tag{29}$$

that, for a purely inertial load (*TL* = 0), coincides with the value introduced in Pasch et al. (1984):

$$\tau' = \sqrt{\frac{J_M}{J_L}} \tag{30}$$

The choice of the optimum transmission ratio allows system acceleration to be maximized (supplying the same motor torque) or to minimize the torque supplied by the motor (at the same acceleration).

For real transmissions, eq.(29) takes the general form:

$$\tau = \tau_{opt,\eta} = \sqrt{\frac{J_M\dot{\omega}_{L,rms}}{T_{L,rms}^{*}}\eta} = \tau_{opt}\sqrt{\eta} \tag{31}$$

showing the dependence of the optimum transmission ratios on the mechanical efficiency η. Eq. (31) shows how the optimization of the performance of the motor-reducer unit through the concept of inertia matching is considerably affected by the mechanical efficiency. In the following (par. 6.3) this effect will be graphically shown.

4.3 Limit on the motor maximum torque

As shown in Fig.3, each motor can supply a maximum torque *TM,max*(ω*M*) that depends on the speed ω*M*. However this relationship cannot be easily described by a simple equation. As a result, it is difficult to express condition (10) as a range of suitable transmission ratios.

Moreover the maximum torque that can be exerted depends on the maximum current supplied by the drive system. For this reason, these conditions will be checked only once the motor and the transmission have been chosen. It has to be:

$$T_{M,max}(\omega_M) \geq \max \left| \frac{\tau T_L}{\eta} + \left(\frac{J_M + J_T}{\tau} + \frac{J_L \tau}{\eta} \right) \dot{\omega}_L \right| \qquad \forall \omega. \tag{32}$$

This test can be easily performed by superimposing the motor torque *TM* (ω*M*) on the motor torque/speed curve.

4.4 Checks

Once the transmission has been chosen, it is important to check the operating conditions imposed by the machine task satisfy the limits imposed by the manufacturers. Main limits concern:

- the maximum achievable speed;
- the maximum torque applicable on the outcoming shaft;
- the nominal torque applicable on the outcoming shaft;
- the maximum permissible acceleration torque during cyclic operation (over 1000/h) using the load factor.

There is no a standard procedure for checking the transmission. Each manufacturer, according to his experience and to the type of transmission produced, generally proposes an empirical procedure to check the right functioning of his products.

These verifications can be carried out by collecting information from catalogs.

5. IDEAL TRANSMISSION

If both the inertia of the transmission and the power losses are negligible, the gearbox can be considered as ideal (*JT* = 0 kgm2, η = 1). In this cases the problem is easier and the equations describing the dynamical behavior of the machine and the corresponding operating conditions can be simplified.

The limit on the maximum achievable speed remains the same (6), since it arises from kinematic relationships, while limits on the root mean square torque (7) and maximum torque (8) are simplified. Inequality (22) can be written as:

$$\frac{T_{M,N}^2}{J_M} \geq \tau^2 \frac{T^{*2}_{L,rms}}{J_M} + J_M \frac{\dot{\omega}^2_{L,rms}}{\tau^2} + 2(T_L^* \dot{\omega}_L)_{mean}. \tag{33}$$

whose solutions are between

$$\tau_{min}, \tau_{max} = \frac{\sqrt{J_M}}{2T^*_{L,rms}} \left[\sqrt{\alpha - \beta + 4\dot{\omega}_{L,rms} T^*_{L,rms}} \pm \sqrt{\alpha - \beta} \right]. \quad (34)$$

Accordingly, the condition on the maximum torque can be expressed as:

$$T_{M,max}(\omega_M) \geq \max \left| \tau T_L + \left(\frac{J_M}{\tau} + J_L \tau \right) \dot{\omega}_L \right| \qquad \forall \omega. \quad (35)$$

5.1 Selection of gearbox and motor

The main steps to select the gear-motor are:

STEP 1: Creation of a database containing all the commercially available motors and reducers useful for the application. For each motor the accelerating factor (αi) must be calculated. Once the database has been completed it can be re-used and updated each time a new motor-reducer unit selection is needed.

STEP 2: Calculation of the load factor β, on the basis of the features of the load ($TL*$).

STEP 3: Preliminary choice of useful motors: all the available motors can be shown on a graph where the accelerating factors of available motors are compared with the load factor. All the motors for which $\alpha < \beta$ can be immediately rejected because they cannot supply sufficient torque, while the others are admitted to the next selection phase. Figure 7 is related to the industrial example discussed in par.5.2 and displays with circles the acceleration factors αi of the analyzed motors, while the horizontal line represents the load factor β.

STEP 4: Identification of the ranges of useful transmission ratios for each motor preliminarily selected in step 3. For these motors a new graph can be produced displaying for each motor the value of the transmission ratios τmax, τmin, τopt and $\tau M,lim$. The graph is generally drawn using a logarithmic scale for the y-axis, so τopt is always the midpoint of the adoptable transmission ratios range. In fact:

$$\tau^2_{opt} = \tau_{min}\tau_{max} \quad \Leftrightarrow \quad \log \tau_{opt} = \frac{\log \tau_{min} + \log \tau_{max}}{2} \quad (36)$$

A motor is acceptable if there is at least a transmission ratio τ for which equations (18, 28) are verified. These motors are highlighted by a vertical line on the graph. Figure 8 is related to the same industrial example and shows the useful transmission ratios for each motor preliminarily selected.

STEP 5: Identification of the useful commercial speed reducers: the speed reducers available are represented by horizontal lines. If one of them intersects the vertical line of a motor, this indicates that the motor can supply the required torque if that specific speed reducer is selected. Table 2 sums up the acceptable combinations of motors and speed reducers for the case shown in figure 8. These motors and reducers are admitted to the final selection phase.

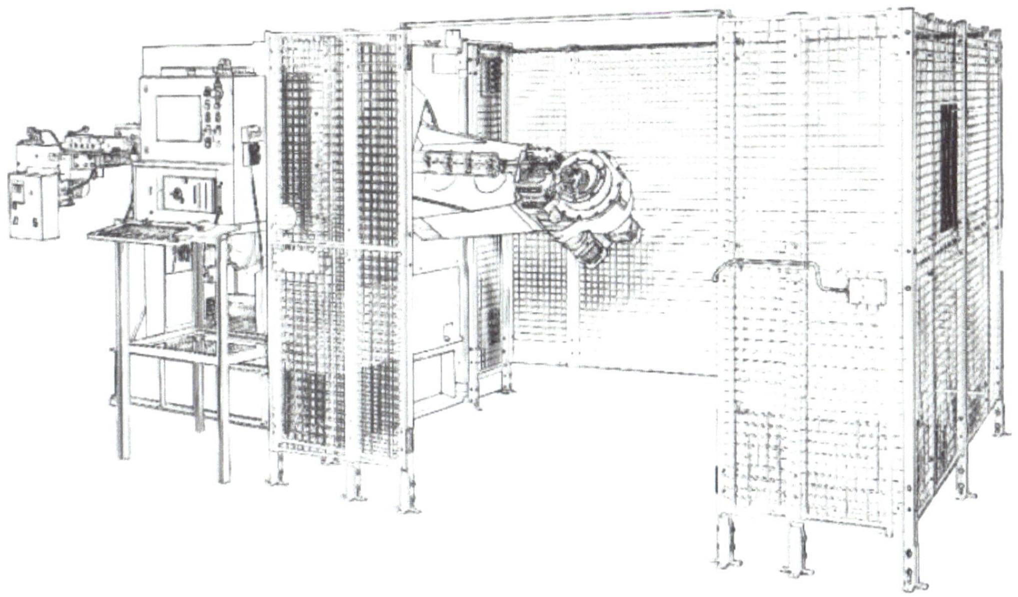

Fig. 5. CNC wire bending machine

STEP 6: Optimization of the selected alternatives: the selection can be completed using different criteria such as economy, overall dimensions, space availability or any other depending on the specific needs.

STEP 7: Checks (see Sec.4.4).

5.2 Example

Fig. 5 shows a CNC wire bending machine. The system automatically performs the task of bending in the plane, or three-dimensionally, a wire (or tape) giving it the desired geometry. The machine operation is simple: semi-finished material is stored in a hank and is gradually unrolled by the unwinding unit. The straightened wire is guided along a conduit to the machine's bending unit, which consists of a rotating arm on which one or more bending heads are mounted. Each head is positioned in space by a rotation of the arm around the axis along which the wire is guided in order to shape it in all directions 6.

The production capacity of the machine is related to the functionality of the heads, while bending productivity depends strongly on arm speed, which allows the heads to reach the position required for bending. The design of the system actuating the rotating arm (selection of motor and speed reducer) is therefore one of the keys to obtaining high performance.

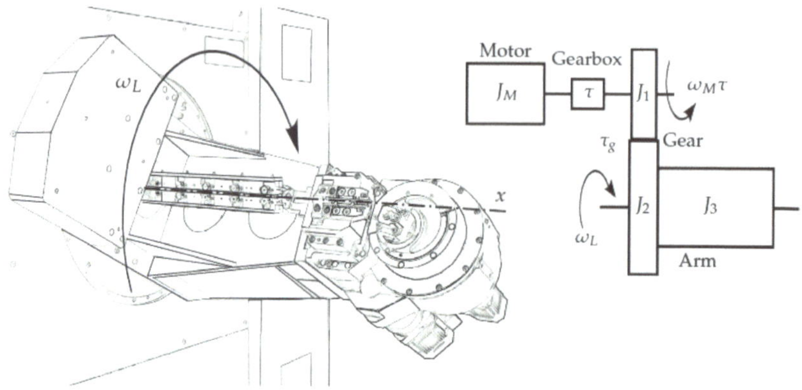

Fig. 6. The moving arm and mechanical system layout

Consider now only the bending unit: the motor, with its moment of inertia JM , is connected through a planetary reducer with transmission ratio τ to a pair of gear wheels that transmits the rotation to the arm.

The pair of gear wheels has a ratio $\tau_2 = 1/5$ which is dictated by the overall dimension of the gearbox and cannot be modified. Putting $J1 = 0.0076$ [kgm2], $J2 = 1.9700$ [kgm2] and $J_3 = 26.5$ [kgm2] respectively as the moment of inertia referred to the axes of rotation of the two wheels and of the arm, the comprehensive moment of inertia referred to the output shaft of the planetary gear is:

$$JL = J1 + (J2 + J3)\tau2_{2} = 1.1464 \text{ [kgm2]}$$

Since the load is purely inertial3, the load factor can easily be calculated as:

$$\beta = 4J_L\dot{\omega}^2_{L,rms} \tag{37}$$

where ω˙ L,rms is a function of the law of motion used. The choice of the law of motion depends on the kind of operation requested and, in the most extreme case, it consists of a rotation of $h = 180°$ in $ta = 0.6$ [s]. After this, a stop of $ts = 0.2$ [s] before the next rotation is normally scheduled.

The value of ω˙ L,rms can be expressed through the mean square acceleration coefficient (ca,rms) using the equation:

$$\dot{\omega}_{L,rms} = c_{a,rms}\frac{h}{t_a^2}\frac{1}{\tau_2}\sqrt{\frac{t_a}{t_a + t_s}}$$

As is known (Van de Straete et al., 1999), the minimum√mean square acceleration law of motion is the cubic equation whose coefficient is $crms = 2\sqrt{3}$. . Moreover, this law of motion has the advantage of higher accelerations, and therefore high inertial torques, corresponding to low velocities. Substituting numerical values in eq.(37) one gets: $\beta = 7.8573 \cdot 10^4$ [W/s].

Considering the same law of motion, maximum acceleration and maximum speed can easily be obtained by:

$$\dot{\omega}_{L,max} = c_a \frac{h}{t_a^2} \frac{1}{\tau_2} \simeq 261.8\ [\mathrm{rad/s^2}]; \quad \omega_{L,max} = c_v \frac{h}{t_a} \frac{1}{\tau_2} \simeq 39.3\ [\mathrm{rad/s}] \tag{38}$$

where ca = 6 and cv = 1.5.

Knowing the load factor β and after selecting the motors and transmissions available from catalogs, the graph shown in Fig.7 can be plotted. Available motors for this application are synchronous sinusoidal brushless motors4. Manufacturer's catalogs give information on motor inertia, maximum and nominal torque. A first selection of suitable motors can be performed. Motors whose accelerating factor αi is lower than the load factor β can be discarded.

For all the accepted motors a new graph can be produced. It displays, for each motor, the corresponding minimum and maximum transmission ratios and the optimum and the minimum kinematic transmission ratios.

They can be obtained using the simplified expression for the purely inertial load case:

$$\tau_{min}, \tau_{max} = \frac{\sqrt{J_M}}{2T^*_{L,rms}} \left[\sqrt{\alpha} \pm \sqrt{\alpha - \beta}\right]. \tag{39}$$

and eq.(29),(17). Commercial transmissions considered for the selection are planetary reducers5.

The graph in Fig.8 shows all the available couplings between the motors and transmissions considered. Three of the eleven motors (M1, M2 e M7) are immediately discarded, since their accelerating factors α are too small compared with the load factor β. Motors M3, M4, M5, and M6 are eliminated because their maximum speed is too low. Suitable motors are M8, M9, M10 and M11. The selection can be completed evaluating the corresponding available commercial speed reducers whose ratio is within the acceptable range. Motor M8 is discarded since no transmission can be coupled to it. Suitable pairings are shown in Tab.2.

Table 2. Combination of suitable motors and speed reducers for the industrial example discussed in par.5.2

Motor	Speed reducer
M9	τ = 1/10, τ = 1/7
M10	τ = 1/5, τ = 1/4
M11	τ = 1/5, τ = 1/4, τ = 1/3

The final selection can be performed using the criterion of cost: the cheapest solution is motor M9 and a reducer with a transmission ratio $\tau = 1/10$. The main features of the selected motor6 and transmission7 are shown in table 3.

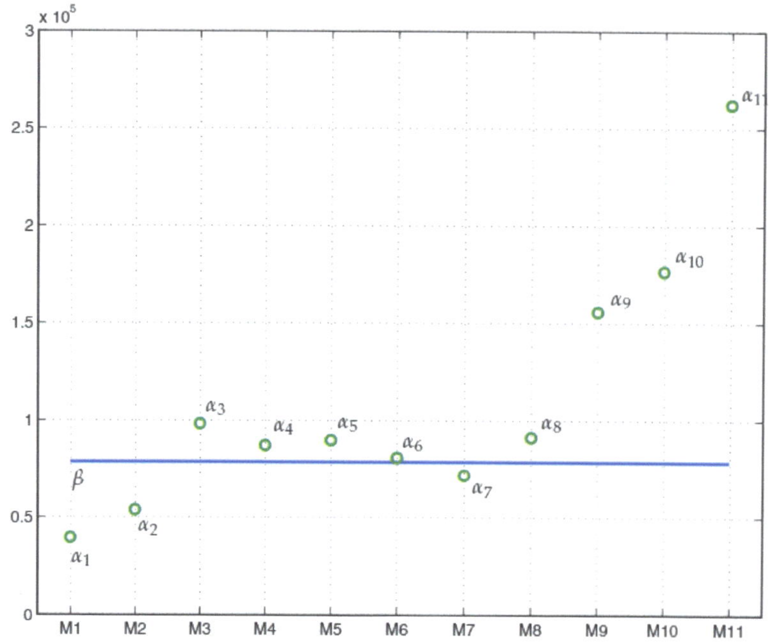

Fig. 7. A first selection of suitable motors

Table 3. Main features of selected motor and transmission

Motor M9	
moment of inertia nominal torque maximum Torque	J_M = 0.0046 [kgm^2] $T_{M,N}$ = 26.7 [Nm] T^{TH} = 132 [Nm]
maximum achievable speed	M,max $\omega_{M,max}$ = 5000 [rpm]
Speed reducer τ क्ष 1/10	
moment of inertia nominal torque	$J_T = 5.8 \cdot 10^{-4}$ [kgm^2] $T_{T,N}$ = 220 [Nm]
maximum Torque	$T_{T,max}$ = 480 [Nm]
maximum endurable speed	$\omega_{T,max}$ = 4000 [rpm]
nominal speed	$\omega_{T,N}$ = 2600 [rpm]
mechanical efficiency	η = 0.97

Figure 9 shows the required motor torque as a function of speed during the working cycle. It has been calculated considering the inertia of both the motor and the gearbox and the mechanical efficiency of the transmission. Since the mechanical efficiency of the speed reducer in backward power flow mode is not available, it is assumed to be equal to that in direct power flow. To verify the condition on the maximum torque reported in eq.(32), the curve has to be contained within the dynamic working field. Note how the maximum torque achieved by the motor is limited by the drive associated with it. From Fig.9 it is possible to check that the condition on the maximum torque is verified.

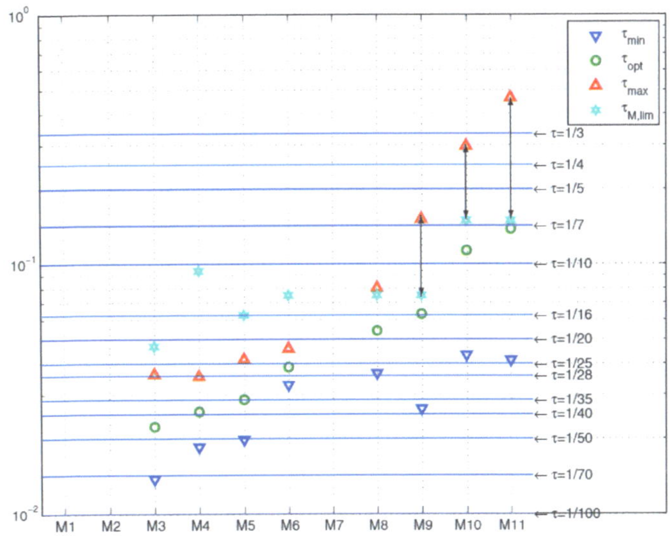

Fig. 8. Overview of available motor-reducer couplings

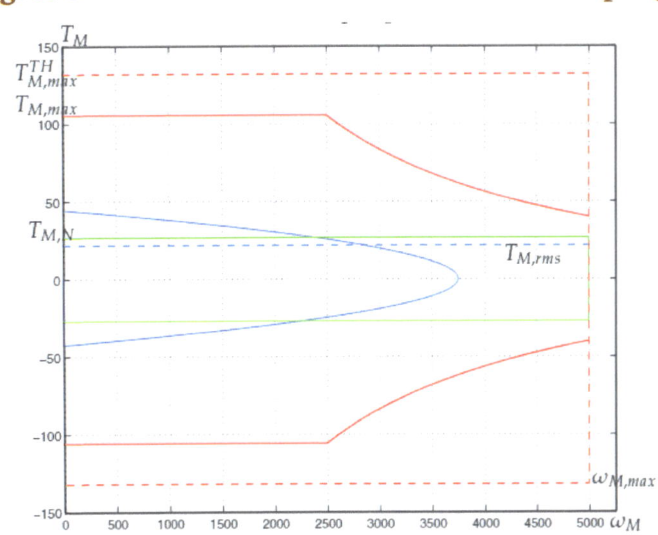

Fig. 9. Checkouts on maximum and nominal torque.

The motor root mean square torque can now be updated, considering the inertia of the transmission and its mechanical efficiency.

Finally, checks should be carried out on the reducers following the manufacturer's guidelines. In this case they mainly consist of verifying that both the maximum and the nominal torque applied to the transmission incoming shaft are lower than the corresponding limits shown in the catalog ($T_{T,max}$, $T_{T,N}$).

$$T_{max} \simeq 300[Nm] < T_{T,max} = 480[Nm]$$

$$T_n \simeq 150[Nm] < T_{T,N} = 220[Nm]$$

In addition, the maximum and the mean angular speed of the incoming shaft have to be lower than the corresponding limits on velocity ($\omega_{T,max}$, $\omega_{T,N}$).

$$n_{max,rid} \simeq 3750[rpm] < \omega_{T,max} = 4000[rpm]$$

$$n_{mean,rid} \simeq 1873[rpm] < \omega_{T,N} = 2600[rpm]$$

The selected motor-transmission pairing satisfies all the checks and provides margins for both the motor (≈ 20% on the nominal torque) and the reducer.

6. REAL TRANSMISSION

For machines working with direct power flow, a decrease of the performance of the transmission that corresponds to an increase in the power dissipation may result in a motor overhead which makes it non longer adequate.

6.1 The mechanical efficiency limit

A motor which is able to perform the task planned in ideal conditions ($\eta = 1$), when coupled with a transmission characterized by poor efficiency, could be discarded. Referring to equation (27), once β is known, a minimum transmission mechanical efficiency exists, for each motor, below which τ_{min} and τ_{max} are undefined. The limit value is called the transmission *mechanical efficiency limit* and it is defined as the ratio between the load factor and the accelerating factor:

$$\eta \geq \eta_{lim} = \frac{\beta}{\alpha} \tag{40}$$

This parameter gives to the designer a fundamental indication: if the task required by the machine is known (and thus the load factor can be calculated), for each motor there is a minimum value of the transmission mechanical efficiency below which the system can not work. This limit is not present in the case of backward power flow functioning.

$$\tau_{min}, \tau_{max} = \begin{cases} \eta\sqrt{J_M}\dfrac{\sqrt{\alpha - \frac{\beta}{\eta} + \frac{4\dot{\omega}_{L,rms}T^*_{L,rms}}{\eta}} \pm \sqrt{\alpha - \frac{\beta}{\eta}}}{2T^*_{L,rms}} & \text{if } \eta \geq \eta_{lim} \\ \text{undefined} & \text{if } \eta < \eta_{lim} \end{cases} \quad (41)$$

6.2 Restriction of the range of useful transmission ratios

For each selectable motor it is possible to graphically represent the trend of both the minimum and maximum transmission ratios. Combining equations (27) and (40), the two functions τmin and τmax are respectively defined as:

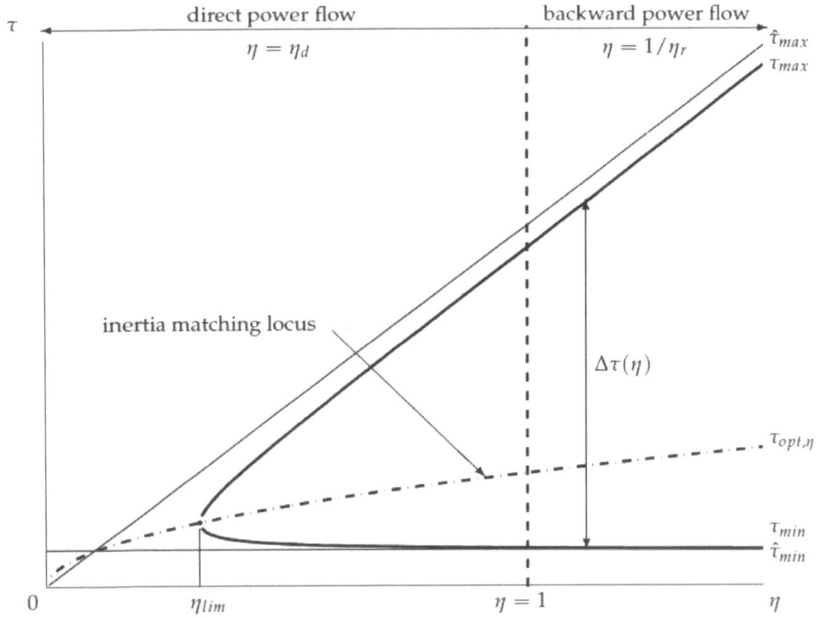

Fig. 10. Trends of τmin and τmax as functions of mechanical efficiency (using the notation introduced in eq. (16)), for a certain motor.

For each motor it is possible to plot τmin and τmax as functions of the mechanical efficiency, highlighting a region in the plane $\eta\tau$ satisfying the condition on the root mean square torque (Fig.10). Depending on the machine functioning mode (direct or backward power flow) the left side or the right side of the graph should be taken into account. On the same graph the trends of the optimum transmission ratio $\tau opt,\eta$ and the breadth $\Delta\tau$ of the range of useful transmission ratios is shown. Note that the range grows monotonically with the difference between the accelerating and load factors. In particular, the range breadth is:

$$\Delta\tau(\eta) = \frac{\sqrt{J_M}}{T^*_{L,rms}}\eta\sqrt{\alpha - \frac{\beta}{\eta}} \quad (42)$$

and decreases appreciably with the transmission mechanical efficiency. However, while the limit on the maximum transmission ratio *τmax* varies considerably, the minimum transmission ratio *τmin* remains almost constant. This behavior is clearly visible in the ητ graph which plots the asymptotes of the two functions (Fig. 10) described respectively by:

$$\hat{\tau}_{max} = \frac{T_{M,N}}{T^*_{L,rms}}\eta \qquad \hat{\tau}_{min} = \frac{J_M}{T_{M,N}}\dot{\omega}_{L,rms} \tag{43}$$

It is interesting to observe that, while τ *max* depends on the reducer, τ *min* depends only on the chosen motor and on the law of motion defined by the task. This is because the transmission ratio τ is so small that the effect of the load is negligible compared to the inertia of the motor.

The power supplied, therefore, is used just to accelerate the motor itself. Note that, for values of η > 1, that is backward power flow functioning, the range of suitable transmission ratios is wider than in the case of direct power flow functioning, which is the most restrictive working mode. For this reason, in all cases where the direction of power flow is not mainly either direct or backward, the first functioning mode can be considered as a precautionary hypothesis on the root mean square torque and the left side of the ητ graph (η < 1) can be used.

6.3 The extra-power rate factor

Inequality (25) can be written as:

$$\alpha \geq \frac{\beta}{\eta} + \gamma(\tau, \eta, J_M) \tag{44}$$

where:

$$\gamma(\tau, \eta, J_M) = \left[\frac{T^*_{L,rms}}{\eta}\left(\frac{\tau}{\sqrt{J_M}}\right) - \dot{\omega}_{L,rms}\left(\frac{\sqrt{J_M}}{\tau}\right)\right]^2 \tag{45}$$

The term γ is called the *extra-power rate factor* and represents the additional power rate that the system requires if the transmission ratio is different from the optimum $(\tau \neq \tau_{opt})$.

Figure 11 shows the trends of the terms of the γ function when the ransmission efficiency changes. Note that, when the transmission ratio is equal to the optimum (τ = *τopt*), the curve γ reaches a minimum. For this value the convexity of the function is small and, even for large variations of the transmission ratio, the extra-power rate factor appears to be contained in eq.(44).

With the mechanical efficiency decreasing, two effects take place: first the optimum transmission ratio decreases, moving on the left of the graph, secondly the convexity of the curve γ is more pronounced and the system is more sensitive to changes in τ with respect to the optimum. For transmissions characterized by poor mechanical efficiency, in the case of the direct power flow mode, the choice of a gear ratio different from the optimum significantly affects the choice of the motor.

6.4 Effect of the transmission inertia

The inclusion of the transmission inevitably changes the moment of inertia of the system. With *JT* as the moment of inertia of the speed reducer, referred to its outgoing shaft, the resistive torque is generally given by eq.(15). Entering this new value in eq.(24), the load factor can be updated. This change makes the system different from that previously studied with the direct consequence that the limit on the mean square torque can no longer be satisfied.

In particular, for the *ith* transmission, characterized by a moment of inertia *JT,i*, the limits on the transmission ratio to satisfy the root mean square torque condition can be expressed as *τmin,i* and *τmax,i*. Let's consider, as example, a machine task characterized by a constant resistant load (*TL* = cost). Figure 12 shows how the range of suitable transmission ratios is reduced when the inertia of the transmission increases. The same reduction can be observed in the ητ graph (Fig. 13) for a purely inertial load. Note that, even for the moment of inertia, there is a limit value beyond which, for a given motor, there is no suitable transmission ratio.

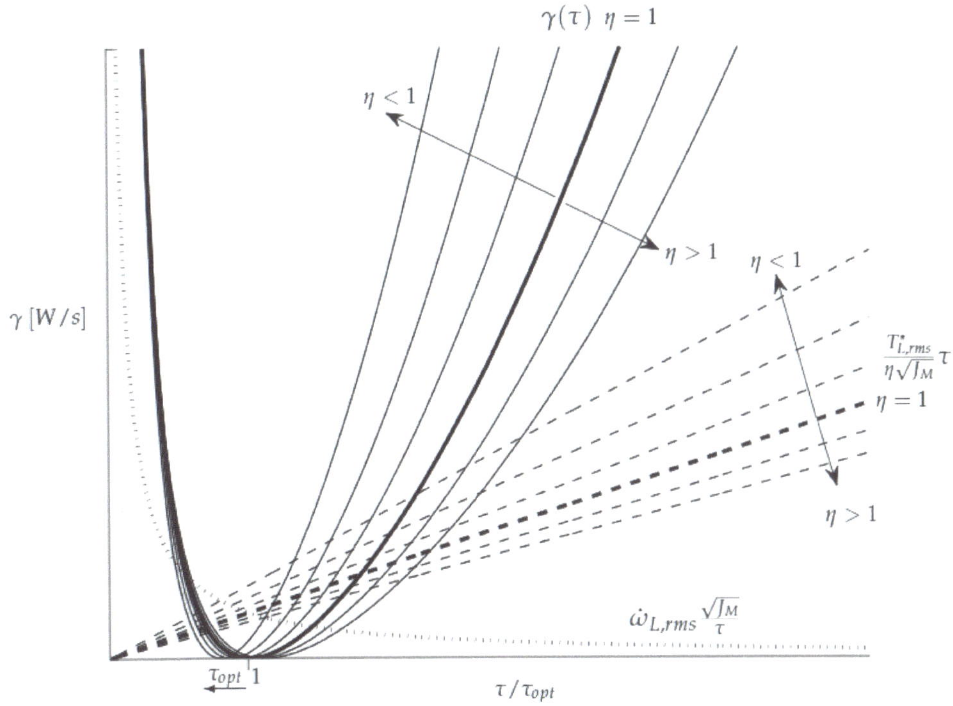

Fig. 11. Variation of the extra-power rate factor as a function of the transmission ratio and mechanical efficiency.

7. GUIDELINES FOR THE MOTOR-REDUCER SELECTION

The theoretical steps presented can be summarized by a series of graphs for evaluating the effect of the transmission on the choice of the motor that make the selection process easy to use.

Firstly, to ensure the condition on the root mean square torque it should be verified that the accelerating factor α of each available motor is greater than the limit β/η. However, since the transmission has not yet been selected and thus its efficiency and inertia are still unknown, it is possible to perform only a first selection of acceptable motors, eliminating those for which $\alpha < \beta$.

For each selectable motor a $\eta\tau$ graph can be plotted (Fig. 14), with the limits on the transmission ratio defined by equations (18), (28). Since these functions depend on the transmission moment of inertia, such limits are plotted for each available transmission.

Available transmissions can be inserted in the $\eta\tau$ graph (there is an example in Fig. 14) using their coordinates (ηdi, τi) and (ηri, τi) which can easily be found in manufacturers' catalogs. Each speed reducer appears twice: to the left of the dashed line for direct power flow, to the right for backward power flow.

Remember that for tasks characterized by mainly backward power flow, only the transmissions on the right half plane should be considered ($\eta > 1$). For all other cases only the transmissions on the left half plane ($\eta < 1$) should be taken into account.

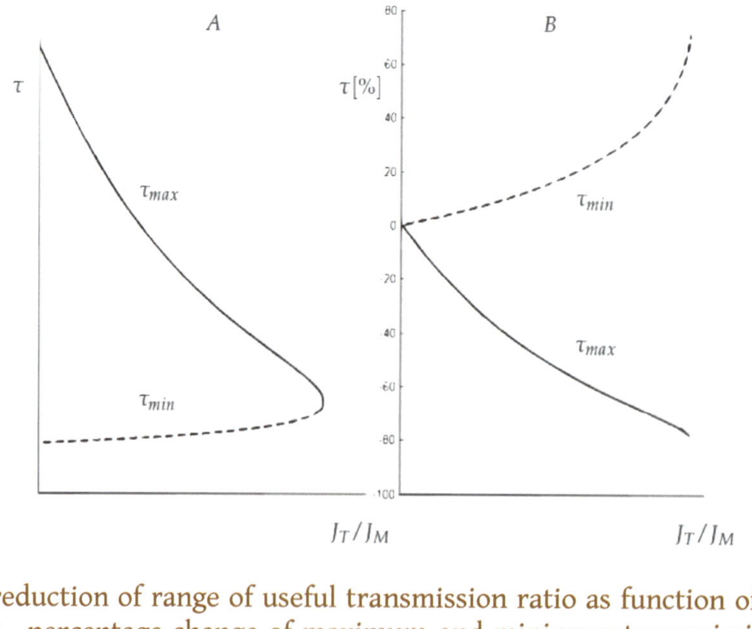

Fig. 12. A - reduction of range of useful transmission ratio as function of *JT /JM* (for *TL* = cost); B - percentage change of maximum and minimum transmission ratios as function of *JT /JM* (for *TL* = cost).

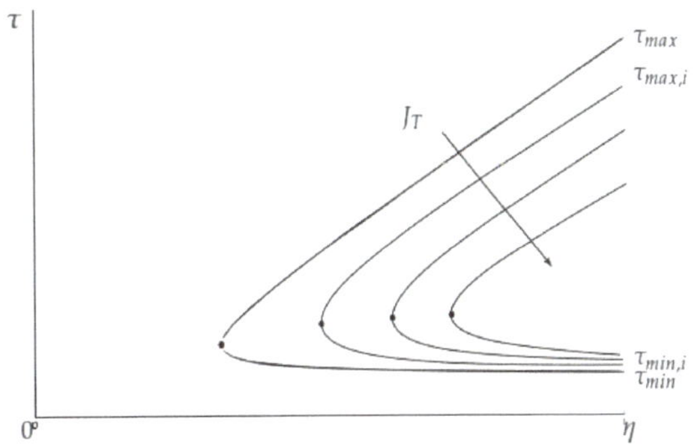

Fig. 13. Effect of the transmission moment of inertia on the ητ graph (for *TL* = cost).

The *ith* speed reducer is acceptable if it lies inside the area limited by the limit transmission ratio (τ*kin*) and the corresponding maximum and minimum ratios (τ*max,i*, τ*min,i*). These ⊙,reducers are highlighted on the graph with the symbol unacceptable ones with ⊗

Table 4 resumes all the alternatives both for the direct and backward power flow modes.

Note that, for the direct power flow mode, transmissions which would be acceptable with normal selection procedures, are now discarded (e.g. transmission T_2 because of insufficient efficiency, reducer T_4 because of its excessively high moment of inertia).

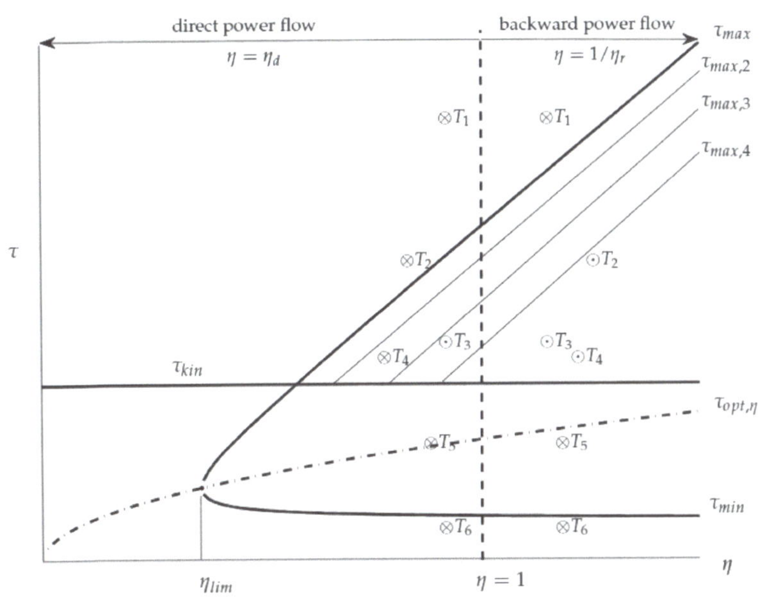

Fig. 14. Graph ητ for the selection of the speed reducer for a specific motor: (⊙, acceptable transmission, ⊗ unacceptable transmission)

Table 4. Overview of acceptable and unacceptable transmissions for the example in Fig. 14

Transmission	Direct P.F.	Cause	Backward P.F.	Cause
T_1	unaccept.	$\tau_1 > \tau_{max}$	unaccept.	$\tau_1 > \tau_{max}$
T_2	unaccept.	$\tau_2 > \tau_{max}$	accept.	—
T_3	accept.	—	accept.	—
T_4	unaccept.	$\tau_4 > \tau_{max,4}$	accept.	—
T_5	unaccept.	$\tau_5 < \tau_{kin}$	unaccept.	$\tau_5 < \tau_{kin}$
T_6	unaccept.	$\tau_6 < \tau_{kin} \wedge \tau_6 < \tau_{min}$	unaccept.	$\tau_6 < \tau_{kin} \wedge \tau_6 < \tau_{min}$

Moreover it is evident that, for the motor considered, there are no acceptable transmissions with a ratio equal to the optimum. More generally it could happen that a motor, while meeting all the constraints mentioned, may have a range of acceptable gear ratios within which there is no reducer commercially available. For this reason it cannot be chosen.

The choice of the motor-reducer unit is made easy by comparing the $\eta\tau$ graphs for each selectable motor.

The resulting graphs give an overview of all the possible pairings of motors and transmissions that satisfy the original conditions. They allow the best solution to be selected from the available alternatives, in terms of cost, weight and dimensions, or other criteria considered important according to the application.

Once the motor and the transmission have been selected and all their mechanical properties are known, the final checks can be performed.

8. CONCISE SUMMARY

The appropriate selection of the motor-reducer unit is a crucial element in automation applications. The selection must consider the mechanical limits of the components, particularly the operational ranges of the drive system and the mechanical characteristics of the gearbox. The research examines the impact of these restrictions on the optimal selection of the motor-reducer unit at a theoretical level and presents a methodology for its selection that facilitates the identification of the most effective combination through a pragmatic approach to the issue.

It delineates the impact of the transmission's mechanical efficiency and inertia on the interaction between the motor and reducer, illustrating their effects on the optimal solution. The technique involves creating a chart that consolidates all necessary information for accurate system sizing, thereby summarizing all potential solutions and facilitating rapid comparison to identify the optimal choice.

9. NOMENCLATURE

Symbol	Description
T_M	motor torque
J_M	motor moment of inertia
$T_{M,N}$	motor nominal torque
$T_{M,max}$	motor maximum torque
$\omega_M, \dot{\omega}_M$	motor angular speed and acceleration
T_L	load torque
J_L	load moment of inertia
T_L^*	generalized load torque
$T_{L^*,rms}$	generalized load root mean square torque
$T_{L,max}$	load maximum torque
$\omega_L, \dot{\omega}_L$	load angular speed and acceleration
$\dot{\omega}_{L,rms}$	load root mean square acceleration
t_a	cycle time
$\tau = \omega_L/\omega_M$	transmission ratio
τ_{opt}	optimal transmission ratio
η	transmission mechanical efficiency
η_d	transmission mechanical efficiency (direct power flow)
η_r	transmission mechanical efficiency (backward power flow)
J_T	transmission moment of inertia
α	accelerating factor
β	load factor
γ	extra-power rate factor
τ_{min}, τ_{max}	minimum and maximum acceptable transmission ratio
τ_{kin}	minimum kinematic transmission ratio
$\omega_{M,max}$	maximum speed achievable by the motor
$\omega_{L,max}$	maximum speed achieved by the load
W_M	motor side power
W_L	load side power

FOOTNOTES

1. In general, the p-norm $\|\cdot\|_p$ of a time function $f(t)$ over the period t_a is defined as:

$$T_L^* = (J_L + J_T)\,\dot{\omega}_L$$

2. For purely inertial load eq.(15) can be written as:

$$T_L^* = (J_L + J_T)\,\dot{\omega}_L$$

Then eq.(21) becomes:

$$\left(\frac{\tau\,(J_L + J_T)}{\eta_d} + \frac{J_M}{\tau}\right)|\dot{\omega}_L| \geq \left(\tau\,(J_L + J_T)\,\eta_r + \frac{J_M}{\tau}\right)|\dot{\omega}_L|$$

and thus:

$$\frac{1}{\eta_d} \geq \eta_r$$

which is always true since both ηd and ηr are defined as positive and smaller than

3. In the selecting phase frictions are not considered.
4. Produced by "Mavilor", *http://www.mavilor.es/.*
5. produced by "Wittenstein", *http://www.wittenstein.it/.*
6. Model Mavilor BLS 144.
7. Model Alpha SP+140.

REFERENCES

1. Pasch, K.A., Seering, W.P., "On the Drive Systems for High-Performance Machines", *Transactions of ASME*, 106, 102-108 (1984).
2. Cusimano, G., "A Procedure for a Suitable Selection of Laws of Motion and Electric Drive Systems Under Inertial Loads", *Mechanism and Machine Theory* 38, 519-533 (2003)
3. Chen, D.Z., Tsai, L.W., "The Generalized Principle of Inertia Match for Geared Robotic Mechanism", *Proc. IEEE International Conference on Robotics and Automation* , 1282-1286 (1991)
4. Van de Straete, H.J, Degezelle, P., De Shutter, J., Belmans, R., "Servo Motor Selection Criterion for Mechatronic Application", *IEEE/ASME Transaction on Mechatronics* 3, 43-50 (1998).
5. Van de Straete, H.J, De Shutter, J., Belmans, R., "An Efficient Procedure for Checking Performance Limits in Servo Drive Selection and Optimization", *IEEE/ASME Transaction on Mechatronics* 4, 378-386 (1999)
6. Van de Straete, H.J, De Shutter, J., Leuven, K.U., "Optimal Variable Transmission Ratio and Trajectory for an Inertial Load With Respect to Servo Motor Size", *Transaction of the ASME* 121, 544-551 (1999)

7. Giberti, H, Cinquemani, S., Legnani, G., "A Practical Approach for the Selection of the Motor-Reducer Unit in Electric Drive Systems" *Mechanics Based Design of Structures and Machines*, Vol. 39, Issue 3, 303-319 (2011)
8. Giberti, H, Cinquemani, S., Legnani, G., "Effects of transmission mechanical characteristics on the choice of a motor-reducer", *Mechatronics* Vol. 20, Issue 5, 604-610 (2010)
9. Giberti, H, Cinquemani, "On Brushless Motors Continuous Duty Power Rate", *ASME 2010 10th Biennial Conference on Engineering Systems Design and Analysis, ESDA 2010*, Istanbul, Turkey, 12-14 July (2010).
10. Legnani, G., Tiboni, M., Adamini, R., "Meccanica degli Azionamenti", Ed. Esculapio, Italy, (2002)
11. Roos, F., Johansson, H., Wikander, J., "Optimal Selection of Motor and Gearhead in Mechatronic Application", *Mechatronics* 16, 63-72 (2006)
12. Magnani P.L., Ruggieri G. Meccanismi per Macchine Automatiche, UTET, Italy (1986) Melchiorri C., Traiettorie per azionamenti elettrici, Esculapio, Italy (2000)

Chapter 6 ELECTRICAL DRIVE SYSTEMS FOR CRANE APPLICATIONS

1. INTRODUCTION: AN OVERVIEW

A crane is a form of machinery primarily employed for lifting heavy loads across various industrial sectors, including metallurgy, paper manufacturing, and cement production. By their construction, cranes are categorized into overhead cranes and gantry cranes. An overhead crane, also referred to as a bridge crane, is a form of crane in which the hook and hoist mechanism operate along a horizontal beam that moves along two widely spaced rails. It is frequently situated within a factory structure and operates along rails affixed to the two elongated walls. A gantry crane resembles an overhead crane, constructed such that the superstructure supporting the trolley is securely supported by two or more legs that move along fixed rails embedded in the floor. Stationary or portable units may be installed either outdoors or indoors. Certain industries, such as port container handling and open storage receptacle operations, necessitate the use of wide span gantry cranes. In outdoor applications, the impact of wind on the drive's performance can be significant (Busschots, 1991). Wind and skew can substantially impact the secure operation of the crane. This will definitively determine the type of crane design (lattice or box type) from a mechanical perspective, as well as influence the selection, dimensions, and control of the crane's electrical motors.

Electrical technology for crane control has experienced a substantial transformation over the past few decades. The transition from the Ward Leonard system to DC drive technology, coupled with the emergence of advanced Insulated Gate Bipolar Transistors (IGBTs) in the 1990s, facilitated the development of AC drives (Backstrand, 1992; Paul et al., 2008). Conventional AC-operated crane engines employ slip ring induction motors, whose rotor windings are connected to a resistor bank in four to five stages via power contactors. Reversal is achieved by altering the phase sequence of the stator supply via line contactors. Braking is accomplished through the process of filling. The primary disadvantage is that the actual pace is dependent on the load. An electronic control system has recently been implemented to continuously regulate the rotor resistor value. Currently, these systems are replaced by frequency converters powered by squirrel-cage induction motors for all categories of motion (Paul et al., 2008). Control concepts based on the application of Programmable Logic Controllers (PLCs) and industrial communication networks (fieldbuses) constitute a standard solution employed in complex applications (Slutej et al., 1999).

Overhead and gantry cranes are commonly employed for the transportation of containers, loading of vehicles, or storage of materials. This form of crane generally involves three distinct movements for the transportation of materials. The initial movement involves the hoist, which elevates and descends the material. The second component is the trolley (cross travel), which enables the hoist to be precisely positioned directly above the material for positioning. The third component is the gantry or bridge movement (longitudinal travel), which enables the entire crane to be repositioned within the working area. In many industrial applications, supplementary motors such as auxiliary hoists, power cable reels, and conveyor belts are frequently required. Therefore, a crane is typically considered sophisticated apparatus.

Depending on the crane capacity, each of the specified motors can be implemented as a multi-motor system. The term multi-motor drive refers to all mechanisms employed within a technological process. If the operation of the drives must be regulated according to the process, based on the controlled speed of each individual drive, the term controlled multi-motor drives is appropriate. For numerous such drives, the mechanical coupling on the load side is typical (Jeftenic et al., 2006; Rockwell, 2000). In crane applications, the coupling of individual motors is achieved through a mechanical transmission device, which is generally designed to be mechanically robust and reliable.

2. POSSIBLE LOAD SHARING CONFIGURATIONS OVERVIEW

Controlled drives are usually fed from the power converter, which is also true for controlled multi-motor drives. The kind, the type and the number of converters used depend on the type of motors, their power ratings, and of the kind of the multi-motor drive. The control and regulation also depend on the type of the multi-motor drive, but also on the type of the converter selected, therefore the selection of the converter and the controller for these drives must be analyzed together. Regarding the power supply of the motor, the following cases are possible (Jeftenic et al., 2006):

- multiple motors fed by a single converter (multiple motors - single converter),
- motors controlled by separate converters (multiple motors - multiple converters).

In crane applications multi-motor drives are used very often and a proportional share of power between motors is required. Load sharing is a term used to describe a system where multiple converters and motors are coupled and used to run one mechanical load (Rockwell, 2000). In the strictest sense, load-sharing means that the amount of torque applied to the load from each motor is prescribed and carried out by each converter and motor set. Therefore, multiple motors and converters powering the same process must contribute its proportional share of power to the driven load.

Multiple motors that are run from a single converter do not load share because torque control of individual motors is not possible. The load distribution, in that case, is influenced only by the correct selection of the torque-speed mechanical characteristic. For the squirrel- cage induction motors, there is no economical method for the adjustment of the mechanical characteristic of the ready-made motors, but this has to be done during the selection. For the slip-ring induction motor, the mechanical characteristic can be adjusted afterwards, with the inclusion of the rotor resistors. Motors that are controlled by separate converters without any interconnection also do not share the load. The lack of interconnection defeats any possible comparison and error signal generation that is required to compensate for the differences in the load that is applied to any single drive and motor set.

Control topologies for load sharing consider the presence of interconnection, i.e. information knowledge about load (motor current or torque). There are three categories of load sharing techniques: common speed reference, torque follower and speed trim follower.

The common speed reference is the simplest the least precise and the less flexible form of load sharing to set up, Fig. 1a). The precision of this control depends on the drives control algorithm, the motor characteristics and the type of load to be controlled.

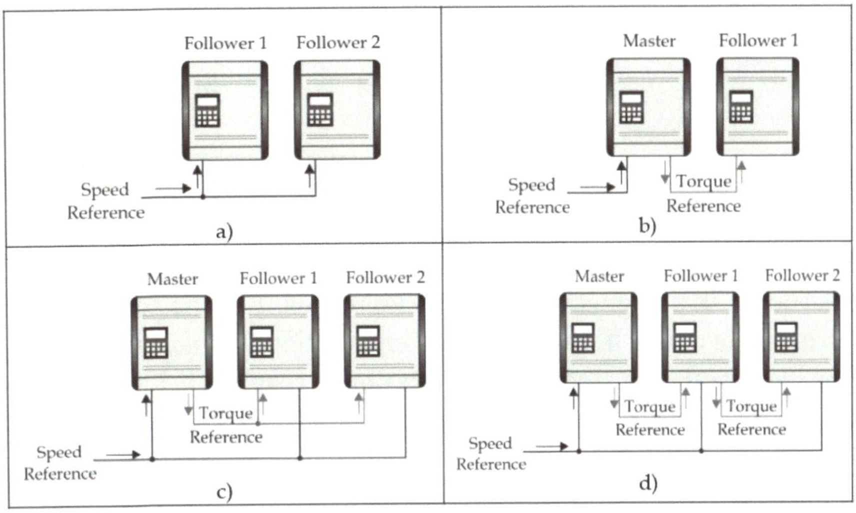

Fig. 1. Load sharing configuration a) Common speed reference, b) Torque follower, c) and d) Speed trim follower.

The torque follower type of load sharing requires the frequency converter to have the capability of operation in "torque mode", Fig. 1b). If speed regulation is required, one of the converters ("master") may be in "speed mode". In speed mode controller provides a torque command at output which can be distributed to the other converters ("slaves" or "torque followers"). The second converter operates in torque regulation mode with the torque reference of the master as command. This torque signal may be scaled to divide load sharing in any desired ratio.

In speed trim follower configuration, Fig. 1.c) and d), all converters are operated in speed regulation mode and receive the same speed reference. The torque reference of the master is sent to the follower converters. Each follower converter compares its own torque reference with that of the master, Fig. 1c). The output of the comparator is an error signal that trims the speed of the follower. Alternative configuration cascades the torque reference comparison, Fig. 1d). The first follower compares the master to its internal value. The second follower compares the foregoing follower to its internal value etc.

2.1 Torque and power requirements for crane drives

Speed control is an essential feature in crane drives. It is required for allowing soft starting and stopping of the travel motions for enabling its correct positioning of load. For the lifting drive the speed control in a wide speed range, from zero to nominal values, is required.

Because of the precision when raising and lowering load, the possibility of working at a very low speed and hold a load in the standstill is required, without using the mechanical brakes.

The torque and power that have to be delivered by the drive may be obtained from the torque versus speed characteristic from the load (so-called mechanical characteristics) and the differential equation of motion, (Belmans et al., 1993; Mitrovic et al., 2011).

The differential equation of motion, describing the behavior of the drive is:

$$J\frac{d\omega}{dt} = T_e - T_l \tag{1}$$

where T_e is the electromagnetic torque of the motor, T_l is the torque of the load, J is the inertia of the drive. If $T_e > T_l$, the system accelerates ($d\omega/dt > 0$), if $T_l/T_e < 0$ it decelerates ($d\omega/dt>0$). The stedy state operation is reached if $T_l = T_e$ and $\omega = const.$

Multiplying equation (1) by the rotating speed ω, yields the power:

$$P_e = \omega T_e = \omega T_l + \omega J\frac{d\omega}{dt} = P_m + P_d \tag{2}$$

This equation shows that the mechanical power $P_e=\omega T_e$, obtained after the electromechanical conversion in the motor, is equal to the power absorbed by the load $P_m=\omega T_l$ only when the speed does not change. Otherwise, the amount corresponding to change in kinetic energy must be added (if the speed increases) or subtracted (if the speed decreases):

$$P_d = \frac{dE_{kin}}{dt} = \omega \frac{Jd\omega}{dt} \tag{3}$$

In the following, the travel and hoist motion of the crane drives will be analyzed.

The mechanical characteristic is given in Fig.2a) for travel motion. Apart from the zone around zero, the torque is constant. The available torque is used for accelerating the system. For a travel in one direction, braking and reversing to full speed in other direction, the speed reference signal is given by top curve of Fig.2b). The torque reference signal is generated by converter (second curve), leading to the machine actual speed. Multiplying the actual and torque reference, yields the actual power (third curve). The peak power is found at the end of the acceleration period. If wind forces are taken into consideration, the torque versus speed curve is shifted horizontally as shown in Fig.2c). The torque and speed reference remain the same, as well as the actual speed. However, the torque reference and the actual power differ, as shown on Fig.2d).

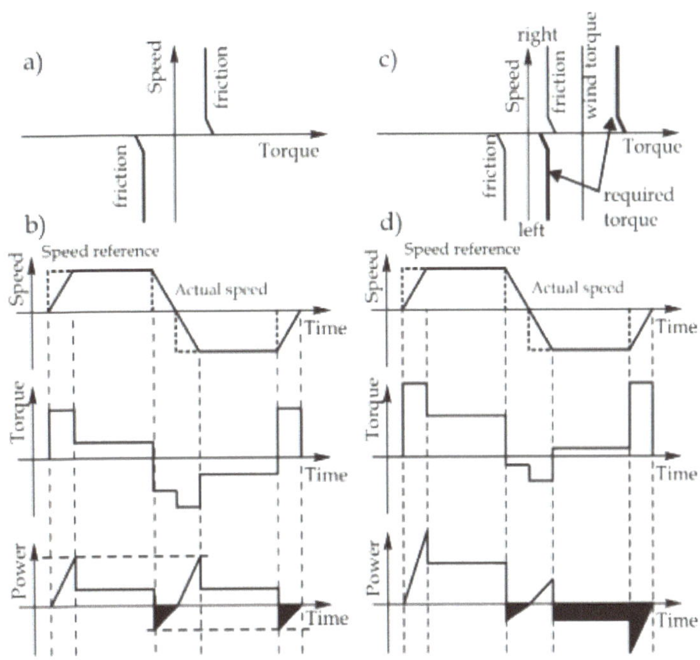

Fig. 2. Power and torque requirements for travel motion, a) and b) without wind influence, c) and d) with wind influence.

During acceleration, kinetic energy is stored in the system. To stop the crane, this energy must be absorbed by the drive. In the indoor situation, this energy is well known and only present for a short period of time. For outdoor applications, the wind forces may become very important. When travelling in the same direction as the wind, the wind drives the crane and a situation may occur, where a continuous electrical braking is required. The drive must be capable of handling this inverse power direction either by consuming the power in a resistor or preferably by feeding it back to the supply.

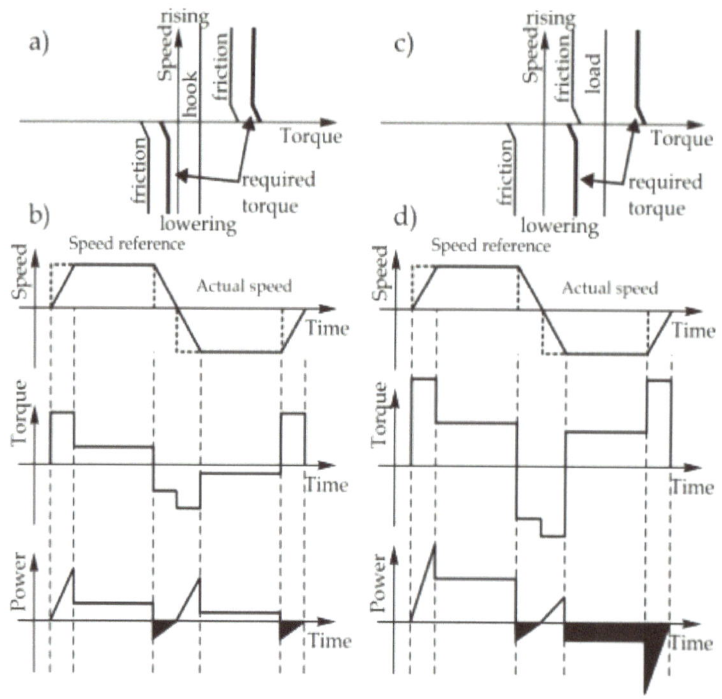

Fig. 3. Power and torque requirements for hoist a) and b) without load, c) and d) with load.

The hoist torque-speed characteristic is shown in Fig.3a) for an unloaded hook. The characteristic resembles the one for the travel motion. However, it is always asymmetric with respect to the vertical axis, due to the gravitation force. This asymmetry becomes more pronounced when the hook is loaded (Fig.3c). For both unloaded and loaded situation, the speed, torque and power are given in Fig.3b) and Fig.4d). Again the amount of braking power is indicated. The worst braking case with a hoist motion, is when sinking a loaded hook. It should be noted that the weight of the hook may be considerable. The hook may be simple, or may consist of several parts to handle the load. For the hoist motion, the speed control is very important in the low speed range: avoiding damage to the load when putting it down and minimizing the stress on the mechanical brakes.

2.2 Frequency converters for induction motor drives

Today's adjustable speed drives (ASD) in the low and mid power range are normally based on the concept of variable voltage, variable frequency (VVVF). Fig.4 shows the basic concept of a single variable speed drive. The three-phase AC supply network is rectified. The DC capacitor, which links the supply rectifier to the inverter, assures that the inverter sees a constant DC voltage from which it generates the required supply voltage and frequency to the motor.

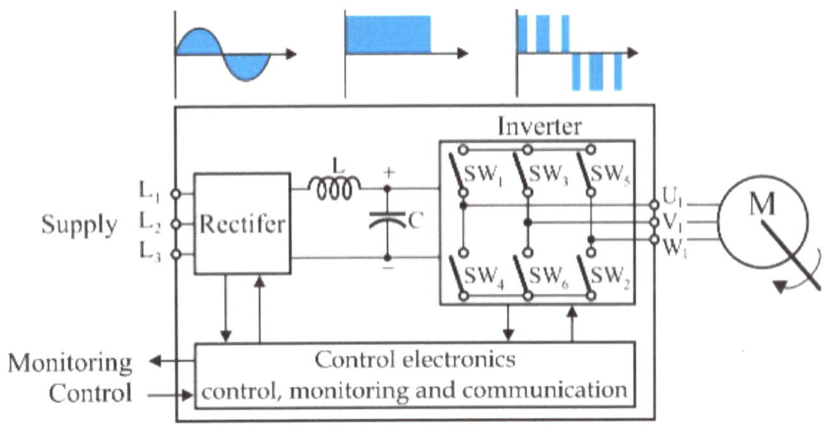

Fig. 4. Basic concept of a variable speed drive.

General classification divides induction motor control schemes into scalar and vector-based methods (Petronijevic et al., 2011). Opposite to scalar control, which allows control of only output voltage magnitude and frequency, the vector-based control methods enable control of instantaneous voltage, current and flux vectors. In numerous industrial applications, such as HVAC (heating, ventilation and air conditioning), fan or pump applications, good dynamic performances are not usually the main control objective. Fig. 5a) illustrates a V/Hz open loop control scheme, where pulse-width modulation (PWM) is realised applying the space vector technique (SVPWM) and output voltage fundamental component amplitude is modified with a voltage drop compensation U_{s0} at low output frequencies. Basic control structure of rotor field oriented control (RFO) is illustrated in Fig. 5b). Two inner PI- controlled current loops for *d* and *q* stator current components are shown, as well as synchronous speed estimator (ω_e, based on reference stator currents components). In its basic version, direct torque control (DTC) consists of a three-level hysteresis comparator for torque control and a two-level hysteresis comparator for flux control as shown in Fig.5.c).

Type of the front end converter, regardless of the control schemes, depends on the power and torque requirements of the drive. Adjustable speed drives in industrial applications are usually characterized by a power flow direction from the AC distribution system to the load (Rashid, 2001). This is, for example, the case of an ASD operating in the motoring mode. In this instance, the active power flows from the DC side to the AC side of the inverter. However, there are an important number of applications in which the motor begins to act as a generator and regenerates energy back into the DC bus of the drive. Moreover, this could be a transient condition as well a normal operating condition. This is known as the regenerative operating mode. For example, these regenerative conditions can occur when quickly decelerating a high inertia load and this can be considered as transient condition. The speed control of a load moving vertically downward (hoist) can be considered as normal operating condition.

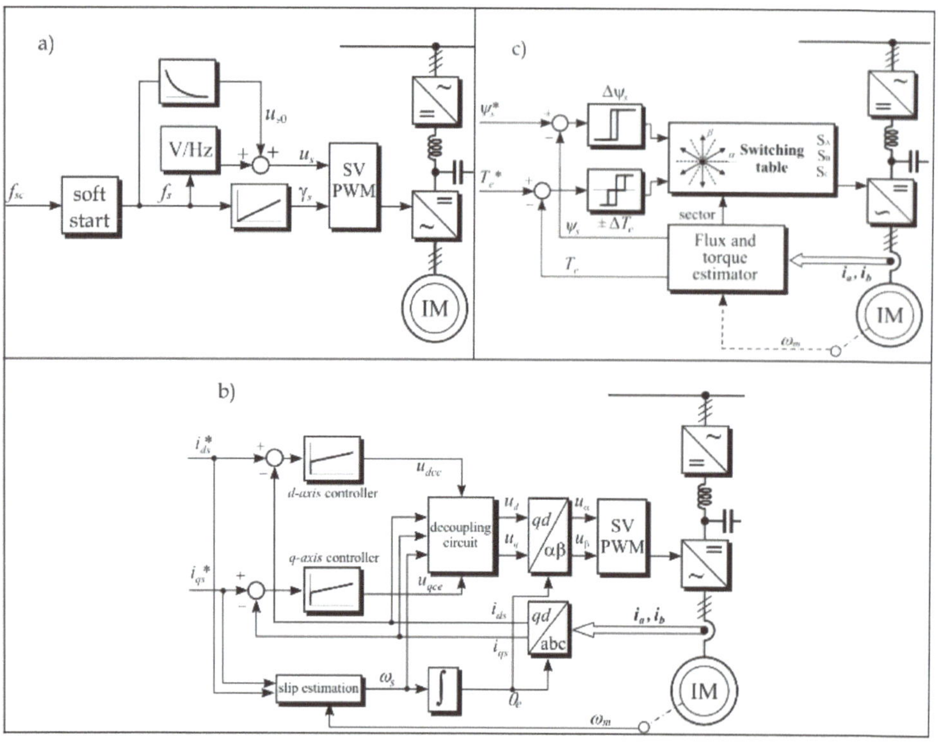

Fig. 5. Control scheme: a) V/Hz, b) vector control, c) DTC control.

Drive applications can be divided into three main categories according to speed and torque. The most common AC drive application is a single quadrant application where speed and torque always have the same direction, i.e. the power flow from inverter to process. The second category is two-quadrant applications where the direction of rotation remains unchanged but the direction of torque can change, i.e. the power flow may be from drive to motor or vice versa. The third category is fully four-quadrant applications where the direction of speed and torque can freely change. These applications are typically elevators, winches and cranes (ABB, 2011).

In order for an AC drive to operate in quadrant II or IV in speed-torque plane, a means must exist to deal with the electrical energy returned to the drive by the motor. The typical pulse width modulated AC drive is not designed for regenerating power back into the three phase supply lines. Electrical energy returned by the motor can cause voltage in the DC link to become excessively high when added to existing supply voltage. Various drive components can be damaged by this excessive voltage. These regenerative conditions can occur when:

- quickly decelerating a high inertia load,
- controlling the speed of a load moving vertically downward (hoist, declining conveyor),
- a sudden drop in load torque occurs,
- the process requires repetitive acceleration and deceleration to a stop,
- controlling the speed of an unwind application.

In standard drives the rectifier is typically a 6-pulse diode rectifier only able to deliver power from the AC network to the DC bus but not vice versa. If the power flow changes as in two or four quadrant applications, the power fed by the process charges the DC capacitors and the DC bus voltage starts to rise. The capacitance is a relatively low value in an AC drive resulting in fast voltage rise, and the components of a frequency converter may only withstand voltage up to a certain specified level.

In order to prevent the DC bus voltage rising excessively, the inverter itself prevents the power flow from process to frequency converter. This is done by limiting the braking torque to keep a constant DC bus voltage level. This operation is called overvoltage control and it is a standard feature of most modern drives. However, this means that the braking profile of the machinery is not done according to the speed ramp specified by the user.

There are two technologies available to prevent the AC drive from reaching the trip level: Dynamic braking and active front end regeneration control. Each technology has its own advantages and disadvantages.

3. DYNAMIC BRAKING

A dynamic brake consists of a chopper and a dynamic brake resistor. Fig.6 shows a simplified dynamic braking schematic. The chopper is the dynamic braking circuitry that senses rising DC bus voltage and shunts the excess energy to the dynamic brake resistor. A chopper contains three significant power components: The chopper transistor is an IGBT. The chopper transistor is either ON or OFF, connecting the dynamic braking resistor to the DC bus and dissipating power, or isolating the resistor from the DC bus. The current rating of the chopper transistor determines the minimum resistance value used for the dynamic braking resistor.

The chopper transistor voltage control regulates the voltage of the DC bus during regeneration. The dynamic braking resistor dissipates the regenerated energy in the form of heat.

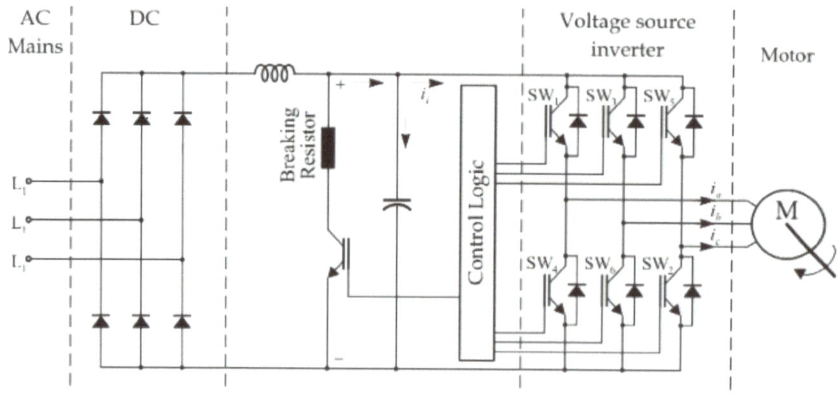

Fig. 6. Voltage source inverter with diode front end rectifier and dynamic brake module.

As a general rule, dynamic braking can be used when the need to dissipate regenerative energy is on an occasional or periodic basis. In general, the motor power rating, speed, torque, and details regarding the regenerative mode of operation will be needed in order to estimate what dynamic braking resistor value is needed. The peak regenerative power of the drive must be calculated in order to determine the maximum resistance value of the dynamic braking resistor.

The peak breaking power required to decelerate the load, according to equation (4) is:

$$P_b = \frac{J\omega_b(\omega_b - \omega_0)}{t_b} \tag{4}$$

where t_b represents total time of deceleration, ω_b and ω_0 initial and final speed in the process of braking.

The value of P_b can now be compared to the drive rating to determine if external braking module is needed. If peak braking power is 10% greater than rated drive power external braking module is recommended. Compare the peak braking power to that of the rated motor power, if the peak braking power is greater than 1.5 time that of the motor, then the deceleration time, needs to be increased so that the drive does not go into current limit.

The peak power dynamic brake resistance value can be calculated as:

$$R_{db} = \frac{V_{dc}^2}{P_b} \tag{5}$$

The choice of the dynamic brake resistance value should be less than the value calculated by equation (5). If a dynamic braking resistance value greater than the ones imposed by the choice of the peak regenerative power is made and applied, the drive can trip off due to transient DC bus overvoltage problems. Once the approximate resistance value of the dynamic braking resistor is determined, the necessary power rating of the dynamic braking resistor can be calculated. The power rating of the dynamic braking resistor is estimated by applying what is known about the drive's motoring and regenerating modes of operation.

To calculate the average power dissipation the braking duty cycle must be determined. The percentage of time during an operating cycle (t_c) when braking occurs (t_b) is duty cycle ($\varepsilon=t_b/t_c$). Assuming the deceleration rate is linear, average power is calculated as follows:

$$P_{av} = \frac{t_b}{t_c}\frac{P_b}{2}\frac{\omega_b + \omega_0}{\omega_b} \tag{6}$$

Steady state power dissipation capacity of dynamic brake resistors must be greater than that average. If the dynamic braking resistor has a large thermodynamic heat capacity, then the resistor element will be able to absorb a large amount of energy without the temperature of the resistor element exceeding the operational temperature rating.

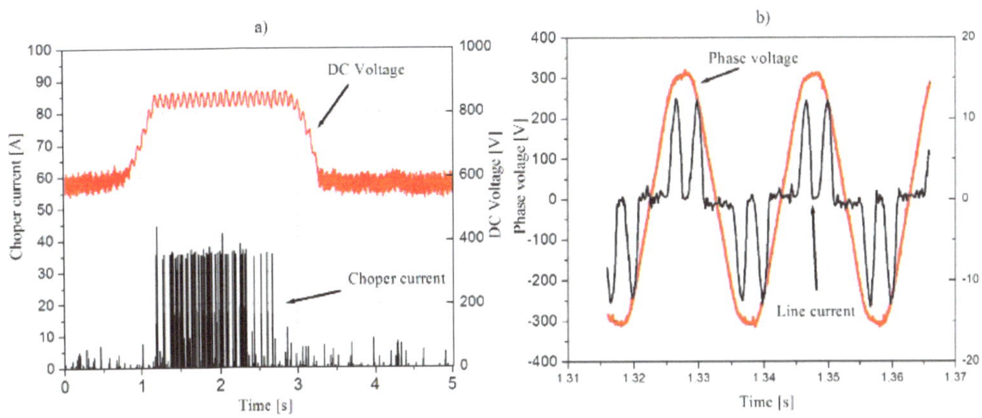

Fig. 7. a) DC voltage and chopper current, b) line voltage and current.

Fig.7a shows the experimental results (DC voltage and chopper current) for the variable frequency drive with braking module in DC link and external braking resistor, under a step change of induction motor load in regenerative regime. Danfoss frequency (series VLT 5000) converter is used in experimental set-up. For the supply voltage of 400 V, DC link voltage is about 540 V. When negative load torque is applied, DC link voltage rises. The chopper transistor voltage control regulates the voltage of the DC bus during regeneration to near 800 V allowing current flow in the resistor. Regenerative energy is then realised into heat. After the end of the regenerative period, DC voltage returns to a value that corresponds to a motor regime. The Fig.7b) shows the line voltage and current at the input of the diode rectifier.

A voltage source PWM inverter with diode front-end rectifier is one of the most common power configurations used in modem variable speed AC drives, (Fig. 6). An uncontrolled diode rectifier has the advantage of being simple, robust, and low cost. However, it allows only unidirectional power flow. Therefore, energy returned from the motor must be dissipated on a power resistor controlled by a chopper connected across the dc link. A further restriction is that the maximum motor output voltage is always less than the supply voltage.

4. ACTIVE FRONT END RECTIFIER

Various alternative circuits can be used to recover the load energy and return it to power supply. One such scheme is shown in Fig. 8 and presents the most popular topology used in ASD. The diode rectifier is replaced with PWM voltage source rectifier. This is already an industrially implemented technology and known as most successful active front end (AFE) solution in ASD if regenerative operation is needed (e.g. for lowering the load in crane) and therefore was chosen by most global companies: Siemens, ABB, and others.

The term Active Front End Inverter refers to the power converter system consisting of the line-side converter with active switches such as IGBTs, the DC link capacitor bank, and the load-side inverter. The line-side converter normally functions as a rectifier. But, during regeneration it can also be operated as an inverter, feeding power back to the line. The line- side converter is popularly referred to as a PWM rectifier in the literature. This is due to the fact that, with active switches, the rectifier can be switched using a suitable pulse width modulation technique.

The PWM rectifier basically operates as a boost chopper with AC voltage at the input, but DC voltage at the output. The intermediate DC-link voltage should be higher than the peak of the supply voltage. The required DC-link voltage needs be maintained constant during rectifier as well as inverter operation of the line side converter. The ripple in DC link voltage can be reduced using an appropriately sized capacitor bank. The AFE inverter topology for a motor drive application, as shown in Fig.8, has two three-phase, two-level PWM converters, one on the line side, and another on the load side. The configuration uses 12 controllable switches. The line-side converter is connected to the utility through inductor. The inductor is needed for boost operation of the line-side converter.

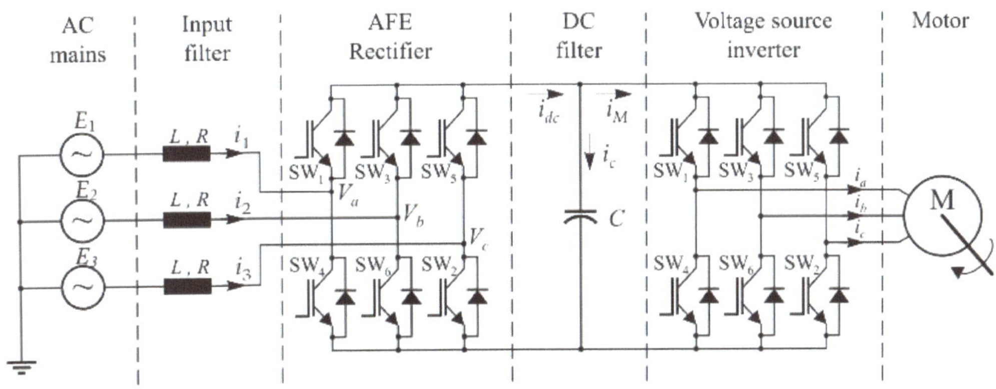

Fig. 8. Active front end inverter topology.

For a constant dc-link voltage, the IGBTs in the line-side converter are switched to produce three-phase PWM voltages at *a*, *b*, and *c* input terminals. The line-side PWM voltages, generated in this way, control the line currents to the desired value. When DC link voltage drops below the reference value, the feed-back diodes carry the capacitor charging currents, and bring the DC-link voltage back to reference value.

The steady state characteristics as well as differential equations describing the dynamics of the front-end rectifier can be obtained independent of an inverter and motor load. This is because the DC-link voltage can be viewed as a voltage source, if V_{dc} is maintained constant for the full operating range. The inverter is thus connected to the voltage source, whose terminal voltage V_{dc}, remains unaffected by any normal inverter and motor operation (Jiuhe et al., 2006).

Furthermore, as shown in Fig.8, the rectifier can also be viewed as connected to the voltage source V_{dc}. Thus, the rectifier is able to control magnitude and phase of PWM voltages V_{abc} irrespective of line voltages E_{123}.

The dynamic equations for each phase can be written as,

$$\begin{bmatrix} E_1 \\ E_2 \\ E_3 \end{bmatrix} = L\frac{d}{dt}\begin{bmatrix} i_1 \\ i_2 \\ i_3 \end{bmatrix} + R\begin{bmatrix} i_1 \\ i_2 \\ i_3 \end{bmatrix} + \begin{bmatrix} V_{a0} \\ V_{b0} \\ V_{c0} \end{bmatrix} \tag{7}$$

In synchronous rotating *d-q* reference frame Equations 8 and 9 represent the dynamic *d-q* model of an active front end inverter in a reference frame rotating at an angular speed of ω.

$$L\frac{di_{qe}}{dt} = E_{qe} - \omega L i_{de} - R i_{qe} - V_{qe} \tag{8}$$

$$L\frac{di_{de}}{dt} = E_{de} + \omega L i_{qe} - R i_{de} - V_{de} \tag{9}$$

The differential equation governing DC link voltage also needs to be added to the above set of system equations to completely define system dynamics:

$$C\frac{dV_{dc}}{dt} = i_{dc} - i_M \quad (10)$$

where, i_{dc} is the total DC link current supplied by the rectifier, while i_M is the load-side DC current which is the result of induction motor operation.

In Equations 8 and 9, the terms E_{qe} and E_{de} are computed from source voltages, E_1, E_2, and E_3. Since line voltages are known, the angular frequency ω, can be easily estimated. The PWM voltages V_{qe} and V_{de} are the two inputs to the system which are generated using the sine- triangle PWM controller. *L* and *R* represent series impedance.

Equations (8 and 9) shows that *d-q* current is related with both coupling voltages ωLi_q and ωLi_d, and main voltage E_d nd E_q, besides the influence of PWM voltage V_{qe} and V_{de}. Voltage V_{qe} and V_{de} are the inputs, controlled in such a way as to generate desired currents. Now define new variables V'_{qe} and V'_{de} such that (Hartani & Miloud, 2010):

$$V_{qe} = -V'_{qe} - \omega Li_{qe} + E_{qe} \quad (11)$$

$$V_{dc} = -V'_{dc} + \omega Li_{dc} + E_{dc} \quad (12)$$

So that the new system dynamic equations become:

$$L\frac{di_{qe}}{dt} = -i_{qe}R + V'_{qe} \quad (13)$$

$$L\frac{di_{se}}{dt} = -i_{dc}R + V'_{dc} \quad (14)$$

We can see from equations that the two axis current are totally decoupled because V'_{qe} and V'_{de} are only related with i_{qe} and i_{de} respectively.

The simple proportional-integral (PI) controllers are adopted in the current and voltage regulation, Fig.9. The control scheme of the PWM rectifier is based on a standard cascaded two-loop control scheme implemented in a *d-q* rotating frame: a fast control loop to control the current in the boost inductors and a much slower control loop to maintain constant dc- link voltage. The reference angle for the synchronous rotating *d-q* frame θ, is calculated, based on the three input phase voltages.

For the current control loop *d-q* synchronously rotating reference frame with the fundamental supply voltage frequency is used (Odavic et al., 2005). The line currents (i_1, i_2, i_3) are measured and transformed to the *d-q* reference frame, Fig.10.

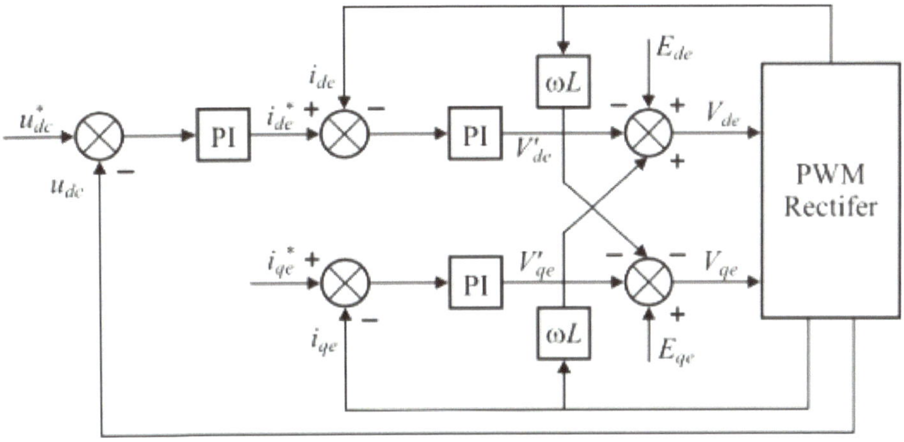

Fig. 9. Decoupled current control of PWM rectifier.

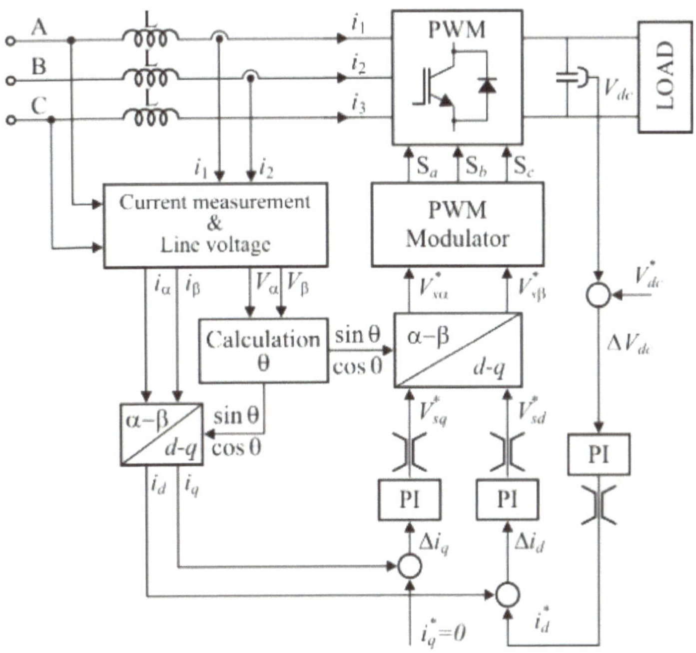

Fig. 10. Simplified block diagram of the AFE.

To get information about the position of the line voltage vector PLL (phase locked loop) is implemented. PI controllers for the *d-q* components of line current are identical and ωL terms are included to eliminate the coupling effect among the *d* and *q* components. Outputs of the line current PI controllers present *d* and *q* components of the voltage across the line inductance. Subtracting this voltage from the supply voltage gives the converter voltage from the AC side that is used to get the modulation signal for proper switching of six switching devices.

The main task of the sinusoidal front end is to operate with the sinusoidal line current; so *d* and *q* components of the line current reference are DC values. Using this approach of control it is possible to control the output voltage of converter as well as the power factor of converter in the same time. To achieve unity power factor the reference of *q* current component need to be set on zero.

Based on analysis, the simulation model of the whole is built using Matlab/Simulink to test the performance of the active front end rectifier. On the load side is the field oriented induction motor drive with topology as shown in Fig.5c). The whole system behavior is simulated as a discrete control system. The AC source is an ideal balanced three phase voltage source with frequency 50 Hz. The phase to phase voltage is 400 V. capacitor in DC link is 4700 μF. The line resistor and line inductance of each phase is 0.1 Ω and 3 mH respectively. The induction motor rated power is 30 kW.

The induction machine is initially running at a constant speed reference (100 rad/s) and under a no load regime. From this situation, we apply a rated load torque during a time interval of 0.5 s and then we remove the load. This case corresponds to a step up and a step down torque perturbation.

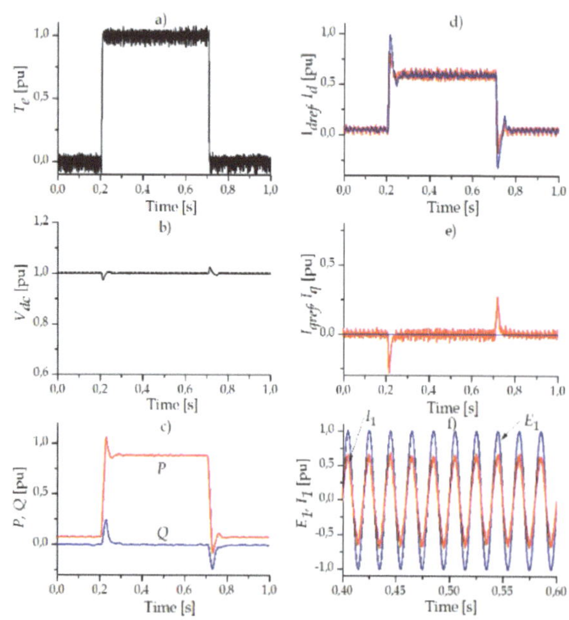

Fig. 11. Behavior of the rectifier under a step change of the induction motor load (motor operation).

The following figures summarize the results of the simulation. Fig. 11 and Fig.12 shows the behavior of the rectifier under a step change of the induction motor load. Fig.11 refers to the motor mode of operation (lifting) and Fig.2 to the generator mode (lowering).

Fig. 11b and Fig. 12b) show the behavior of the DC link voltage of the PWM rectifier in response to a step change in the load. Fig. 11c) and Fig. 12c) presents the active and reactive power on the line side of converter. Since that the q component of the set point current is I_{qref}=0 (Fig.11e and Fig.12e) there is only *d* current component (Fig.11d and Fig.12d), reactive power is zero. Fig.11f) and Fig.12f) show the behavior of steady state voltage and current delivered by the source when the line side converter works in the rectifier and regenerative mode. The input current is highly sinusoidal and keeps in phase with the voltage, reaching a unity power factor.

The simulation results show that the rectifier has excellent dynamic behavior and following advantages:

- Power flow between the motor and the mains supply is possible in both directions, and so this makes the drive more efficient than when a braking resistor is used. As well as recovering energy during deceleration, it is possible to recover energy from overhauling load such as a crane when the load is being lowered.
- Good quality input current waveforms are possible with unity power factor.
- It is possible to boost the DC link voltage to a level that is higher than would be possible with a simple diode rectifier.

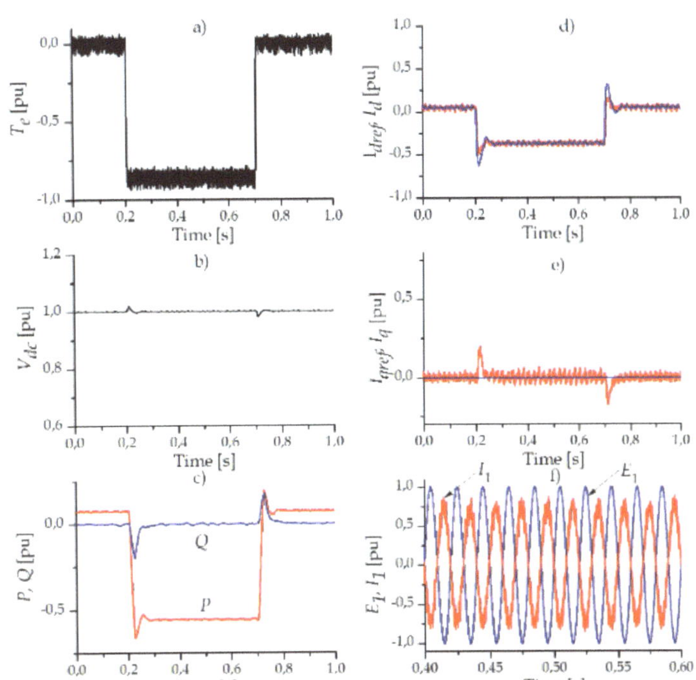

Fig. 12. Behavior of the rectifier under a step change of the induction motor load (generator operation).

5. CASE STUDY 1: DERRICK CRANE

The experimental behavior analysis of some drives is considered in a derrick crane, which serves for load handling in many industry branches. The main task in adjustable speed drives design is a safe, multi-axis movement that allows material handling throughout the working area. The derrick crane with following technical details has been taken for experimentation with adjustable frequency drive:

- Main hoist load capacity: 60 t;
- Auxiliary load capacity: 12.5 t;
- Main hoist height: 46 m;
- Auxiliary hoist height: 49 m;
- Jib boom radius: 0-82.5°;
- Length of runway rail path: 350 m
- Working conditions: outdoor.

Fig. 13. Derrick crane with indicated drives.

Using AFE rectifier/regenerative unit on common DC bus, six groups of inverter-motor combinations are supplied:

- hoist motion with 2x55 kW vector control inverter, the motor is a six pole, 2x45 kW,
- auxiliary hoist motion with a 55 kW vector control inverter, supplying a four pole, 45 kW motor,

- jib motion with 2x55 kW vector control inverter, the motor is a six pole, 2x45 kW,
- travel motion with 3x7.5 kW vector control inverter, the motor is six pole, 3x7.5 kW,
- motor driven cable reel with 3 kW vector control inverter supplying 3 kW motor,
- auxiliary drives with 8x1.1 kW and motors (8x1.1 kW).

The rating of the AFE rectifier/regenerative unit output at cosφ=1 and 400 V supply voltage is 177 kW. This is far less than the sum of the ratings of the individual invertors, being 300 kW. In the Fig.13 crane with indicated drives is shown.

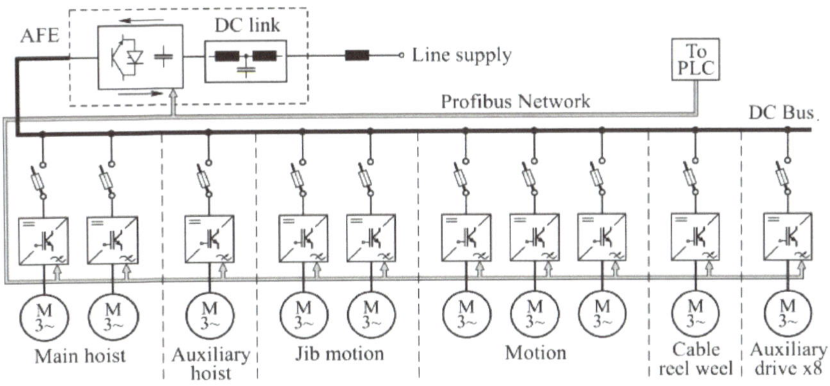

Fig. 14. Single line power circuit topology.

Fig. 14 shows the power circuit topology of the derrick crane using AFEs at the input side. The AFE is connected upstream to the standard frequency inverter and consists of three components:

- Active Infeed Converter.
- Line Filter Module (EMC filter, line contactor and charging circuit).
- Line Filter Choke

Power flows from the line through the input transformer and the input reactance into AFE, creating a common DC bus. The inverters take energy from the common DC bus to control the induction motors for the different movements.

Fig.15a) shows a hoist movement with the 30% of full load and Fig.15b) for auxiliary hoist with an unloaded hook measured in similar conditions. In Fig.15a), curve 1 gives the actual speed signal (reference speed signal is given at 100% from the crane driver joystick command). Curves 2, 3 and 4 show the torque, power and motor current, respectively. After an acceleration period (ending at 5 s), a constant torque is delivered. This transition in torque level coincides with reaching the prescribed speed. At 17.5 s, the speed reference signal is made zero (stop command). The driving torque becomes zero and the system decelerates due to gravity. After 20 s, zero speed is reached, and the drive has to deliver the torque required for holding the load before closing the mechanical brake. After that the same the same measurements were performed during lowering. Due to high friction losses, torque was required to start the decent. After short initial period, only the dynamic friction is present, yielding a small driving torque. After the acceleration, the power flow is reversed and the drive lowers the load at a constant speed. From 43 s on, the system is braked with maximum torque, until standstill.

The same measurements were performed during lowering and hoist movement with the 30% of full load, Fig. 16a). The reference speed was 25, 50, 75 and 100% of rated speed. As on the Fig.15 actual speed, motor torque, power and current are shown. After that for jib-boom motion the same signal was record as shown on Fig.16b). In both figure regenerative periods during the lowering, at any reference speed, can be seen.

From the Fig.15 and Fog.16, can be seen periods when the energy recovery occurs at the point of load lowering. It is very important to point out that the AFE topology allows for fully regenerative operation, which is quite important for crane application.

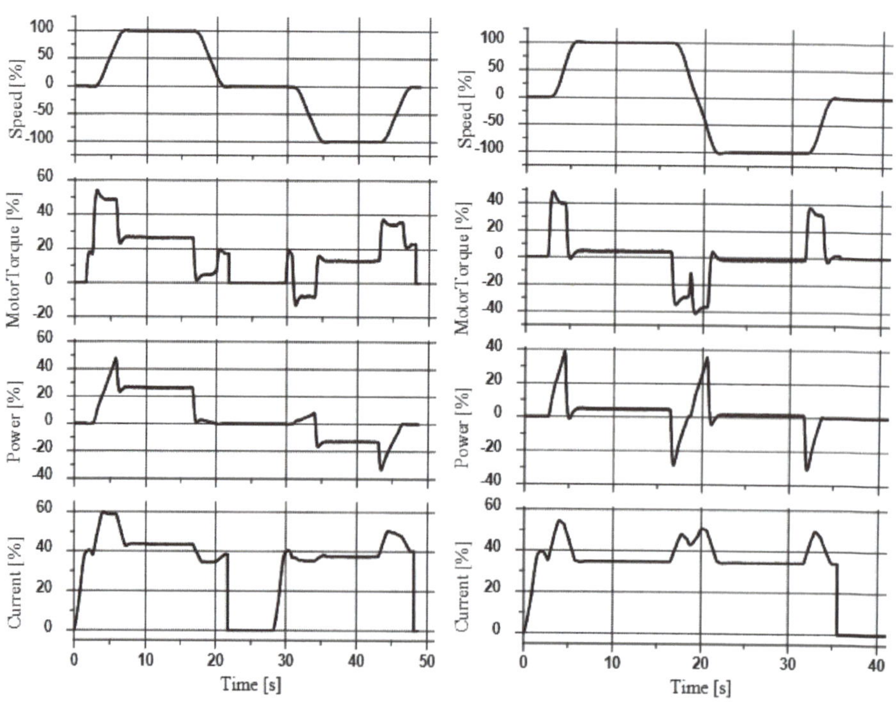

Fig. 15. Measured pattern of the a) hoist motion, b) auxiliary hoist.

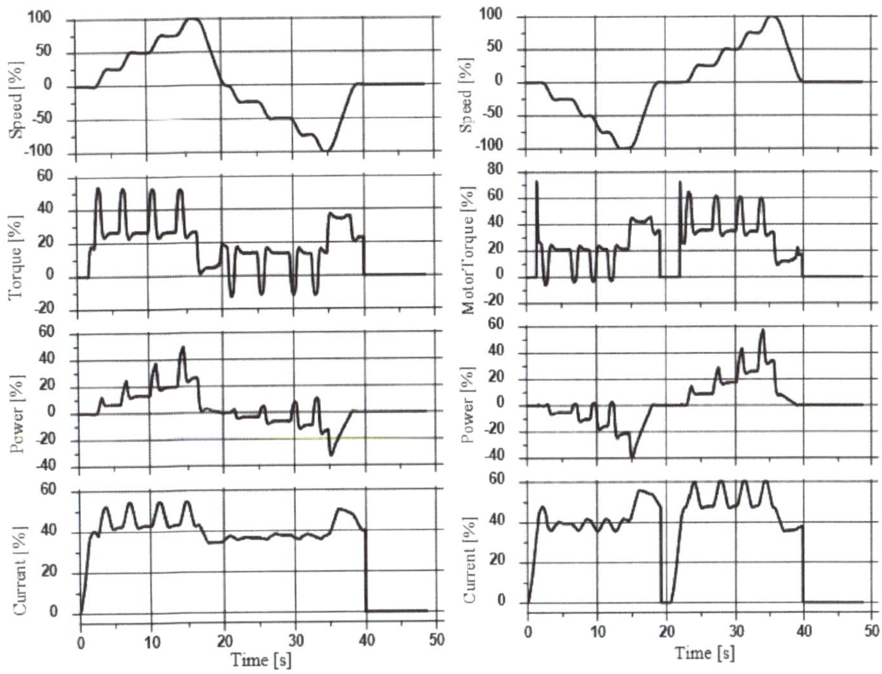

Fig. 16. Measured pattern at 25, 50, 75 and 100% of rated speed a) hoist motion, b) jib motion.

Different tests have been performed on the system to show some of the capabilities in the AFE inverter system. The measurements are done at steady-state operation. During experiments, the DC link voltage is boosted to 650 V. The first test is rectifier system operation when the induction machine operates as motor during lifting of the load, Fig.17a), and second test is regenerative operation during lowering of the load, Fig.17.b). Both figures show the measured line currents, line voltages and DC voltage. It can be observed a high stationary performance both in motor and generator operation. The line current is nearly a sine wave with unity power factor while DC voltage is unchanged.

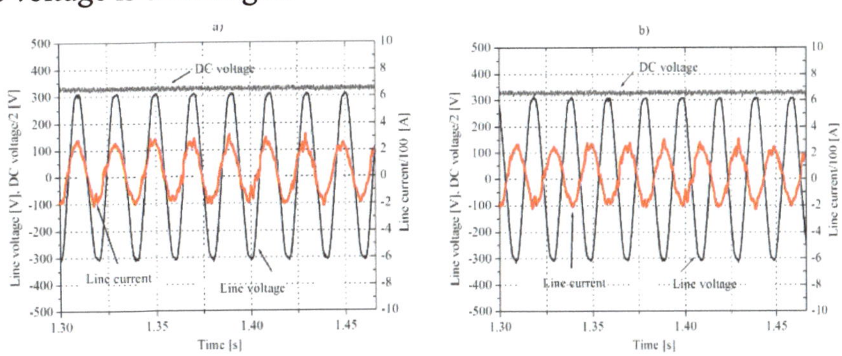

Fig. 17. Waveforms under steady-state operation: Line voltage, line current and dc link voltage a) motor operation, b) generator operation.

6. CASE STUDY 2: WIDE SPAN GANTRY CRANE

The experimental behavior analysis of some drives is considered on the example of crane with wide span, which in sugar factory serves for continuous transport of sugar beet from the reception position to the factory storage.

The crane with following details has been taken for experimentation with adjustable frequency drive:

- Handling capacity: 500 t/h;
- Gantry span: 64.5 m;
- Runway rail path: 300 m;
- Hoist height: 18 m;
- Working conditions: outdoor.

Gantry crane for sugar beet storage is designed from the following functional parts:

1. Gantry drive (16 m/min) with four induction motors of 5.5 kW, two per leg.
2. System conveyor belts (2 m/s) with "battered" (30kW), horizontal (30kW) and "butterfly" conveyor (11kW).
3. Trolley drive (12 m/min) with four motors of 1.1kW.
4. "Butterfly" hoist (3 kW).
5. Motor driven cable reel (1.1 kW).
6. Decentralized crane control system with appropriate PLC, Profibus communication between converters and other intelligent devices (encoders, operator panels etc.).

In the Fig.18 gantry crane with indicated drives is shown. All motors are three phase fed by frequency converters.

Certainly, the most complicated is the gantry drive, from following reasons:

- that is multi motor drive which consists of two motors on each side,
- the span is wide,
- construction is lattice, therefore it is elastic,
- plant is located outdoor so the influence of the wind may be considerable,
- the length of runway rail path is 300 m.

Basic requirements set in front of this drive are: equal load distribution between motors located on the same side, as well as skew elimination between fixed and free gantry leg.

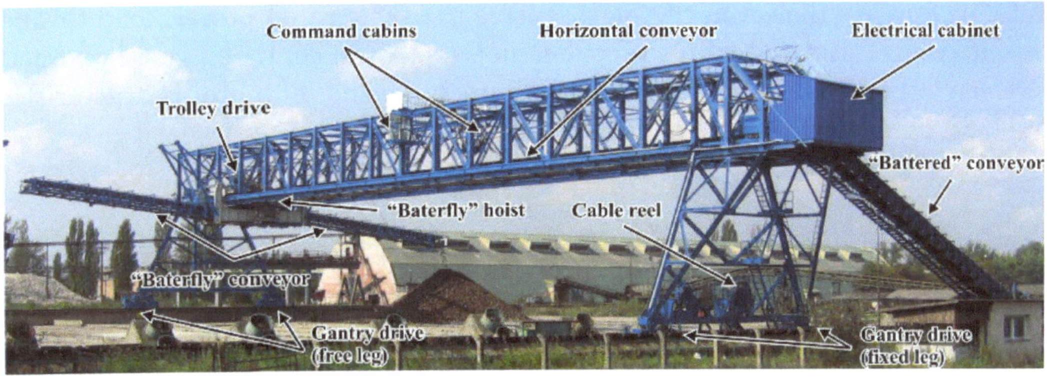

Fig. 18. Gantry crane with indicated drives.

6.1 Load sharing

Although the motors have the same power, there are few necessary reasons to do the load distribution: different wheel diameter, unequal adhesion, geometrical imperfection of the construction, slipping of the pinion wheel due to wet or frozen rails. Load distribution is resolved by using speed trim load sharing configuration, Fig.1c). Load distribution controller is realized by PLC.

In the Fig.19 the principle block scheme for load distribution between two rail coupled induction motors (IM_1 and IM_2) fed by frequency converters (FC1 and FC2) is shown. Starting point at design of load sharing controller is that the less loaded motor should accelerate in order to take over the part of load from the more loaded motor. Information about the load can be obtained in different ways. The easiest one is by motor current. Modern converters used in drives, enable to obtain information about the motor torque in percentage in relation to rated torque.

As we can see in the Fig.19, the speed reference of only one motor (n*2) is updated in relation to the main speed reference (n*=n*1). Reference correction Δn^* is proportional to the difference of estimated electromagnetic torque ($\Delta T_e = T_{e1} - \Delta T_{e2}$). Proportional gain of load sharing regulators is denoted as K_{LS}.

In order to ensure the stabile operation of the motors during the large external disturbances, especially at low speed when estimation of electromagnetic torque in speed sensorless drives looses on accuracy, it is necessary to limit the correction value Δn^*, as shown in Fig.19.

For the purpose of suggested algorithm verification the trolley load sharing is analyzed. Because of the short distance between left and right side the skew may be neglected. Trolley drive consists of four motors, two on each side (IM_1-IM_2 on left and IM_3-IM_4 on right side). Frequency converters are set on speed sensorless vector control mode. Motors have the common reference speed. In the Fig.20a) motors torque without load distribution is shown. At reference speed, in steady state, we can see that even the motors have the same rated power, load torques are different. Estimated motor torque is not applied in control algorithm. Speed between left and right side is different because it depends of motor characteristics and load, as shown in Fig.20a).

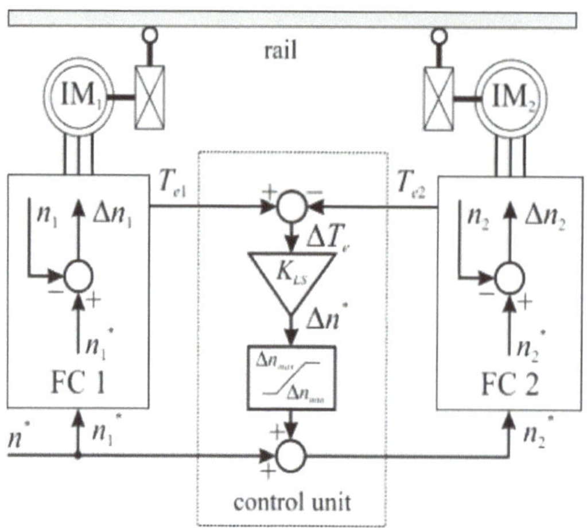

Fig. 19. The principle of load sharing based on estimated torque.

Effect of load sharing is shown in Fig.20b). The approximately equal motors torque on the same leg can be easily seen. Used system enables that speed of every motor is regulated, but also the load difference is controlled. In this way the load difference is being maintained on the desired accuracy.

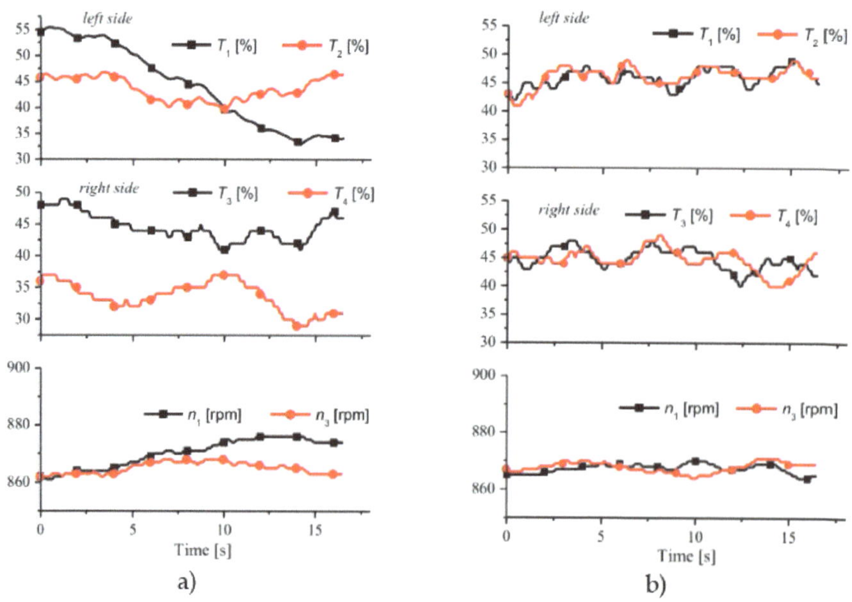

Fig. 20. Motors torque and speed: a) without load sharing; b) with load sharing.

Depending on the purpose of drives and needed accuracy of maintaining load distribution, load controller can be with only proportional effect, but also with proportional integrated effect. In our case only proportional controller with K_{LS}=1 p.u. is used. Output from the load controller is restricted on only several percentages of maxsimum speed reference (in our example $\Delta n_{min\text{-}max}$=2%). That is quite enough to provide necessary load regulation and not to "break" the drive speed regulation by too big effect on the speed reference. This solution can be applied for all kinds of multi motor drives on cranes.

6.2 Skew elimination

On most rails mounted wide span gantry cranes skewing problem is associated with poor rail conditions, uneven wheel wear, wind influence, wheel slippage or unequal load conditions when the trolley is operating at one end of the crane bridge. The skewing of the crane can cause excessive wheel abrasion and stress, especially to the wheel flanges. It can also produce horizontal or lateral forces that can result in unusual stresses to the crane runway beams and building structure. This often results in differing diameters of drive wheels, which subsequently cause the crane to skew.

The crane construction consists of opposed pairs of end truck assemblies (left hand side is named as free leg and right hand side is named as fixed leg). These are movable along a track and a long transverse support member between the end truck assemblies. Each end truck assembly includes two sets of trolleys and an upper load bar laterally interconnects the two sets of trolleys.

The hardware for skew elimination consists of a PLC with Field-bus communication, two absolute multi-turn encoders, two proximity sensors and four frequency converters for motor supply of trolley drives, as shown in Fig.21). On each end truck, one of the converters is master and the other one is slave. The master-slave references distribution is modified according to the load sharing principle as shown in Fig.1c).

The main devices for skew tracking are two absolute encoders (E_1 and E_2) installed on a special, non-tractive wheel (so called free wheel), in order to avoid slipping. Encoders measure the traveled distance, and absolute position is transferred to the anti-skew control subsystem in PLC, as shown in Fig.21). The fixed leg frequency converter (FC_1) is set as a master for gantry drive skew elimination algorithm; while in this case, the speed reference for the frequency converter on the free leg (FC_2) is modified with the anti-skew controller output.

The control scheme for skew elimination between the master and slave motor of gantry drive is shown in detail in Fig.22). As it can be seen, we propose a simple proportional (P) controller acting as an additional, outer correction loop, which supplies speed control loop. The speed reference of one motor (n*2) is updated in relation to the main speed reference ($n^*=n^*1$) with reference correction value Δn^*. If the encoder position difference ΔE related to maximum allowed skew is known, the controller gain K_{SC} can be calculated, (Mitrovic et al., 2010).

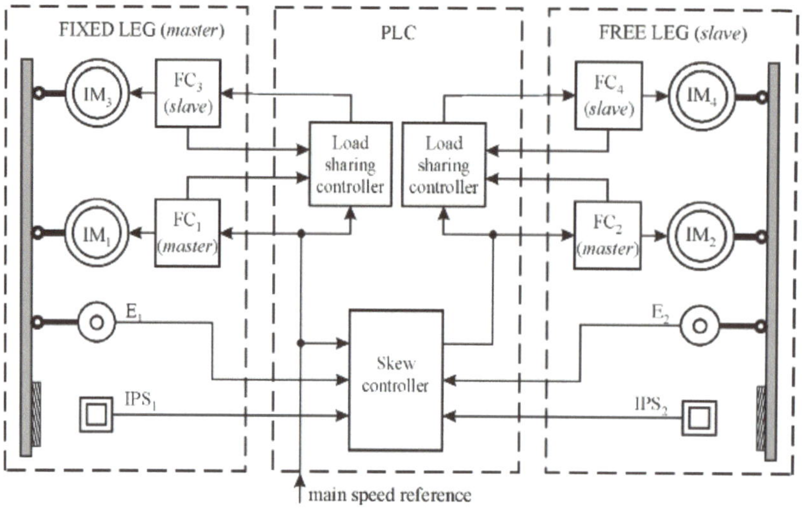

Fig. 21. Block scheme of gantry drive.

In order to ensure the stabile and safe operation of the motors during the large external disturbances and at low speed, when estimation of electromagnetic torque in speed sensor- less drives looses on accuracy, it is necessary to limit the correction value Δn^*.

A reliable operation (even in terms of key components failure - for example encoders) requests an additional external disturbance compensator (EDC) which includes several pairs of position bars (or markers, M) and inductive proximity sensors (IPS). The EDC takes into account all external influences on the position difference of the two encoders: the free wheels diameter difference and an accidental wheel and encoder joint slipping.

The proximity sensors are fitted on the end truck holders, while the position bars are equidistantly mounted along the rails. During the crane movement, proximity sensors detect the moment when the fixed (or free) leg passes above the markers and so register the crane actual skew. Now when both legs are positioned on the markers, absolute encoders measure the trajectory difference, as shown in Fig.22). In fact, this difference is the real skew (s) of the crane, determined at each crossing over the markers. If the difference is greater than the length of the markers (l_m) that means the crane skew is bigger than allowed. For this reason it is required that the length of markers matches the allowed skew of the crane. The distance between successive markers (l_{ms}) depends on the length of marker and maximum expected liner speed difference between the legs.

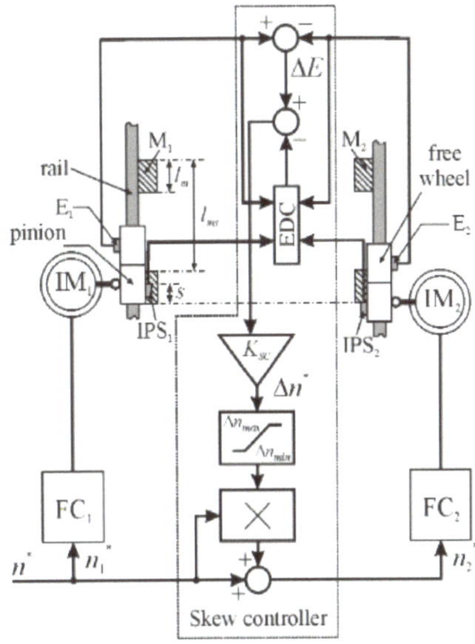

Fig. 22. The principle block diagram of skew controller.

The limited number of the necessary input data for the calculation and design of the skew controller allows quick adjustment of parameters and the choice of EDC components, and the proposed algorithm makes it suitable for industrial applications.

In the analyzed example, the loads of fixed and free legs are different, partly because of asymmetry of gantry, but mostly because of the trolley moving along the gantry. The calculated critical skew of gantry structure is 100 cm, but during normal operation the maximum allowed skew is 50 cm. A preview of gantry drive parameters, controller design and set-up values are taken from reference (Mitrovic et al., 2010).

At the beginning, we analyzed the behavior of gantry drives without a skew controller and the main results are shown in Fig.23a).

In this case, the load-sharing controllers for the fixed and free gantry leg are applied. Three working sections are noticeable: crane acceleration, steady state operation, and crane deceleration. The encoder measures the motor speed, while torque is estimated from the frequency converters. The measured data are collected in PLC SCADA system. The observed variables are master motor (IM_1) speed n_1, speed difference n_1-n_2 between the master motor (IM_1) on the fixed leg and the master motor (IM_2) on the free leg, torque differences between motors on the same leg (IM_1-IM_3, IM_2-IM_4) and the value of actual skew (*s*). In this case, as the skew is not controlled, it can be seen the increase of the value. During the crane skew, motors (IM_1 and IM_3) on the fixed leg are more loaded than the motors (IM_2 and IM_4) on the free leg. In addition to that, the effects of the load-sharing controller can be noticed because the motors on the same leg share loads approximately, i.e. torque difference oscillates about zero value.

The next experiment was performed including the skew controller and under the similar operational regimes as in the previous case: acceleration, steady state operation and deceleration. The experimental results are shown in Fig.23b). During the crane acceleration/deceleration, due to different loads between the fixed and free leg temporarily skew can be observed. The skew controller action eliminates this start-up disturbance in a few seconds. Simultaneously, with the action of a skew regulator, load-sharing controllers provide motor loading in proportion to their rated power. At constant speed operation, the trolley moves between the fixed and free leg, which causes additional differences in loads, but the proposed controller successfully compensates for these disturbances. In the case of crane deceleration, it can be seen that the characteristic case of the free leg stopping is postponed in order to complete the elimination of skew and for the fine position adjustment.

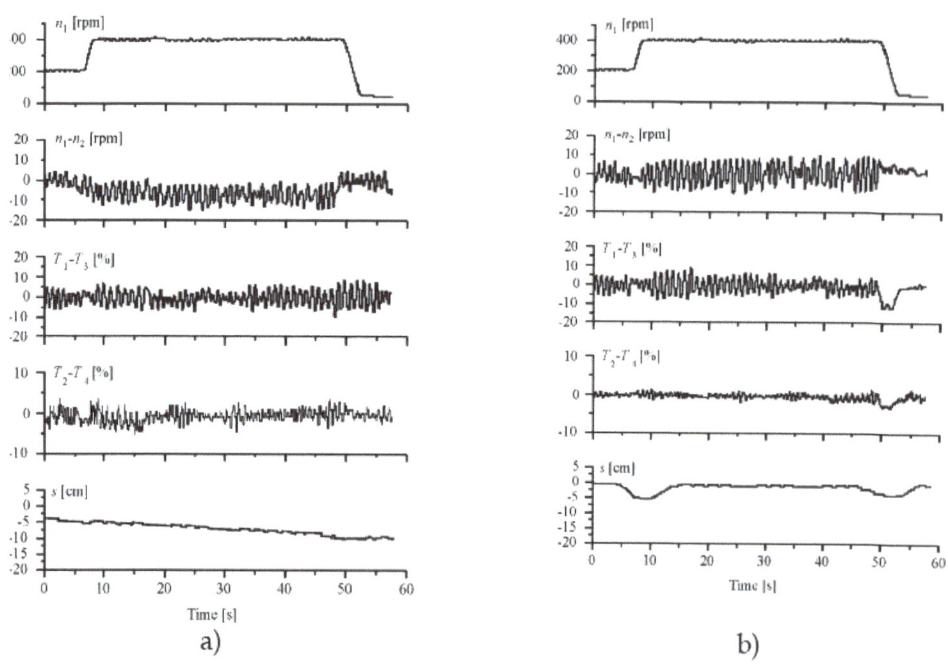

Fig. 23. Behavior of gantry drives: a) without skew controller; b) with skew controller.

7. CONCISE SUMMARY

The utilization of squirrel cage induction motors powered by frequency converters (also referred to as adjustable speed drives) has become the established standard for contemporary crane drives. However, the conventional inverter configuration cannot be employed for certain drives, primarily due to regenerative operation, which may be intermittent (such as during long or transverse travel) or continuous (such as during lowering). The detailed specifications and analyses of the power and torque requirements for such devices are provided. From the perspective of the necessary power, crane motors are frequently designed as multi-motor systems. An essential issue in this case is the load distribution between the motors proportional to their power ratings, which can be addressed by implementing modern converters within a master-follower configuration.

This chapter delineates the solutions that are frequently employed in contemporary crane propulsion systems. In the event of a lackadaisical recovery, dynamic braking is employed. If continuous regeneration occurs, an active front-end rectifier capable of returning energy to the supply network is employed. The following two case studies have been selected. Case study 1 is representative because the AFE is employed, which, in addition to enabling power recovery, also functions to supply all devices on the common DC bus. Case study 2 concerns gantry cranes characterized by extensive spans. Solutions to two issues that arise in these varieties of cranes are presented as follows: Distribution of burden among multiple motors and the skew issue arising from an extended span.

REFERENCES

1. ABB, Technical guide No.8, Electrical Braking, 2011.
2. Backstrand, J.E. (1992). The Application of Adjustable Frequency Drives to Electric Overhead Cranes, *Industry Applications Society Annual Meeting*, 1992, Conf. Rec.1992 IEEE 4-9 Oct. 1992, vol.2, pp.1986 – 1991.
3. Belmans R., Bisschots F. & Timmer R. (1993). Practical Design Considerations for Braking Problems in Overhead Crane Drives, *Conf. Rec. IEEE-IAS*, Vol.1, pp. 473-479.
4. Busschots F., Belmans R. & Geysen W. (1991). Application of Field Oriented Control in Crane Drives, *Proc. IEEE-IAS, Annual Meeting, Dearborn, Michigan, USA*, September 28-October 4, 1991, pp. 347-353.
5. Hartani, K. & Miloud, Y. (2010). Control Strategy for Three Phase Voltage Source PWM Rectifier Based on the Space Vector Modulation, *Advances in Electrical and Computer Eng.*, Vol.10, Issue 3, pp. 61-65.
6. Jeftenic B., Bebic M. & Statkic S. (2006). Controlled Multi-motor Drives, *Intern. Symp. SPEEDAM* 2006, Taormina (Sicily) - Italy, 23-26 May 2006 , pp. 1392-1398.
7. Jiuhe, W., Hongren, Y., Jinlong Z & Huade, L. (2006). Study on Power Decoupling Control of Three Phase Voltage Source PWM Rectifiers, *Power Electronics and Motion Control Conference*, 2006.

8. Mitrovic N., Petronijevic M., Kostic V. & Bankovic B. (2011). Active Front End Converter in Common DC Bus Multidrive Application, XLVI *Proc. of Inter. Conf. ICEST* 2011, Serbia, Nis, Vol.3, pp. 989-992, 2011.
9. Mitrovic, N., Kostic, V., Petronijevic, M. & Jeftenić, B. (2010). Practical Implementation of Load Sharing and Anti Skew Controllers for Wide Span Gantry Crane Drives", *JME*, Vol. 56, no. 3, pp. 207-216, 2010.
10. Odavic M., Jakopovic Z. & Kolonic F. (2005). Sinusoidal Active Front End under the Condition of Supply Distortion, *AUTOMATIKA* 46(2005), 3–4, pp.135–141, 2005.
11. Paul, A. K., Banerje, I., Snatra, B.K. & Neogi, N. (2008). Application of AC Motors and Drives in Steel Industries, XV *Natinal Power System Conference*, Bombay, Dec.2008, pp.159- 163.
12. Petronijevic M., Veselic B., Mitrovic N., Kostic V. & Jeftenic, B. (2011). Comparative Study of Unsymmetrical Voltage Sag Effects on Adjustable Speed Induction Motor Drives, *Electric Power Applications, IET* , vol.5, no.5, pp.432-442, May 2011
13. Rashid H., *Power Electronics Handbook*, Academic Press, San Diego, 2001.
14. Rockwell Automation (2000), Load Sharing for the 1336 PLUS II AC Drive, *Publication number* 1336E-WP001A-EN-P, June 2000.
15. Slutej, A., Kolonic, F. & Jakopovic, Z. (1999). The New Crane Motion Control Concept with Integrated Drive Controller for Engineered Crane Application, ISIE'99, *Proc.of the IEEE International Symposium*, Volume 3, 1999, pp.1458 – 1461.

Chapter 7

ANISOTROPIC MECHANICAL CHARACTERISTICS OF ABS COMPONENTS MANUFACTURED VIA FUSED DEPOSITION MODELING

1. INTRODUCTION: AN OVERVIEW

Layered manufacturing (LM) techniques have historically been employed primarily for rapid prototyping (RP), with the main objective of producing models for visualization, design validation, and the assessment of kinematic functionality in developing assemblies throughout the product realization process (Caulfield et al., 2007). Without requiring any tooling or fixturing, LM enables the computer-controlled manufacturing of components in a single configuration directly from a digital solid model. These attributes have demonstrated advantages in achieving the goal of decreasing the duration required to complete the product development process (Chua et al., 2005).

A variety of LM processes are currently accessible in the market, such as stereolithography (SLA), fused deposition modeling (FDM), selective laser sintering (SLS), and three-dimensional printing (3DP), all of which are additive manufacturing techniques that share significant similarities (Upcraft & Fletcher, 2003). For each of these procedures, the object design is initially modeled as a solid within a computer-aided design (CAD) software and subsequently exported in a tessellated format as an STL file. This faceted model is subsequently transferred into the appropriate LM machine software, where it is mathematically sectioned into a series of parallel cross-sections or layers. The software generates a traversal path for each slice, including detailed instructions for constructing any required infrastructure to support overhanging segments of the slice. The tangible component is subsequently manufactured, beginning with the lowest layer, by progressively constructing each model slice atop the previously fabricated layer. This additive layering process is therefore capable of manufacturing components with intricate geometrical shapes in a single configuration, eliminating the necessity for tooling, human intervention, or supervision.

In recent years, layered manufacturing processes have evolved from rapid prototyping techniques to rapid manufacturing methods, with the current aim of producing complete components suitable for potential end-use in products (Caulfield et al., 2007). LM is particularly advantageous for the production of specialized, low-volume items such as replacement components for larger systems. This trend emphasizes the importance of a comprehensive comprehension of the related mechanical properties and the subsequent behavior of components manufactured through layered techniques. The base material must not only exhibit durability but also possess mechanical properties of the stratified components that are adequate to withstand in-service loading and operational demands, while remaining reasonably comparable to parts manufactured through more conventional techniques.

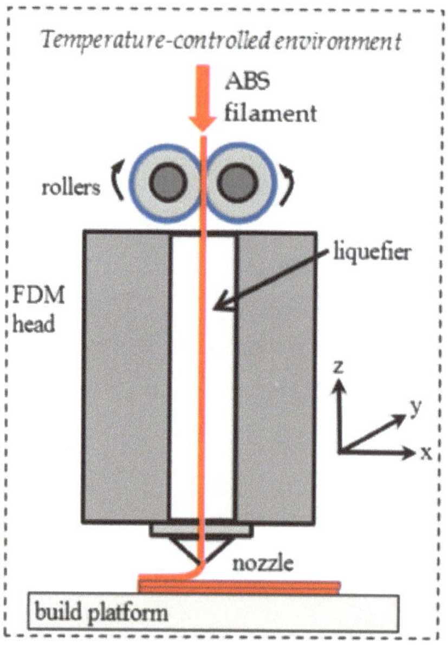

Fig. 1. Schematic of the FDM process

Fused deposition modeling (FDM) by Stratasys Inc. is a type of layered manufacturing technology that creates components with intricate geometries through the deposition of extruded materials, such as robust acrylonitrile butadiene styrene (ABS) plastic (Figure 1). In this procedure, the build material initially exists in the form of a flexible filament. The feedstock filament is subsequently partially dissolved and extruded through a heated nozzle within a temperature-controlled build environment. The material is extruded as a thin layer onto the previously constructed model layer on the build platform, following a specified two-dimensional (x-y) pattern (Sun et al., 2008). The deposited material cools, solidifies, and adheres to the adjacent material. Following the complete deposition of each layer, the build platform advances downward along the z-axis by a distance equivalent to the filament height (layer thickness), and the subsequent layer is then deposited atop the previous one.

In recent years, multiple researchers have specifically examined the anisotropic properties of FDM components. Rodriguez et al. (2001) examined the tensile strength and elastic modulus of FDM specimens with different mesostructures in comparison to the properties of the ABS monofilament substrate. They established that the tensile strength was maximized in components where the fibers were aligned with the direction of the tension force. Ahn et al. (2002) conducted a factorial experiment to assess the influence of model temperature, bead width, raster orientation, air gap, and ABS color on the tensile and compressive strengths of FDM parts. It was concluded that both air gap and raster orientation exerted significant influence on the resultant tensile strength, whereas these factors did not impact the compressive strength. Their findings comprise a collection of recommended guidelines for the design of FDM components. A comparable study was conducted by Sood et al. (2010), examining the effects of various factors including layer thickness, build orientation, raster angle, raster breadth, and air gap. These researchers employed a central composite experimental design and examined the functional relationship between process parameters and specimen strength through response surface methodology. Their findings suggest that the evaluated factors affect both the mesostructural arrangement of the manufactured component and the bonding and distortion characteristics within it. Based on this study, Sood et al. (2011) further investigated the influence of the same five process parameters on the subsequent compressive strength of test specimens. Their research offers valuable insights into the intricate relationship between compressive stress and these parameters and formulates a statistically validated predictive equation. Results highlight the significance of fiber-to-fiber bond strength and the management of distortion throughout the manufacturing process. Lee et al. (2005) determined that layer thickness, raster angle, and air gap affect the elastic performance of compliant ABS prototypes produced through fused deposition. A study conducted by Es Said et al. (2000) examined the influence of raster orientation on the alignment of polymer molecules along the deposition direction during fabrication. These researchers examined the matter of volumetric contraction and raster orientation in relation to tensile, flexural, and impact strengths. Lee et al. (2007) investigated the compressive strength of layered components in relation to construct orientation. They established that the axial FDM specimens exhibit higher compressive strength compared to the transverse specimens. The preceding studies demonstrate the directional dependence or anisotropy of the mechanical properties of FDM components, attributable to mesostructure and fiber-to-fiber bond strength, and offer valuable insights and recommendations concerning key process parameters and the formulation of component build guidelines.

The objective of this project is to quantitatively assess the potential of fused deposition modeling to completely develop into a rapid manufacturing solution. The aim of the project is to establish an understanding of how the mechanical properties of FDM components depend on raster orientation and to evaluate their ability to sustain structural integrity under service loads. The study draws upon the insights of prior researchers and further investigates the influence of fiber orientation on various key mechanical properties of ABS components manufactured through fused deposition modeling. This study employs FDM build recommendations from prior research, along with the predefined machine default parameters, to concentrate the analysis specifically on the critical issue of fiber or raster orientation, namely the direction of the polymer beads (roads) in relation to the loading direction of the component. The tensile, compressive, flexural, impact, and fatigue strength characteristics of FDM specimens are analyzed, assessed, and contextualized in comparison with the properties of injection-molded ABS components.

2. EXPERIMENTAL PROCEDURE

2.1 Materials

All of the FDM specimens tested and analyzed in this study were acrylonitrile butadiene styrene (ABS). ABS is a carbon chain copolymer belonging to styrene ter-polymer chemical family. It is a common thermoplastic that is formed by dissolving butadiene-styrene copolymer in a mixture of acrylonitrile and styrene monomers, and then polymerizing the monomers with free radial initiators (Odian, 2004). The result is a long chain of polybutadiene crisscrossed with shorter chains of poly(styrene-co-acrylonitrile). The advantage of ABS is that it combines the strength and rigidity of the acrylonitrile and styrene polymers with the toughness of the polybutadiene rubber. The proportions can vary from 15 to 35% acrylonitrile, 5 to 30% butadiene, and 40 to 60% styrene. In this study, the resulting composition was 90-100% acrylonitrile/butadiene/styrene resin, with 0-2% mineral oil, 0-2% tallow, and 0-2% wax.

2.2 Specimen preparation and equipment

This project included five different mechanical tests: tension, compression, flexural (3-point bend), impact, and tension-tension fatigue. Three unique specimen designs were required. The tension, flexural, and fatigue specimens were all thin rectangular slabs (Figure 2a) fabricated to be 190.5 mm long, 12.7 mm wide, and 2.6 mm thick in accordance with ASTM D3039 (ASTM, 1998), ASTM D790 (ASTM, 2007), and ASTM D3479 (ASTM, 2007a) standards respectively. Compression specimens were cylindrical and fabricated with dimensions conforming to the ASTM D695 standard (ASTM, 1996). Each cylinder was 25.4 mm long and 12.7 mm diameter (Figure 2b). Impact specimens were fabricated with dimensions conforming to the ASTM D256 standard (ASTM, 2010). The geometry was a v-notched rectangular block of 63.5 mm long, 25.4 mm wide, and 25.4 mm thick (Figure 2c). The v- notch was modeled within the computer solid model of the specimen and was produced directly on the FDM machine.

Fig. 2. Specimen geometries associated with each test

All FDM specimens were fabricated with a Stratasys Vantage-*i* machine. Solid models were first created using Pro/Engineer® software, and then tessellated and exported in STL format. Digital models were then sliced using the Vantage machine's Insight software, and layer extrusion tool paths were generated, i.e. raster patterns used to fill interior regions of each layer, to represent the four different fiber orientations studied in each test.

Layered specimens were all fabricated in a build orientation that aligned the minimum part dimension with the z-axis of the machine, i.e. perpendicular to the build platform. In this orientation, five to ten replicate specimens were built with each of four different raster patterns relative to the part loading direction, for each of the five different tests completed. The four raster orientations included: (a) longitudinal or 0º, i.e. rasters aligned with long dimension of the specimen, (b) diagonal or 45º, i.e. rasters at 45º to the long dimension of the specimen, (c) transverse or 90º, i.e. rasters perpendicular to long dimension of the specimen, and (d) default or +45º/-45º criss-cross, i.e. representing the machine's default raster orientation (Figure 3).

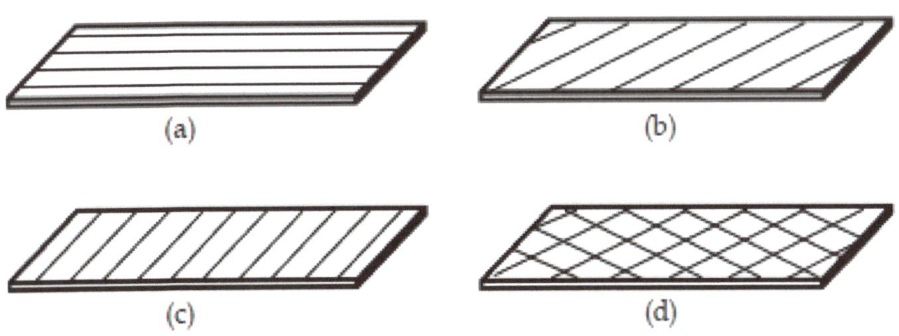

Fig. 3. Four different raster orientations investigated

All FDM spec/imens were built while holding all other machine process settings at the recommended or default values displayed in Table 1.

Table 1. Fixed FDM process settings

Factor	Value/Level
Air gap	0.0 mm
Nozzle	T12
Road width	0.3048 mm
Slice height	0.1778 mm
Part interior fill style	Solid normal
Part fill style	Perimeter/raster
Liquefier temperature	320 ºC
Envelope temperature	80 ºC

In order to measure the reference strength and behaviour of the ABS filament material, for comparisons with the layered parts, additional specimens were fabricated by injection molding for the same five tests. Aluminium molds for each of the three previously described geometries (Figure 2) were designed using Pro/Engineer® software, and manufactured on a Haas VF-1 CNC machining center. Mold cavity dimensions were the same as those described for the FDM specimens, with slight increases to compensate for the shrinkage of molded ABS at approximately 0.005 cm/cm. Parting lines and runners were located with an effort to avoid potential stress concentrations or anomalies in the resulting specimens that might affect test results. All molded specimens were fabricated from the same material as the layered models by feeding the FDM-ABS filament into a polymer granulator and cutting it into pellets of 3-5 mm in length. The pellets were then fed into the hopper of a Morgan Press G-100T injection molding machine. Molding parameters were set to the recommended values for ABS plastic, including nozzle temperature of 270 °C, mold preheat temperature of 120° C, clamping force of 71 kN (16,000 lb), and injection pressure of 41 MPa (6000 psi). Ten replicate specimens were molded for each of the five tests.

Tensile, compressive, flexural, and tension-fatigue tests were performed on an Instron model 3366 dual column uniaxial material testing with .057 micron displacement precision and up to 0.001 N force accuracy. The machine has a 10kN load force capacity. Impact strength was studied on a TMI impact tester. Resulting fracture surfaces were subsequently prepared by gold sputtering and analyzed with a JSM 500-type JEOL Scanning Electron Microscope (SEM).

3. RESULTS AND DISCUSSION

3.1 Tension testing

The tension specimens were thin rectangular slabs made in compliance with ASTM D3039 (ASTM, 1998). Fabrication utilized the FDM T12-nozzle, providing an individual layer or slice height of 0.1778 mm. The specimen thickness of 2.6 mm was subsequently achieved with a total of 15 layers (Figure 4).

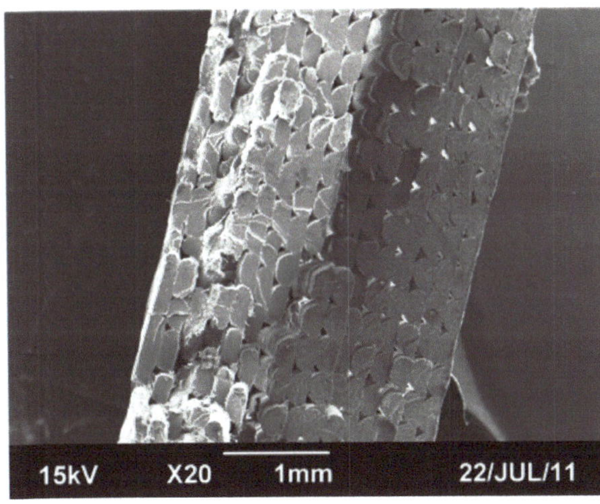

Fig. 4. SEM image displaying 15 layer thickness

A summary of the tension test results for the four raster orientations is displayed in Table 2. The mean ultimate and yield strengths (0.2% offset) were largest for the longitudinal (0°) raster orientation, 25.72 and 25.51 MPa respectively, and weakest for the transverse (90°) raster orientation, 14.56 and 14.35 MPa respectively. The mean ultimate strength of the 90° specimens represented only 56.23% of that of the 0° raster specimens, followed by the 45º specimens at 61.45% and the +45º/-45º specimens at 74.09%.

Table 2. Tension test results

Raster Orientation	*Mean Yield Strength* (MPa), *Std Dev*	*Mean Ultimate* Strength (MPa), *Std Dev*	*Mean Effective* Modulus (MPa), *Std Dev*
Longitudinal (0°)	25.51, *0.73*	25.72, *0.91*	987.80, *19.98*
Diagonal (45°)	15.68, *0.27*	16.22, *0.27*	741.78, *20.28*
Transverse (90°)	14.35, *0.08*	14.56, *0.05*	738.77, *7.91*
Default (+45°/-45°)	18.90, *0.53*	19.36, *0.39*	768.01, *33.31*

A one-way analysis of the variance (ANOVA) was completed in order to consider the equivalence of the population means for the four raster orientations. The results, appearing in Table 3, include a calculated F-test statistic of $F(3,16) = 490.98$ and a *p*-value of 0.0001, indicating a significant difference between some or all of the mean ultimate strength (UTS) values associated with the four raster orientations at a level of significance of $\alpha = 0.05$. The coefficient of determination associated with this analysis was R2 = 0.9886.

Table 3. ANOVA results comparing mean ultimate tensile strengths of 4 raster orientations

Source	DF	SS	MS	F	P
Raster Angle	3	363.991	121.330	460.98	0.0001
Error	16	4.211	0.263		
Total	19	368.202			

Further analysis in the form of post hoc comparisons was performed to determine which raster orientations differed in mean UTS. Tukey's method (Montgomery, 2009), creating a set of 95% simultaneous confidence intervals for the difference between each pair of means, indicated that the difference was significant for all pairwise comparisons of mean UTS values (Table 4). The difference between the mean UTS of the longitudinal rasters (25.72) and that of the transverse rasters (14.56) was the most significant. These results confirm that raster orientation has a significant effect on the tensile strengths of the FDM specimens. Tensile strength is thus verified to be affected by the directional processing and subsequent directionality of the polymer molecules, signifying an anisotropic property.

Table 4. Post hoc Tukey HSD multiple comparisons of mean tensile strengths

Raster Orientation (i)	Raster Orientation (j)	Difference of Mean UTS (i-j)	95% Confidence Interval	
			Lower Bound	Upper Bound
Longitudinal	45-Degree	9.50*	8.579	10.437
	Transverse	11.16*	10.235	12.093
	Default	6.36*	5.438	7.296
45-Degree	Transverse	1.66*	0.727	2.585
	Default	-3.14*	-4.07	-2.212
Transverse	Default	-4.8*	-5.726	-3.868

* *Mean difference is significant at the 0.05 level*

The quantitative data analysis was followed by detailed physical inspection of the specimens at both macro and microscopic levels. Macroscopically, the fracture patterns of the specimens varied somewhat as a function of the raster orientation of the two- dimensional layers and the resulting weakest path for crack propagation (Figure 5). The 90º specimens failed in the transverse direction and the 45º specimens failed along the 45º line. The 0º specimens failed primarily in the transverse direction, although there was some fiber pullout and delamination intermittently evident as well. The +45º/-45º specimens broke at intersecting fracture paths along ±45º, resulting in a saw-tooth fracture pattern across the specimen width. It is likely that fracture paths controlled by weak interlayer bonding are affected by the residual stresses that result from the volumetric shrinkage of the polymer layers during solidification and cooling. In addition, interlayer porosity and air gaps serve to reduce the actual load-bearing area across the layers, providing an easy fracture path.

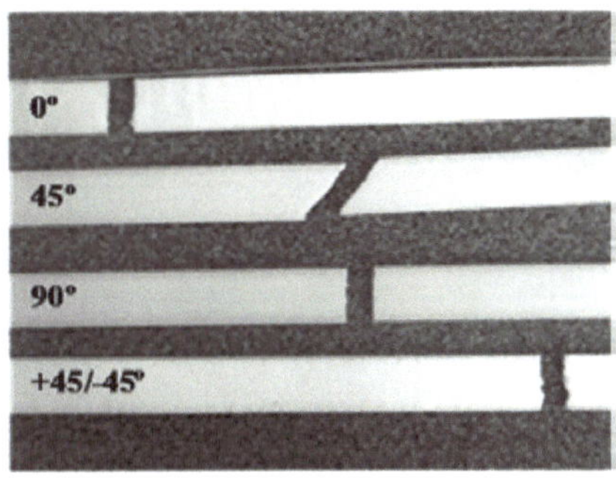

Fig. 5. Failure modes of the specimens with each of the four raster orientations

In specimens with the longitudinal (0°) raster orientation, the molecules tend to align along the stress axis direction. This produces the strongest individual two-dimensional layers subjected to tension loading. During the testing of these specimens, stress whitening due to craze formation and growth was observed to develop prior to reaching the yield stress. Failure occurred at whitened areas from plastic deformation where some evidence of localized fiber delamination was observed. The fracture surfaces were further analyzed with a scanning electron microscope, and displayed failure that was predominantly brittle in nature with localized micro-shearing on each fiber face (Figure 6). The tensile strength of these specimens is thus more heavily dependent on the strength of the ABS monofilament than specimens with fibers running at orientations other than 0° with the stress axis (Rodriguez et al., 2001).

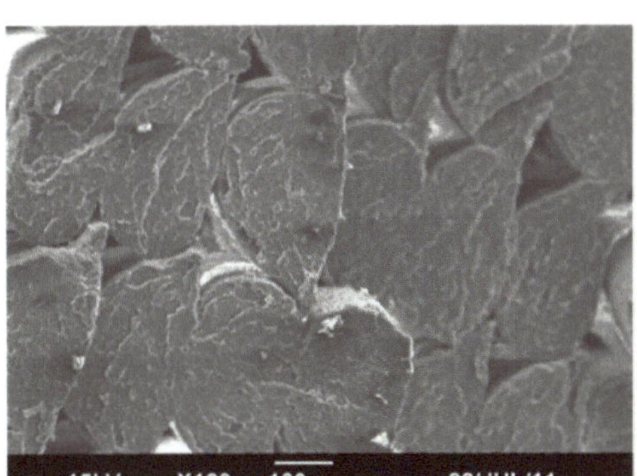

Fig. 6. SEM image of fracture surface of a 0° raster specimen

In contrast, the specimens with transverse (90°) raster orientations did not display obvious crazing during testing, and failure occurred predominantly at the weak interface between layered ABS fibers (Figure 7). These specimens experienced brittle interface fracture. Weak interlayer bonding or some amount of interlayer porosity was evident in the failure of many of the specimens with raster orientations other than 0°, and appeared to be the cause of layer delamination along the fiber orientation during loading. The tensile strength of these specimens depended much more heavily upon the fiber-to-fiber fusion and any air gap resulting between the fibers, as opposed to the strength of the fibers themselves.

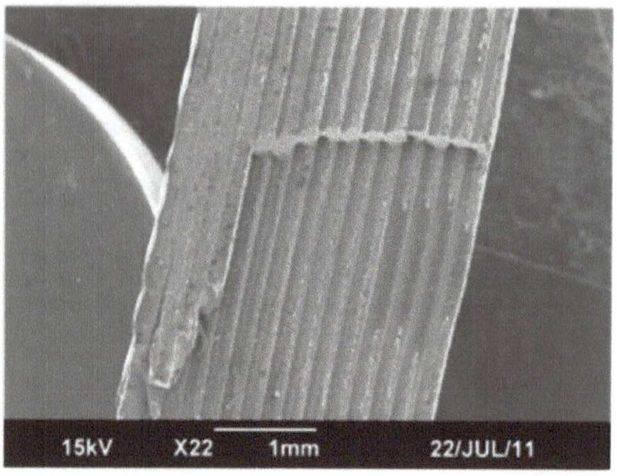

Fig. 7. SEM image of fracture surface of a 90° raster specimen

Under microscopic examination, the +45°/-45° specimens displayed multiple failures of individual raster fibers in both shear and tension (figure 8a). Failure occurred by the pulling and eventual rupturing of individual fibers whereby the material separated at a +45°/-45° angle relative to the tensile load, creating the saw-tooth like appearance evident at the macro-scale (figure 5). Failure of the 45° raster specimens was similar to that of the default +45°/-45° in that it was a brittle shear failure on each of the individual fibers at the microscopic level, as each raster was pulled in tension and failed at 45 degrees relative to the loading axis (figure 8b). Macroscopically, the samples also displayed a characteristic shear failure along the 45° line with the tensile load (figure 5).

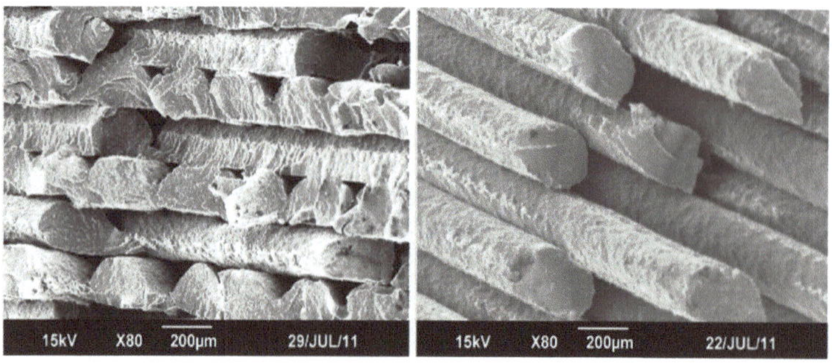

Fig. 8. SEM image of fracture surfaces of (a) +45°/-45° and (b) 45° raster specimen

The air gap that forms during fabrication and remains present between fibers of the FDM specimens is a significant factor in considering tensile ultimate and yield strengths and comparing these properties to those of injection molded ABS specimens. Although the FDM machine setting indicated a desired air gap of 0.0 mm, the fiber geometry inherently causes the presence of triangular air voids as seen in the SEM image in Figure 9. These voids influence the effective tensile strengths and effective elastic moduli of the FDM parts by decreasing the physical cross-sectional area of material specimens. This is in part why the injection molded specimens displayed tensile strengths greater than that of any of the FDM parts, achieving a mean yield and ultimate tensile strength of 26.95 and 27.12 MPa respectively. The mean UTS achieved by the 0° raster specimens was closest to that of the injection molded specimens, representing 94.8% of its value.

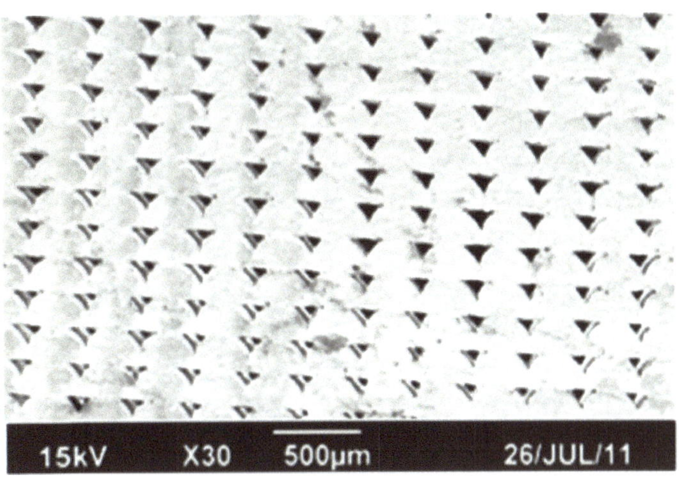

Fig. 9. SEM image of air voids seen on fractured 0º raster specimen

3.2 Compression testing

A summary of the compression test results, as displayed in Table 5, shows strengths that are higher than those obtained in tension testing. Higher compressive strengths are often observed in polymers, and specifically for bulk ABS materials (Ahn et al., 2002). In this study, the average tensile yield strength for FDM specimens was 56% of the average compressive yield strength for FDM specimens. The mean compressive ultimate and yield strengths (0.2% offset) were found to be the largest for the 90° raster orientation, 34.69 and 29.48 MPa respectively. The 45° raster specimens displayed the smallest mean yield strength, 24.46 MPa, representing 82.97% of the yield strength of the 90° raster specimens. The mean yield strength for the injection molded specimens was 35.50 MPa, a value higher than any of the FDM specimens tested. The 90° raster specimens performed the most closely to the injection molded parts, achieving a mean yield strength that was 83.0% of that of the molded specimens.

Table 5. Compression test results

Raster Orientation	Mean Yield Strength (MPa), Std Dev	Mean Ultimate Strength (MPa), Std Dev	Mean Effective Modulus
Longitudinal (0°)	28.83, 1.16	32.32, 0.58	402.64, 3.64
Diagonal (45°)	24.46, 0.30	33.43, 0.20	417.20 10.06
Transverse (90°)	29.48, 0.75	34.69, 0.99	382.21, 10.31
Default (+45°/-45°)	28.14, 0.64	34.57, 0.86	410.44, 11.23

Although mean ultimate strengths are provided in Table 5, it is typically difficult to pinpoint the instance of rupture in compression loading. Most plastics do not exhibit rapid fracture in compression and the focus is therefore on measuring the compressive yield stress at the point of permanent yield on the stress-strain curve (Riley et al., 2006). Although many of the FDM specimens failed by separation between layers, resulting in two or three distinct pieces, yield stresses were analyzed for consistency and to allow for comparisons with that of the injection molded specimens. As a result, a one-way ANOVA was conducted to compare the effect of raster orientation on mean compressive yield strengths in 0º, 45º, 90º, and +45º/-45º conditions. The test indicated that raster orientation had a significant effect on mean compressive yield strength at the $p < 0.05$ level for the four conditions, $F(3, 16) = 31.25$, $p = 0.0001$. Post hoc comparisons using the Tukey HSD test indicated that this significance was limited to and specifically in regard to comparisons with the 45º diagonal condition. The mean yield strength for the 45º raster orientation (24.46 MPa) was significantly different (lower) than that of the other three raster orientations (Table 6), while all other paired comparisons indicated statistically insignificant differences in the mean yield strengths.

Table 6. Post hoc Tukey HSD multiple comparisons of mean yield compressive strengths

Raster Orientation (*i*)	Raster Orientation (*j*)	Difference of Mean Yield Strength (*i-j*)	95% Confidence Interval	
			Lower Bound	Upper Bound
Longitudinal	45-Degree	4.365*	2.739	5.991
	Transverse	-0.653	-2.279	0.973
	Default	0.683	-0.943	2.309
45-Degree	Transverse	-5.018*	-6.644	-3.392
	Default	-3.682*	-5.308	-2.056
Transverse	Default	1.336	-0.290	2.962

* *Mean difference is significant at the 0.05 level*

Inspection of the failed compression specimens provided additional evidence that the 45º raster specimens were significantly weaker in compression than the other raster orientations. The specimens ultimately separated into two or three pieces, following the displacement of the cylinder's top relative to its bottom, as seen in Figure 10. This distortion occurred as a result of the shearing or sliding along the 45º rasters as the specimens was subjected to an axial compressive load. The other three raster orientations displayed less distortion prior to failure and had mean compressive yield strengths that were significantly larger than that of the 45º raster specimens.

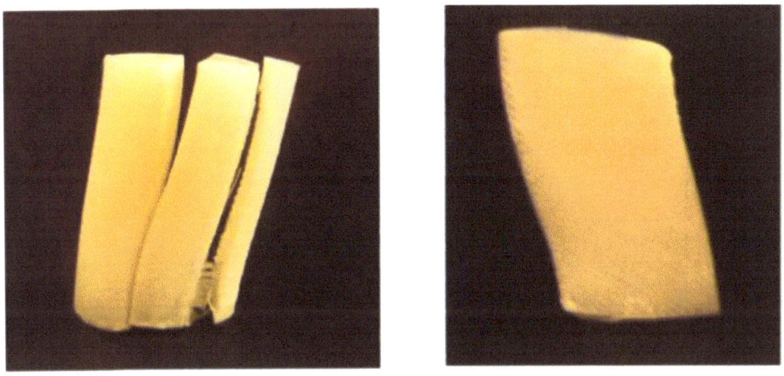

Fig. 10. Photos of failed 45º raster specimens under compression loading

3.3 Flexural testing

Three-point bend tests were completed in order to study the flexural properties of the FDM specimens. Following preliminary testing, a three-inch gage length or span between the outermost points was used for all tests (Figure 11). Flexural strengths were found to be greater than tensile strengths for each raster orientation because the specimens are subjected to both tensile and compressive stresses during bending (Riley et al., 2006). In addition, the three-point test configuration results in the measurement of the maximum strength at the outermost fiber of the beam specimens.

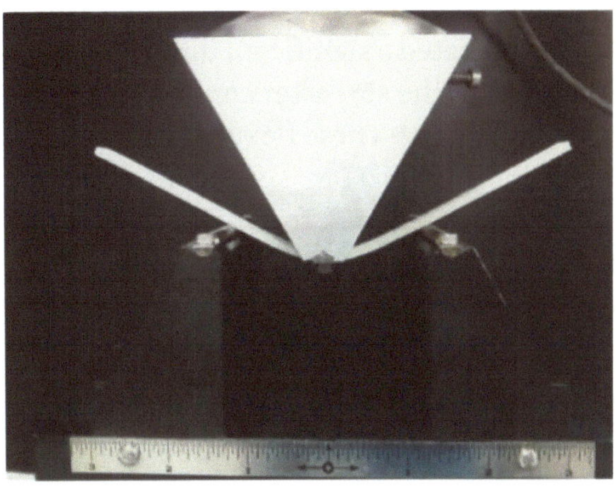

Fig. 11. Three-point bend test configuration

The results of the three-point bend flexural tests are displayed in Table 7. The mean ultimate strength value is the highest for the 0° fiber orientation (38.1 MPa), as was the case during tensile testing. The +45°/-45° orientation had the next highest flexural strength values, followed by 45° and then 90° (23.3 MPa) orientations. Consequently, the flexural test results in Table 7 display the same trend as the tensile test results in Table 2. The flexural strength of the 90° raster specimen was only 60.9% of that of the 0° specimen.

Table 7. Flexural test results

Raster Orientation	*Mean Yield Strength (MPa) Std Dev,*	*Mean Ultimate Strength (MPa)*	*Mean Effective Modulus (MPa), Std Dev*
Longitudinal (0°)	34.2, 2.6	38.1, 2.3	1549.0, 327.3
Diagonal (45°)	21.3, 0.2	25.7, 0.6	1250.0, 36.1
Transverse (90°)	20.8, 0.9	23.3, 1.6	1269.7, 149.6
Default (+45°/-45°)	26.5, 0.7	32.2, 0.5	1438.6, 34.7

Similar to compression testing, however, not all of the specimens ruptured at failure. The 0° raster specimens and the +45°/-45° specimens never fractured, warranting further analysis to be based upon yield rather than ultimate strengths for consistency. A one-way ANOVA was completed and determined that raster orientation had a significant effect on mean flexural yield strengths at the $p < 0.05$ level for the four conditions, with $F(3, 16) = 96.44$ and $p = 0.0001$ (Table 8). The coefficient of determination associated with this analysis was $R2 = 0.9476$.

Table 8. One-way ANOVA results for flexural testing

Source	DF	SS	MS	F	P
Raster Angle	3	582.55	194.18	96.44	0.0001
Error	16	32.22	2.01		
Total	19	614.77			

Post hoc analysis further indicated a significant difference between all paired mean comparisons other than that of the 45º raster condition (21.3 MPa) in comparison to the 90º raster condition (20.8 MPa). These flexural strength results further confirm that the raster orientation of the FDM specimens contributes to directionally dependent performance. The specimen fracture patterns for the 45º and the 90° specimens were similar to those described for the tensile testing. In contract, the 0º and the +45°/-45° specimens never fractured during three-point bend testing, but retained some degree of permanent deformation.

Examination of the fracture surfaces of those specimens that broke into two pieces, i.e. the 45° raster specimens and the 90° specimens, revealed that failure initiated on the side of the part that was under tension loading. As fracture began, the specimen initially remained together by unbroken fibers on portion of it that was in compression. Crack propagation along load direction was erratic and not uniform. This is apparent in Figure 12, which displays clusters of fibers that have bent and then ruptured individually in a catastrophically brittle manner.

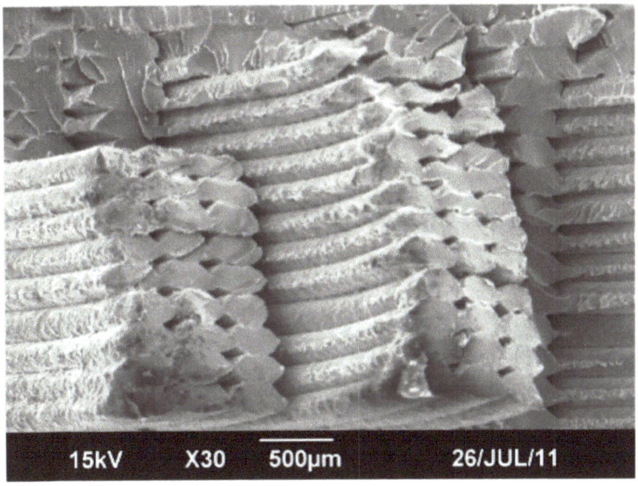

Fig. 12. SEM image of the fracture surface of a 45° raster specimen after flexural loading

Specimens with 0° raster orientation will have fibers that are able to offer more resistance to bending because they are parallel to the bending plane. There is more fiber length over which the load can be distributed. As the raster angle increases to 45° or 90°, the fiber inclination relative to the plane of bending produces rasters with smaller lengths. This results in a net decrease in the ability of the specimen to resist the load. This effect is observed in Figure 13 where the 90° raster specimen shows little evidence of bending.

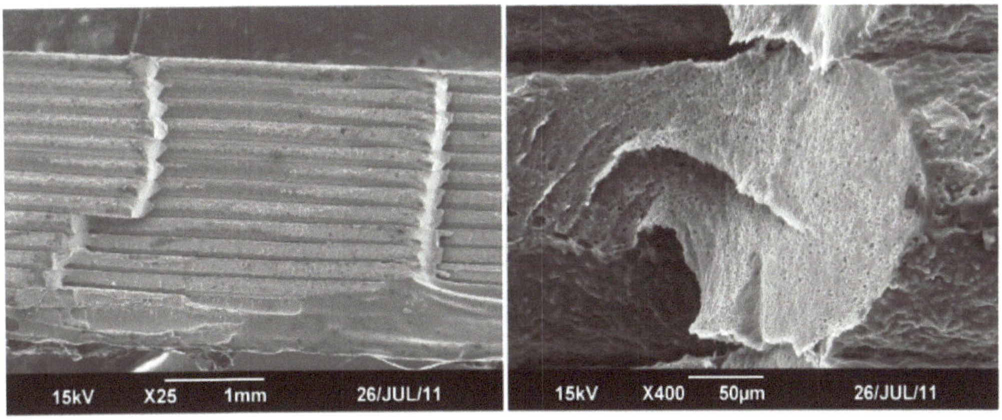

Fig. 13. SEM image of fractured 90° flexural specimen magnified (a) 25X and (b) 400X

The bottom of the specimen shows a large flat area initially affected by the failure of several rasters, from which the crack then splits to the right and left and eventually climbs, as seen in the shear failure of individual rasters in layers. Upon closer microscopic examination, it was observed that individual rasters showed shear failure with a clearly defined exaggerated shear lip on the top of each fiber; something that was not observed in the analysis of 90° raster specimens that failed in tension testing (Figure 7). In both the 45° and the 90° raster specimens, there is little localized plastic deformation before failure initiates and fibers begin to break.

3.4 Impact testing

The impact study utilized an Izod test configuration with a notched specimen held as a vertical cantilevered beam as shown in Figure 14. In this position, the material was subjected to a load in the form of an impact blow from a weighted pendulum hammer striking the notched side of the specimen. The test measures the impact energy or notch toughness, and the results are expressed in energy absorbed per unit of thickness at the notch in units of J/cm. The impact energy absorbed by the specimen during failure is measured by calculating the change in the potential energy of the hammer. The change in potential energy is proportional to difference in the height of the hammer from its initial position to the maximum height achieved after impact.

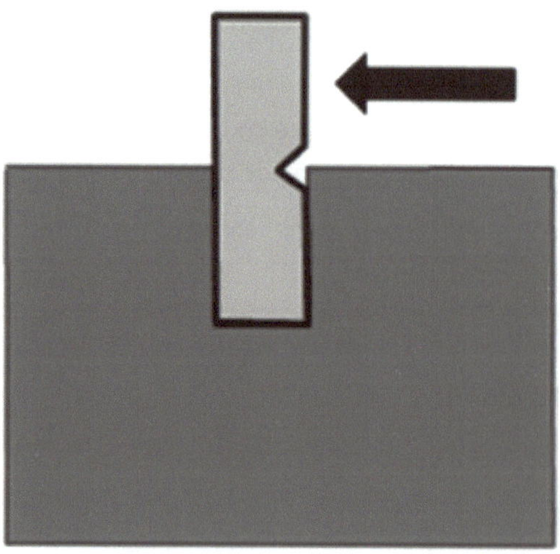

Fig. 14. Izod impact test configuration

Impact tests were completed on 10 specimens with each of the four raster orientations. The mean impact energy results are displayed in Table 9. The absorbed energy was the highest for the longitudinal (0°) fiber orientation (2.989 J/cm) and the lowest for the transverse (90°) orientations (1.599 J/cm). The 45º and +45º/-45º default specimens broke with mean impact resistances between those of the 0° and 90° specimens.

Table 9. Impact test results

Raster Orientation	Mean Impact Energy (J/cm)	Standard Deviation	Fracture Type
Longitudinal (0º)	2.991	0.103	Hinged
Diagonal (45º)	2.339	0.483	Hinged & Complete
Transverse (90º)	1.599	0.014	Complete
Default (+45º/-45º)	2.514	0.338	Hinged & Complete

The relative impact strengths of the four raster orientations correlated well with the tensile strength results. In addition, the variation of the impact strengths was smallest for the transverse orientation (Figure 15), coinciding with the variation of the tensile test results.

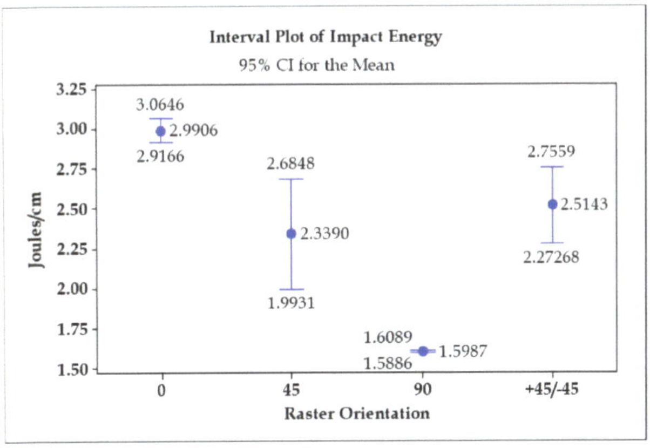

Fig. 15. Interval plot of impact test results

A one-way ANOVA was conducted to compare the effect of raster orientation on mean impact energies in 0º, 45º, 90º, and +45º/-45º conditions. There was a significant effect of raster orientation on impact energies at the $p < 0.05$ level for the four conditions, with F(3, 36) = 37.23, p = 0.0001. Post hoc comparisons using indicated that the mean impact energy for the 45º diagonal condition (2.339 J/cm) did not significantly differ from that of the +45º/- 45º condition (2.514 J/cm), applying a 95% confidence interval (Table 10). All other paired comparisons indicated statistically significant differences in mean impact energies. These results suggest that impact strength has anisotropic characteristics.

Table 10. Post hoc Tukey HSD multiple comparisons of mean impact energies

Raster Orientation (*i*)	Raster Orientation (*j*)	Difference of Mean Impact Energy (*i-j*)	95% Confidence Interval	
			Lower Bound	Upper Bound
Longitudinal	45-Degree	0.652*	0.291	1.012
	Transverse	1.392*	1.031	1.753
	Default	0.473*	0.116	0.837

45-Degree	Transverse	0.740*	0.379	1.101
	Default	-0.175	-0.536	0.184
Transverse	Default	-0.916*	-1.276	-0.555

** Mean difference is significant at the 0.05 level*

Inspection indicated that the fracture patterns of the specimens varied as a function of the raster orientations. The longitudinal (0º) and transverse (90º) pieces fractured along a path oriented 90º to the length of the specimen. The fracture surface appeared smooth along the layers of each of the transverse specimens, all of which experienced clean and complete separation of the specimen into two discrete pieces. Weak interfaces parallel to the crack front affected the transverse raster specimens by providing a straightforward path for crack propagation, resulting in the least amount of energy absorption. Failure initiated from the side on the notch where the energy absorption was greatest, thus causing the rasters to fracture longitudinally along the length of the fiber through several layers until the energy adsorption decreased significantly and the sample broken in half (Figure 16a). Upon closer examination, it can be seen that individual fibers plastically deformed and twisted at the ends upon catastrophic failure, whereby halves of the samples separated (Figure 16b).

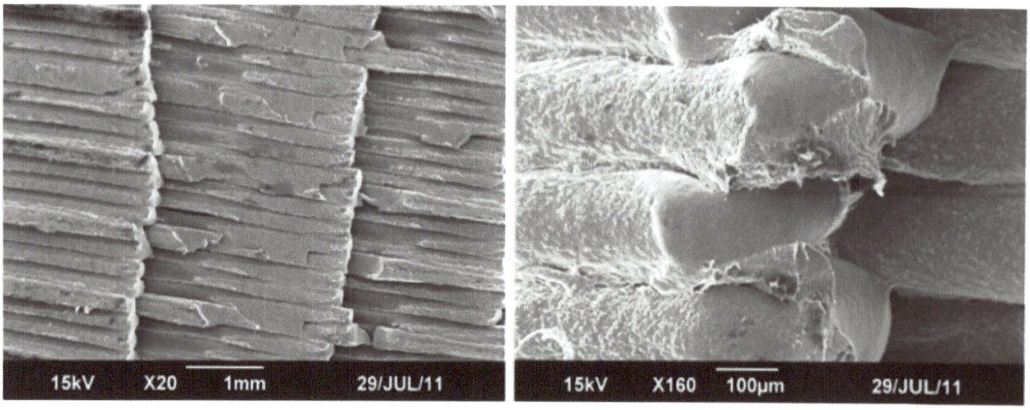

(a) (b)

Fig. 16. SEM image of fractured 90° impact specimen magnified (a) 20X and (b) 160X

The fracture surfaces of the 0° raster specimens were macroscopically rougher, in contrast, displaying slightly jagged edges of individual broken fibers along the transverse fracture plane. Interlayer delamination was also evident and resulted in hinged specimens at failure. Figure 17 displays a 0° specimen where the slightly varied fiber length is evident, along with the formation of a hinge. The longitudinal specimens had the highest mean impact strength, correlating well with the results of three-point bend tests.

Fig. 17. SEM image of impact fractured 0° raster specimen

Weak interfaces running at ±45° to the crack front require relatively high amounts of energy as a result of a mixed fracture mode. As a result, a rougher texture was evident for the fracture surfaces of the default raster specimens, and several hinged as shown on the bottom of Figure 18a. The fracture patter of the 45° specimens, in contrast, fractured more consistently along the 45° maximum shear plane as shown on the top of Figure 18a. The macroscopic roughness of the fracture surface was the result of the varying fiber lengths, evident in the SEM image of Figure 18b.

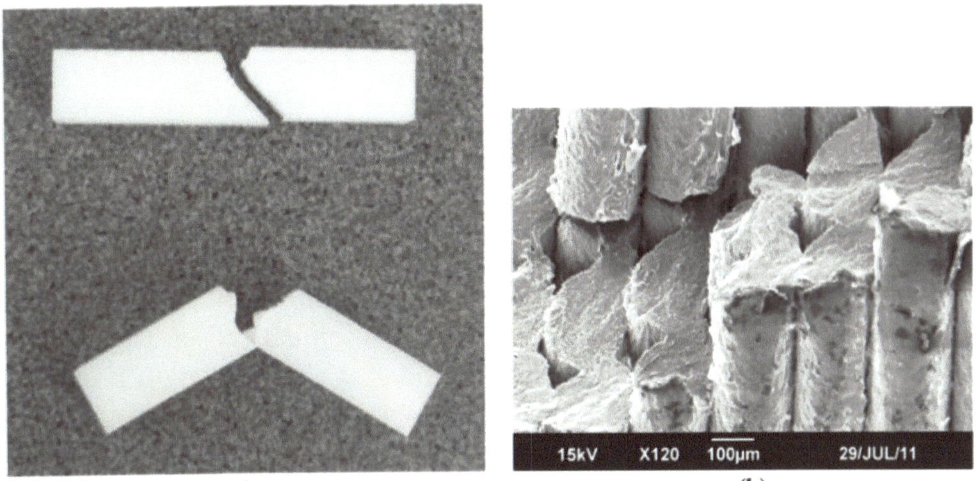

(a) (b)

Fig. 18. (a) Fracture patterns of 45° & +45°/-45° specimens; (b) SEM image of impact fractured 45° raster specimen

3.5 Tension-tension fatigue testing

In analyzing the feasibility of fused deposition modeling to fully evolve into a rapid manufacturing tool, it is important to assess the fatigue properties of FDM specimens and their dependence upon raster orientation. A comprehensive fatigue study is warranted and is currently underway by the authors. At this time, however, a pilot fatigue study utilizing a tension-tension loading configuration has been completed. The fatigue tests utilized specimens originally with natural undamaged surfaces. The maximum load of the fatigue cycles was set to 70% of the mean failure load as determined in the static tensile tests for each specific raster orientation. The minimum cycle load was set as 1/10 of the maximum load, and the resulting stress ratio R for the fatigue tests was 0.1. The value of 70% of the mean failure load was selected following a preliminary study focused on determining a reasonable compromise between the occurrences of excessively long fatigue lives and being too close to the static strength.

All fatigue tests were performed at room temperature. To eliminate any heating effects due to considerable strains, the fatigue loading was applied at a low frequency, 0.25 Hz. The specimen surface was observed during testing and the number of cycles to final failure was determined. The test results appear in Table 11. The 45º raster orientation fractured with the smallest mean number of cycles to failure (1312), representing only 26.7% of that of the +45º/-45º raster orientation (4916).

Table 11. Tension-tension fatigue test results

Raster Orientation	Mean No Cycles to Failure	Standard Deviation
Longitudinal (0º)	4557	694
Diagonal (45º)	1312	211
Transverse (90º)	1616	195
Default (+45º/-45º)	4916	150

A one-way ANOVA was completed in order to consider the equivalence of the mean number of cycles to failure for the four raster orientations. The results, appearing in Table 12, provide a calculated F-test statistic of $F(3,16) = 124.95$ and a *p*-value of 0.0001, indicating a significant difference between some or all of the mean cycle values associated with the four raster orientations at a level of significance of $\alpha = 0.05$. The resulting coefficient of determination associated with this analysis was R2 = 0.9591.

Table 12. One-way ANOVA results for tension-tension fatigue testing

Source	DF	SS	MS	F	P
Raster Angle	3	54093034	18031011	124.95	0.0001
Error	16	2308959	144310		
Total	19	56401993			

Tukey post hoc comparisons indicated that the difference between the mean number of cycles to failure was not significant for the pairwise comparisons of 0° with +45°/-45°, or 45° with 90° raster specimens. All other pairwise comparisons were significantly different.

These results confirm that certain raster orientations have a significant effect on the tension- fatigue properties of the FDM specimens.

The failure modes of the specimens were similar to those for static tension testing (Figure 5), except that several of the 0° raster specimens fractured with a more uneven and almost toothed appearance during fatigue testing. This is shown in Figure 19a where the clusters of rasters have broken at various fiber lengths showing an erratic crack path most probably driven by the areas of weakest fiber bonds and voids between fibers. This SEM image also shows the smooth, brittle, tensile failure on each individual raster face.

The fracture surfaces of the +45°/-45° raster specimens, in contract, showed a mixed mode repeated failure of individual fibers by shearing and tension (Figure 19b). Upon close examination of the individual raster faces, failure initiation sites can be observed at multiple locations. In areas of closely bonded clusters of fibers, "river patterns" can be observed and are believed to occur at large crack growth rates. At the same time, patterns resembling "fish scales" are observed and are often an indication of small crack growth rates. This change of the pattern indicates the existence of a dynamic transition of failure mode.

Fig. 19. SEM images of fatigue fractured specimens with: (a) 0° rasters (b) +45°/-45° rasters

There was some level of correlation between the tension-tension fatigue results and the static tension test results. While the 0° raster orientation achieved the maximum tensile strength, the +45°/-45° specimens survived the most fatigue cycles to failure on average. However, the mean number of cycles to failure for the

0° raster orientation was not found to be statistically different than the +45°/-45° specimens at a level of significance of $\alpha = 0.05$.

Despite being limited to a pilot investigation, the results demonstrate that the orientation of polymer molecules, along with the existence of air gaps and porosity, leads to anisotropic behavior of FDM specimens subjected to tension-fatigue stress.

4. CONCISE SUMMARY

The mechanical characteristics of ABS specimens produced using fused deposition modeling exhibit anisotropic behavior, greatly affected by the orientation of the stacked rasters and the consequent directionality of the polymer molecules. The existence of air gaps and the amount of air voids between the rasters or fibers also affects the strength and effective moduli for all tests conducted in this study.

Tension tests indicate that the ultimate and yield strengths are the largest for the 0° raster orientation, followed by the +45°/-45°, 45°, and 90° orientations in descending order. The differences between mean ultimate tensile strengths are significant for all pairwise comparisons of different raster orientations. Fracture paths are affected by the directionality of the polymer molecules and the strength of individual layers. The longitudinal specimens benefit from the alignment of molecules along the stress axis.

The compression test data indicates that the 45º raster specimens are significantly weaker in compression than the other raster orientations, and they distort prior to failure as a result of shearing along the raster axes. The other three raster orientations have mean yield strengths that are significantly larger than that of the 45º raster specimens, and that are statistically equal to each other at a level of significance of $\alpha = 0.05$.

The results of both three-point bend and impact tests correlate well with tension test results, again indicating that the yield strengths are the largest for the 0° raster orientation, followed by the +45°/-45°, 45°, and 90° orientations in descending order. The 0° rasters offer the most resistance to bending due to the largest effective raster lengths. As raster angle increases, the effective length and associated flexural and impact strengths decrease. Mean flexural and impact strengths are significantly affected by raster orientations, with the pairwise comparison of 45º and 90º rasters as the only one with no statistical difference.

Preliminary tension-tension fatigue tests indicate anisotropic behaviour on the basis of raster orientations. The difference between the mean number of cycles to failure was statistically significant for all pairwise comparisons other than 0° with +45°/-45°, and 45° with 90° raster specimens. Failure modes are similar to those seen in static tension tests.

The outcomes of this experiment are essential in determining the optimal raster orientation for FDM components based on their anticipated in-service loading. The results are valuable for benchmarking future analytical or computational models of FDM strength or stiffness in relation to void density. Ongoing study encompasses a comprehensive fatigue examination of FDM specimens with diverse raster orientations.

REFERENCES

1. Ahn, S., Montero, M., Odell, D., Roundy, S. & Wright, P. (2002), Anisotropic Material Properties of Fused Deposition Modeling ABS. *Rapid Prototyping Journal*, Vol. 8, No. 4, pp. 248 –257, ISSN 1355-2546
2. ASTM Standard D256. (2010). Standard Test Methods for Determining the Izod Pendulum Impact Resistance of Plastics. *ASTM International*, West Conshohocken, Pennsylvania, DOI: 10.1520/D0256-10, Available from: <www.astm.org>
3. ASTM Standard D695. (1996). Standard Test Method for Compressive Properties of Rigid Plastics. *ASTM International*, West Conshohocken, Pennsylvania, DOI: 10.1520/D0695-10, Available from: <www.astm.org>
4. ASTM Standard D790. (2010). Standard Test Methods for Flexural Properties of Unreinforced and Reinforced Plastics and Electrical Insulating Materials. *ASTM International*, West Conshohocken, Pennsylvania, DOI: 10.1520/D0790-10, Available from: <www.astm.org>
5. ASTM Standard D3039/D3039M – 08. (2008). Standard Test Method for Tensile Properties of Polymer Matrix Composite Materials. *ASTM International*, West Conshohocken, Pennsylvania, DOI: 10.1520/D3039_D3039M-08, Available from: <www.astm.org>
6. ASTM Standard D3479. (2007). Standard Test Method for Tension-Tension Fatigue of Polymer Matrix Composite Materials. *ASTM International*, West Conshohocken, Pennsylvania, DOI: 10.1520/D3479M-96R07, Available from: <www.astm.org>
7. Caulfield, B., McHugh, P. & Lohfeld, S. (2007). Dependence of mechanical properties of polyamide components on build parameters in the SLS process. *Journal of Materials Processing Technology*, Vol. 182, pp. 477–488, ISSN 0924-0136
8. Chua, C., Feng, C., Lee, C. & Ang G. (2005). Rapid investment casting: direct and indirect approaches via model maker II. *International Journal of Advanced Manufacturing Technology*, Vol. 25, pp. 11–25, ISSN
9. Es Said, O., Foyos, J., Noorani, R., Mendelson, M., Marloth, R. & Pregger, B. (2000). Effect of layer orientation on mechanical properties of rapid prototyped

samples. *Materials and Manufacturing Processes*, Vol. 15, No. 1, pp. 107–22, ISSN 1532-2475.

10. Lee, B., Abdullah, J. & Khan, Z. (2005). Optimization of rapid prototyping parameters for production of flexible ABS object. *Journal of Materials Processing Technology*, Vol. 169, pp.54–61, ISSN 0924-0136
11. Lee, C., Kim, S., Kim, H. & Ahn, S. (2007). Measurement of anisotropic compressive strength of rapid prototyping parts. *Journal of Materials Processing Technology*, Vol. 187–188, pp. 627–630, ISSN 0924-0136
12. Montgomery, D. (2009). *Design and Analysis of Experiments* (7th Edition), John Wiley & Sons, ISBN 978-0-470-12866-4, Hoboken, New Jersey
13. Odian, G. (2004). *Principles of Polymerization*(4th Edition), John Wiley & Sons, ISBN 978-0- 471-27400-1, Hoboken, New Jersey
14. Riley, W., Sturges, L. & Morris, D. (2006). *Mechanics of Materials* (6th Edition), John Wiley & Sons, ISBN 978-0-471-70511-6, Hoboken, New Jersey
15. Rodriguez, J., Thomas, J. & Renaud, J. (2001). Mechanical Behavior of Acrylonitrile Butadiene Styrene (ABS) Fused Deposition Materials. Experimental Investigation. *Rapid Prototyping Journal*, Vol. 7, No. 3, pp. 148-158, ISSN 1355-2546
16. Rodriguez, J., Thomas, J. & Renaud, J. (2003). Mechanical behavior of acrylonitrile butadiene styrene fused deposition materials modeling. *Rapid Prototyping Journal*, Vol. 9, No. 4, pp. 219-230, ISSN1355-2546
17. Sood A., Ohdar R. & Mahapatra, S. (2010). Parametric appraisal of mechanical property of fused deposition modelling processed parts. *Materials & Design*, Vol. 31, No. 1, pp. 287–95, ISSN 0261-3069
18. Sood, A., Ohdar, R. & Mahapatra, S. (2011). Experimental investigation and empirical modeling of FDM process for compressive strength improvement. *Journal of Advanced Research*, DOI:10.1016/j.jare.2011.05.001, ISSN 2090-1232
19. Sun, Q., Rizvi, G., Bellehumeur, C. & Gu, P. (2008). Effect of processing conditions on the bonding quality of FDM polymer filaments. *Rapid Prototyping Journal*, Vol. 14, No. 2, pp. 72 – 80, ISSN 1355-2546
20. Upcraft, S. & Fletcher, R. (2003). The rapid prototyping technologies. *Assembly Automation*, Vol.23, No.4, pp. 318–330, ISSN 0144-5154

8
Chapter

DESIGN AND ASSESSMENT OF SELF-EXPANDING STENTS APPROPRIATE FOR VARIOUS CLINICAL PRESENTATIONS BASED ON MECHANICAL ENGINEERING PRINCIPLES

1. INTRODUCTION: AN OVERVIEW

Atherosclerosis is a significant disease that causes circulatory dysfunction, affecting big and medium-sized arteries. When cholesterol is present in high concentrations in the blood, it damages the intima, prompting a white blood cell, specifically a monocyte, to infiltrate the intima and transform into a foam cell. Smooth muscle cells move from the media to the intima, where they proliferate extensively. Consequently, cholesterol and other fatty substances build in the intima. Atherosclerosis has emerged as a significant issue in aging developed nations. Consequently, interventions against atherosclerosis have gained significance. Numerous medical treatments exist for atherosclerosis; however, stent implantation has garnered significant interest as a less invasive method for arterial stenotic lesions associated with coronary atherosclerosis and arteriosclerosis obliterans, among others. A stent is a cylindrical, tube-shaped medical device that can constantly expand a stenotic lesion in a blood artery. Two types of stent expansion methods are available for consideration. One is a self-expanding variant that autonomously expands upon being removed from the catheter sheath. Another form is balloon-expandable, which requires forcible expansion via a balloon catheter. The self-expanding stent maintains its expansion to the predetermined diameter at the stenotic lesion, ensuring long-term patency of the arterial wall. The primary focus of the current investigation is the self-expanding kind.

Recently, the significant issue of instent restenosis has emerged in a blood channel containing an implanted stent. In-stent restenosis occurs due to neointimal thickening in the blood artery, resulting from smooth muscle cell hyperplasia. The proliferation of smooth muscle cells is induced by a mechanical stimulation from the stent to the arterial wall. The drug-eluting stent (DES) including immunosuppressive drugs is currently employed in clinical practice to address this issue (Morice et al., 2002). The drug-eluting stent (DES) is more efficacious in inhibiting the occurrence of restenosis compared to a bare metal stent (BMS). Nonetheless, it may be asserted that the DES does not enhance life expectancy or prevent myocardial infarction (Babapulle et al., 2004; Kastrati et al., 2007; Lagerqvist et al., 2007). Reports indicate that the DES may adversely affect life prognosis, although the BMS does not (Nordmann et al., 2006). The prolonged use of immunosuppressive drugs has been associated with depletion. When utilizing a DES, the significant issue of drug-induced adverse effects must also be taken into account. Numerous reports have indicated the utilization of a drug-eluting stent (DES) to avert in-stent restenosis. Nevertheless, limited research has been conducted to avoid in-stent restenosis through the design and modification of bare metal stents (BMS). The majority of studies have been conducted to enhance or optimize the BMS. The morphology, position, and mechanical characteristics of a stenotic lesion are contingent upon the individual patient. Optimization, which produces a singular stent configuration, is not invariably the most advantageous for the patient. Therefore, it is essential to create a stent configuration tailored to each patient. Employing an appropriate stent can minimize the likelihood of in-stent restenosis. Nevertheless, no research has attempted to tailor a stent configuration to address the specific symptoms of each patient.

Two objectives were established to provide a bare metal stent with a reduced risk of in-stent restenosis. The primary aim of our research is to develop a mechanism for designing a stent tailored to each patient's symptoms. The second objective is to design a mechanism for selecting an appropriate stent from commercially available options based on their mechanical qualities. This chapter delineates the design and selection methodologies for a stent tailored to the patient's symptoms, grounded in mechanical engineering principles.

2. METHOD TO SELECT STENT SUITABLE FOR CLINICAL MANIFESTATION BASED ON EVALUATION OF STENT RIGIDITIES

It is important to evaluate mechanical properties of a stent for selecting one suitable for patient's condition. There have been many studies that have evaluated mechanical properties of a stent. (Duda et al., 2000) reported that the important properties of a stent include acceptable weight, stiffness in its radial direction, ease of insertion into the blood vessel, and radiation transmittance capability.

They then proposed a method to evaluate stent and its performances. (Mori & Saito, 2005) performed a four-point bending test using a stainless steel stent to assess flexural rigidity for each different stent structure. (Carnelli et al., 2010) performed two mechanical tests on six kinds of carotid stents. They carried out a four-point bending test to assess flexibility of the stent. Their method of bending test was similar to that by (Mori & Saito, 2005). They also conducted a three-point compression test to measure the radial stiffness of the stent. Based on measurement results, they considered the relation between geometrical features of the stent and its mechanical properties. The results of these results provide a medical doctor the important information. If there exists a selection method of stent by using these evaluation results efficiently, the doctor can select a suitable stent easily in a large proportion of cases. However, there has been no study that has tried to propose the method to select a suitable stent by efficiently using evaluated mechanical properties. In this section, we introduce a method to select a stent suitable for patient's symptom based on mechanical properties of the stent.

2.1 Target stents

Four kinds of commercially available stents, which are already in clinical use, are evaluatedin addition to two types of SENDAI stents having different diameters. One of the evaluated stent is Protege® GPS™ (ev3 Endovascular, Inc., Plymouth, Minnesota, U.S.). The Protege® GPS™ has been developed for a bile duct. Zilver® (COOK MEDICAL Inc., Bloomington, Indiana, U.S.) and JOSTENT® SelfX (Abbott Vascular Devices, Redwood City, California, U.S.) are also biliary stents. Bard® Luminexx™(C. R. Bard, Inc., Murray Hill, New Jersey, U.S.) has been developed as a vascular stent. All six kinds of stents are self- expanding stents made of NiTi shape memory alloy, namely Nitinol. Name, diameter, and length of each target stent are summarized in Table 1.

Table 1. Dimensions of target stents

Stent	Code	Diameter (mm)	Length (mm)
SENDAI	SD10		80
Protege® GPS™ 10	GPS	10	80
Zilver®	ZIL		80
SENDAI	SD8		80
JOSTENT® SelfX	JSX	8	60
Bard® Luminexx™	BLU		100

2.2 Radial compression test and stent stiffness in radial direction

The stent stiffness in radial direction was measured by using the radial compression test machine designed by reference to the method proposed by (Duda et al., 2000).

A stent is mounted on the polytetrafluoroethylene stage with slit and wrapped in a sheet. As illustrated in Fig. 1, one end of the sheet is fixed, and the other end is pulled by the linear actuator (ESMC-A2; ORIENTAL MOTOR Co., LTD., Tokyo, Japan). By applying tensile force to the sheet, the stent is compressed in its radial direction. This tensile force can be measured by using the load cell (LUR-A-200NSA1, load rated capacity: 200 N; KYOWA ELECTRONIC INSTRUMENTS CO., LTD., Tokyo, Japan). In addition, the reduction of the stent diameter is measured by using the LED displacement sensor (Z4WV; OMRON Corporation, Tokyo, Japan). The sheet to wrap the stent consists of a polyethylene film 50 μm thick and a polyethylene terephthalate (PET) film 12 μm thick. Test temperature is 34°C ± 1 .

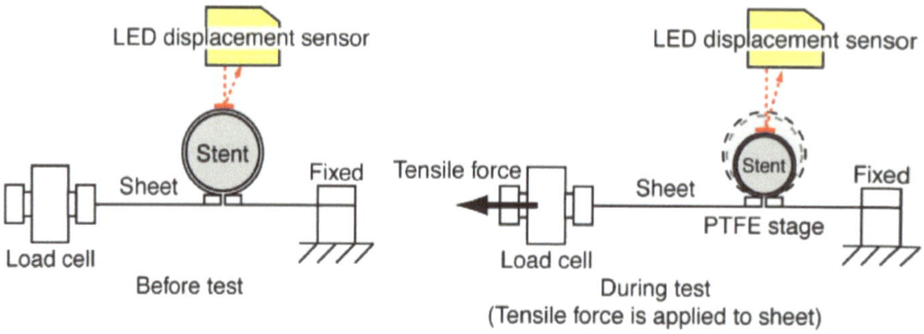

Fig. 1. Schematic view of the measuring method

(Duda et al., 2000) defined two kinds of forces for evaluating the scaffolding property of a self-expanding stent. One of the defined forces is chronic outward force, which is necessary to subtract 1 mm from a stent diameter. The other, namely radial resistive force, is needed to subtract 1 mm from a stent diameter. (Yoshino et al., 2008) defined the radial stiffness based on the radial pressure exerted on a stent for evaluating the scaffolding property. Based on these evaluation indicators, the stent stiffness in radial direction is defined as follows.

$$K_{p,f} = \frac{2\pi F}{\Delta r_s l_s} \tag{1}$$

Here, F is the tensile force measured by using load cell, l_s is the stent length, and Δr_s is the radius reduction of the stent.

The stent stiffness in radial direction was obtained from the measurement result by using equation (1). Figure 2 shows the comparison of the stent stiffness with each stent. For stent diameter of 10 mm, Protege® GPS™ has the highest stent stiffness in radial direction. On the other hand, for the stent diameter of 8 mm, the stent stiffness of Bard® Luminexx™ is the highest. Note that there is a difference of the stent stiffness in each stent.

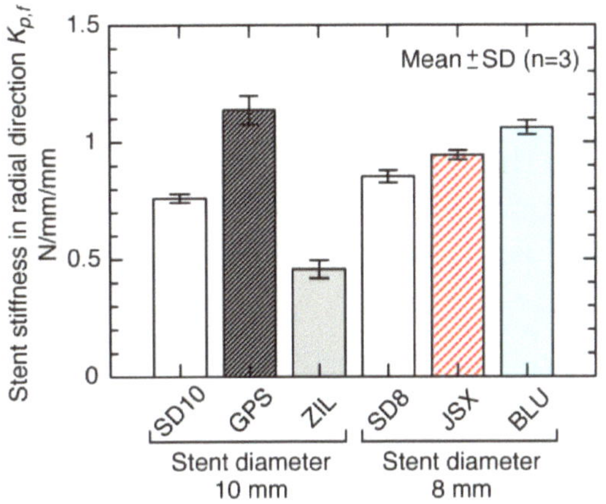

Fig. 2. Stent stiffness in radial direction of each target stent

2.3 Bending test and flexural rigidity

The flexural rigidity was measured by using the designed four-point bending test machine.

As illustrated in Fig. 3, stent is bent by using two pin indenters and a pair of support pins. The load to bend a stent and deflection of the stent are measured by using the micro load capacity load cell (LTS-1KA, load rated capacity: 10 N; KYOWA ELECTRONIC INSTRUMENTS CO., LTD.) and contact displacement transducers (Head: AT-110, Amplifier: AT-210, measurement range: ±5 mm, measuring force: 0.28 N; Keyence Corporation, Osaka, Japan). Here, the interval between support pins l_{sup} and that between pin indenters l_{ind} are set for 40 mm and 12 mm, respectively. In addition, the diameter of these pins is 3 mm. Test temperature is 35 °C ± 1°.

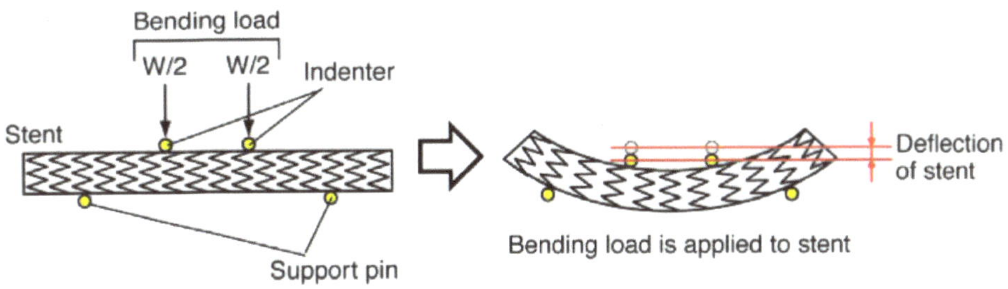

Fig. 3. Schematic view of the bending test

On the four-point bending test, shear force does not act on a stent between pin indenters. This enables us to apply a uniform bending moment to a stent. When considering the stent deformation as the problem of a simply supported beam, the differential equation of the deflection curve at the loading point is presented as follow.

$$\delta = \frac{\left(l_{sup} + 2l_{ind}\right)\left(l_{sup} - l_{ind}\right)^2}{48EI} W \quad (2)$$

Here, W is bending load. The product of Young's modulus E of the stent material and the moment of inertia of cross sectional area of the stent I exactly denotes the flexural rigidity K_b of the stent. Therefore, the flexural rigidity of the stent is derived from equation (2) as follow.

$$K_b = \frac{\left(l_{sup} + 2l_{ind}\right)\left(l_{sup} - l_{ind}\right)^2}{48} \frac{W}{\delta} \quad (3)$$

The flexural rigidity was obtained from measurement results by using equation (3). Figure 4 shows the comparison of the flexural rigidity with each stent. For the stent diameter of 10 mm, Protege® GPS™ has the highest flexural rigidity. On the other hand, the flexural rigidity of Bard® Luminexx™ is the highest in the stents with 8 mm diameter.

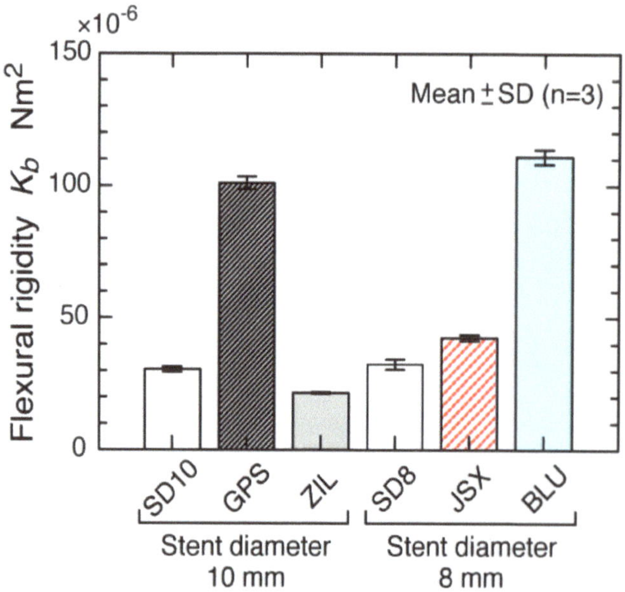

Fig. 4. Flexural rigidity of each target stent

2.4 Method to select stent suitable for clinical manifestation

Figure 5 shows flow of the proposed method to select a suitable stent based on mechanical properties. As preparation for selecting a suitable stent, the map of stent rigidity is made. The selection method is described below according to the flow illustrated in Fig. 5.

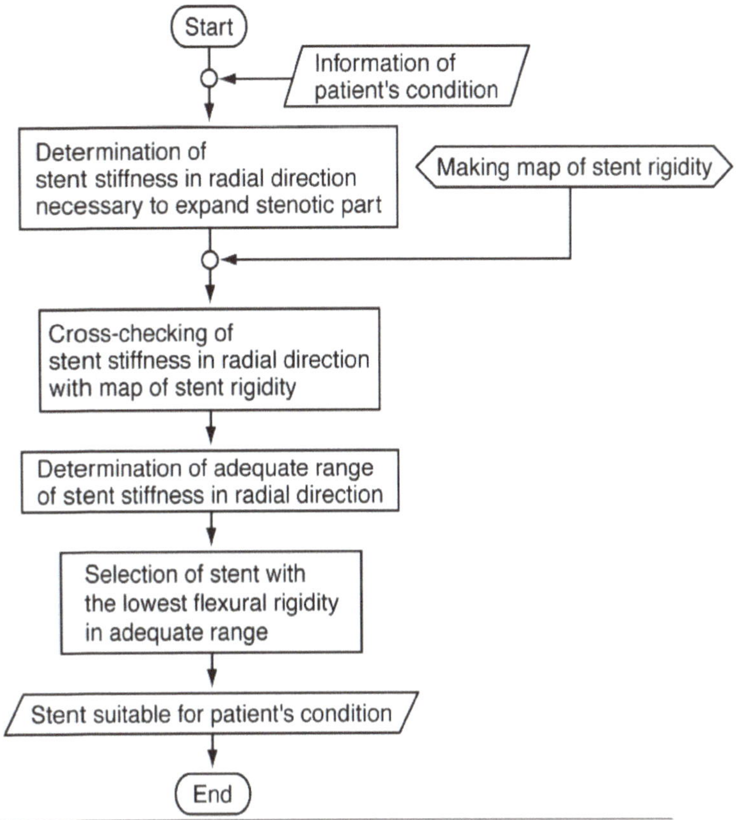

Fig. 5. Flow of selecting stent suitable for clinical manifestation

Step 1. Determination of stent stiffness in radial direction necessary to expand stenotic part: First, the stent stiffness in radial direction necessary to expand stenotic part is determined based on information of the patient's symptom. Requirements are the stent diameter d_s, the outer diameter D_o, the inner diameter D_i, and the least inner diameter D_l of the stenotic part. The pressure strain elastic modulus $E_{p,vl}$ of the diseased blood vessel is also needed. Other important information is the percentage by which to improve the blood flow level, namely the target diameter D_t after treatment. Given all the values listed above, the necessary radial stiffness can be obtained by the following equation (see in Section 6 for details).

$$K_p^* = 2E_{p,vl}\frac{D_t - D_l}{D_o\left(d_s - D_t\right)} \tag{4}$$

Equation (4) is derived from the radial pressure necessary to expand the stenotic part in the blood vessel. In this chapter, the radial force necessary to expand the stenotic part is considered as a standard. Therefore, K^* of equation (4) is converted into the necessary stent stiffness $K_{p,f}^*$ using the circumferential length of the vascular inside wall after treatment as follow.

$$K^{*}_{p,f} = 2\pi E_{p,vl} \frac{D_t(D_t - D_l)}{D_o(d_s - D_t)} \tag{5}$$

With the symptom information assumed as shown in Table 2, the necessary stent stiffness $K_{p,f}^*$ can be calculated using equation (5) as $K_{p,f}^*$ = 0.87 N/mm/mm (d_s = 8 mm), and $K_{p,f}^*$ = 0.48 N/mm/mm (d_s = 10 mm).

Table 2. Information of symptom assumed for selecting of stent

Artery	Carotid artery
Outer diameter, D_o (mm)	6.82
Inner diameter, D_i (mm)	5.60
The least inner diameter of stenotic part, D_l (mm)	2.80
Rate of stenosis by ECTS method (%)	50
Pressure strain elastic modulus of diseased artery $E_{p,vl}$ (MPa)	0.145
Target inner diameter after treatment, D_t (mm)	5.60

Step 2. Cross-checking of stent stiffness in radial direction with map of stent rigidity: The calculated $K_{p,f}^*$ values are plotted onto the map of stent rigidity, and indicated by broken lines presented in Fig. 6. It is difficult that the stent stiffness of a commercially available stent matches the calculated $K_{p,f}^*$ value.

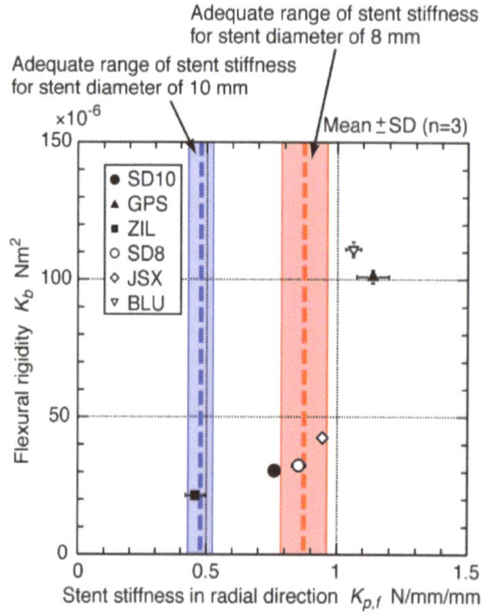

Fig. 6. Selection of suitable stent using map of stent rigidity

Step 3. Determination of adequate range of stent stiffness in radial direction: As described above, there exist few stents that have the stent stiffness value equal to the calculated K^* value. Therefore, the necessary stent stiffness K *p,f* is widened to the extent adequate to expand the *p,f* stenotic part in the blood vessel. The doctor should normally determine this adequate range of the stent stiffness. In this case, it is determined that the range of ±10 % for the necessary stent stiffness K^* is adequate. The shaded areas shown in Fig. 6 are the setup adequate ranges.

Step 4. Selection of stent with the lowest flexural rigidity in adequate range: If the stent, which is in the adequate range, is selected, it can expand the stenotic part in the blood vessel sufficiently. The stent sometimes has too high flexural rigidity for the lesion because of selection only in terms of the stent stiffness in radial direction. When selecting a stent, its flexural rigidity should be considered. But we have no basis of the flexural rigidity for selecting a stent. Therefore, it is decided to select the stent that has the lower flexural rigidity than any other stent being in the adequate range. For the assumed symptom, SENDAI (SD8, d_s = 8 mm) and Zilver® (ZIL, d_s = 10 mm) are most suitable.

In this section, we introduced the method to select a stent suitable for the patient's symptom based on mechanical properties of the stent. It is considered that the selection method can help doctors greatly in clinical sites. Commercially available stents are targeted for this selection method. There are limitations to selecting a suitable stent using this method. Therefore, a novel stent has to be designed for providing the stent more suitable for the patient's symptom. The method to design more suitable stent will be described in the following sections.

3. DESIGN SUPPORT SYSTEM FOR SELF-EXPANDING STENTS

A design support system for a self-expanding stent using CAD and CAE is introduced in this section. This support system can improve the efficiency of the suitable stent design.

3.1 Design variables of SENDAI stent

Figure 7 shows a two-dimensional diagram of a SENDAI stent. Each wire section is constructed from 12 loosely curved S-shaped wires. The strut section of the stent connects them using three bridge wires. On the two-dimensional shape of the SENDAI stent in Fig. 7, the design variables that might affect the mechanical properties of the stent are set. Here, l_w and l_b are the length of the wire and the bridge wire along the axial direction, θ_w and θ_b are the angle of the wire and the bridge wire to the axial direction, t_w and t_b are line element width of the wire and the bridge wire, and r_i and r_o are the inner radius and the outer radius of the wire end part. Every wire is structured in an arc shape. When the design variables l_w, θ_w, and others are given, the arc shape is determined and wire section is constructed through laying out these arc shapes continuously. Furthermore, the number of wires n_w, the number of bridge wires n_b, and the thickness of the tube material t_s are also design variables.

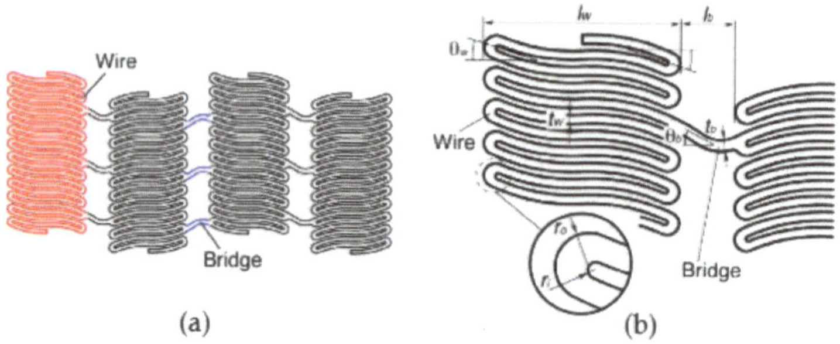

Fig. 7. Two-dimensional diagram of SENDAI stent (a) and main design variables for SENDAI stent (b)

3.2 Framework of design support system

Figure 8 shows the design support system for a self-expanding stent. The left-hand side of the figure shows the production process of the SENDAI stent. It has three production stages: a 'manufacture' stage, during which the NC data are created based on the two-dimensional diagram to manufacture the initial stent shape while the initial stent is manufactured by using laser processing; an 'expansion' stage, during which the initial stent is forcibly expanded in the radial direction by inserting a tapered rod into the stent as it is given shape- memory treatment; and an 'evaluation' stage, during which the performance of the expanded stent is tested.

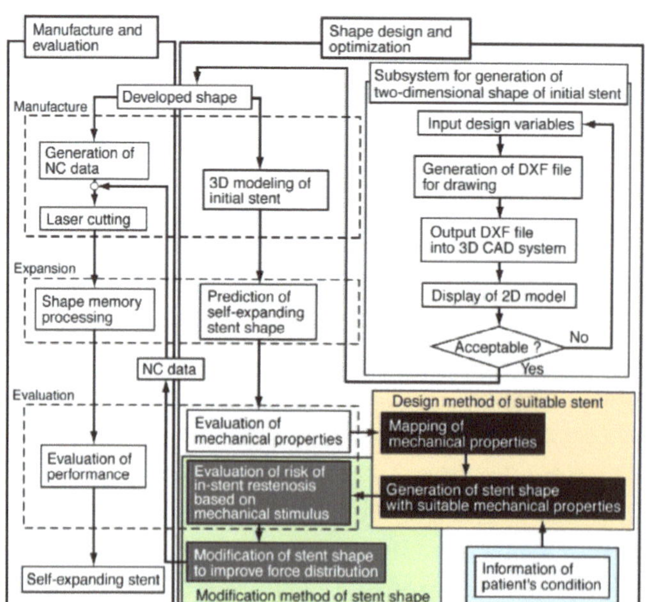

Fig. 8. Design support system for self-expanding stent. The left-hand side shows the production process of the SENDAI stent. The right-hand side shows the flow of the proposed design support system.

The right-hand side of Fig. 8 shows the flow of the shape design for self-expanding stents being proposed. A three-dimensional model of the initial stent manufactured by using laser processing is created from the two-dimensional shape by using 3D CAD. Then, by dividing into finite elements, the finite element model representing the initial stent is created based on the 3D CAD model, and the expanded stent shape is predicted by applying an expansion analysis using the finite element method for large deformation. Based on this prediction, a rigidity analysis is conducted using a non-linear finite element method. The mechanical properties of the stent are evaluated from the results. This process corresponds to the actual production process of the 'manufacture,' 'expansion,' and 'evaluation' stages.

This support system has design method of a self-expanding stent suitable for the patient's symptom based on mechanical properties of a stent. This method is available to introduce into the existing design support system described above, and has two stages which are the design and modification methods. In the first stage, a stent shape with mechanical properties suitable for the patient's symptom is determined and designed. In the second stage, to modify the stent shape in consideration of the risk of in-stent restenosis realizes designing the stent shape more suitable for the patient's symptom. The risk of in-stent restenosis is evaluated based on a mechanical stimulus to a vascular wall by insertion of a stent. These two stages of the design method will hereinafter be described in more detail (see in Section 6).

After the two-stage design method, the design support system ends with the generation of the NC data of the designed stent necessary for moving onto the actual production process. For change in a stent shape, a subsystem for generating the two-dimensional shape of the initial stent was introduced, and this is available for changing the two-dimensional shape flexibly. It can also be used to generate a two-dimensional diagram of the stent in the first place.

4. MECHANICAL PROPERTIES OF STENT

It is important to obtain sensitivities of mechanical properties of a stent to design variables when designing the stent suitable for patient's symptom. In this section, important mechanical properties, namely, radial stiffness, flexural rigidity, and shear rigidity, are evaluated and the maps of their sensitivities are made. The expanded stent shape is used for evaluation of mechanical properties. The expanded shape is predicted from the expansion analysis (see in (Yoshino & Inoue, 2010) for details).

We must know the relationship between the mechanical properties of a stent and the design variables in order to design a stent with specific properties. The mechanical properties of the SENDAI stent, such as radial stiffness, flexural rigidity, and shear rigidity, were evaluated, and their sensitivities to the design variables were also defined, as shown in Fig. 9 (Yoshino & Inoue, 2010). The wire length along the axial direction and the wire width were selected as design variables. Isolines on the maps of mechanical properties are very important for proposing designs. The isoline is plotted onto the maps based on the required mechanical property, and design variables of the proposed design are determined from the isoline. In addition, we assumed the mechanical properties of the stent material as illustrated in Fig. 10: Young's modulus of 28 GPa, and Poisson's ratio of 0.3.

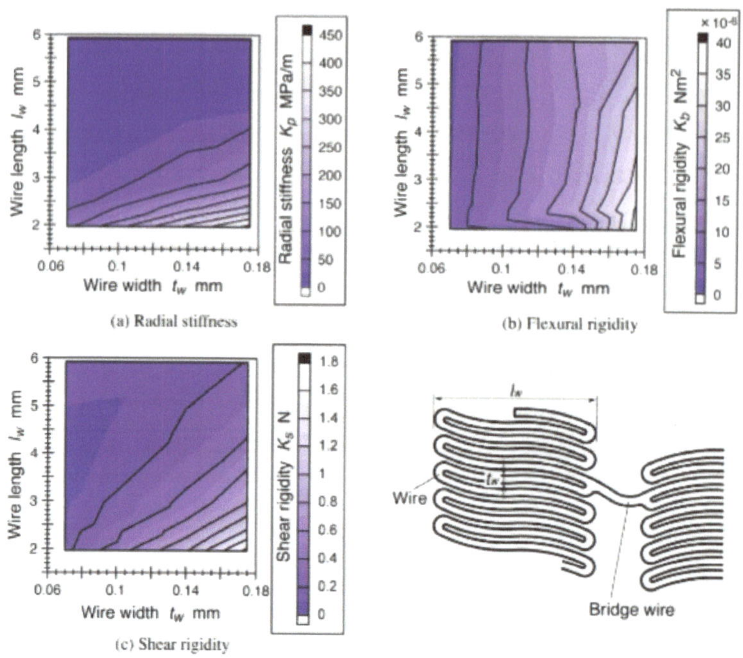

Fig. 9. Sensitivities of the mechanical properties of the SENDAI stent 6 mm diameter to thedesign variables: (a) radial stiffness, (b) flexural rigidity, and (c) shear rigidity

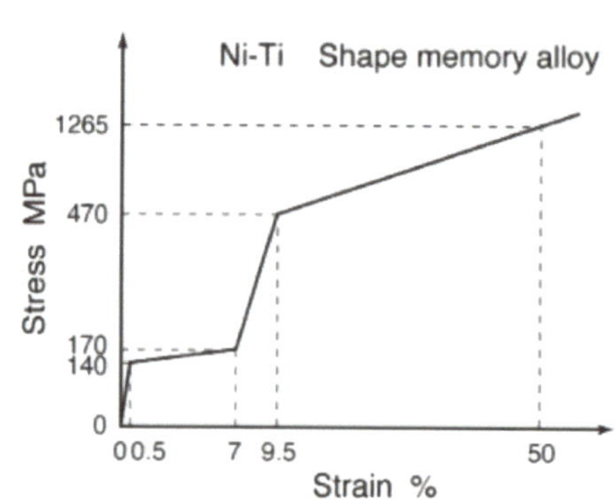

Fig. 10. Assumed stress-strain relationship for the stent material

5. ESTIMATION OF FORCE ON VASCULAR WALL CAUSED BY INSERTION OF SELF-EXPANDING STENT

The forces on a blood vessel are classified into two categories: internal pressure caused by the expansion of the stent and the force resulting from the straightening of the blood vessel, which is a phenomenon whereby a curved blood vessel is straightened by the stent. Straightening of blood vessels often occurs in cases involving the use of a closed-cell stent. As a result, a problem occurs in that the straightening easily encourages kinking of the blood vessel at the flexural area distant from the stented lesion (Tamakawa et al., 2008).

In this section, focusing on a method by which to improve the force distribution on the vascular wall according to the symptoms of the patient, we introduce a method to compute the distribution of the contact force between the stent and the blood vessel under the assumption that the stent is inserted into a straight blood vessel. In the method, the stent and the blood vessel are simplified as axisymmetrical models. Then a method for calculating the distribution of the straightening force on the vascular wall is introduced.

5.1 Computation of contact force distribution on vascular wall caused by expansion of stents

The expansion of a stent in a blood vessel induces pressure on the contact surfaces of the stent and the blood vessel. Analysis of pressure or contact force based on the complicated shape of stent is a difficult and time-consuming task. Therefore, a simplified calculation method using the axisymmetrical models is presented as shown in Fig. 11.

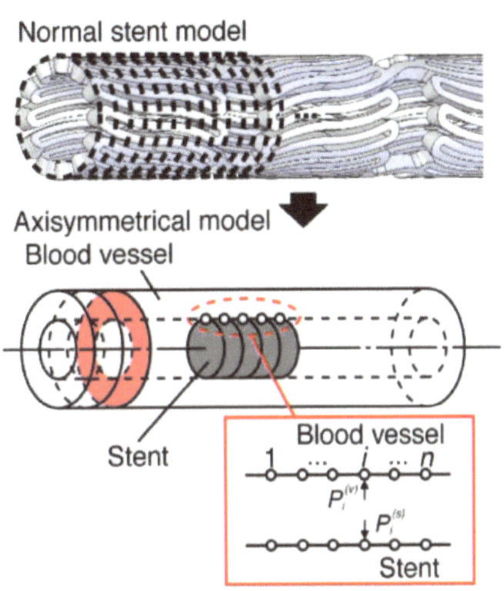

Fig. 11. Model used to compute the contact force between the stent and vascular wall. Thestent and blood vessel are simplified to axisymmetrical models.

The stent is modeled by a number of rings, indicated by broken lines in Fig. 11. The wire section is represented by 11 rings. The blood vessel is similarly modeled by rings, where the ring intervals are the same as those of the stent. In addition, it is assumed that the blood vessel is much longer than the stent. It is also assumed that these rings deform axisymmetrically due to the uniformly distributed radial force.

Next, let us consider ring i (i = 1, 2, ..., n) in contact along the surfaces of stent and vascular wall. When a unit radial force is applied to ring i of model m (m = s, v, which correspond to the stent and the blood vessel, respectively), the radial displacement at ring j, denoted as $r_{ji}(m)$, is calculated using the finite element method. The influence matrix $[C(m)]$ is defined in terms of as $r_{ji}(m)$ as follows:

$$\left[C^{(m)}\right]=\left[\left\{r_1^{(m)}\right\},\left\{r_2^{(m)}\right\},\ldots,\left\{r_n^{(m)}\right\}\right] \tag{6}$$

where

$$\left\{r_i^{(m)}\right\}=\left(r_{1i}^{(m)},r_{2i}^{(m)},\ldots,r_{ni}^{(m)}\right)^{\mathrm{T}} \tag{7}$$

The radial displacement $\{r(m)\}=(r_1(m), r_2(m), \ldots, r_n(m))T$ due to unknown contact force $\{P(m)\}=(P_1(m), P_2(m), \ldots, P_n(m))T$ is given as follows:

$$\left[C^{(m)}\right]\left\{P^{(m)}\right\}=\left\{r^{(m)}\right\} \tag{8}$$

Therefore, the following expressions, namely, the equilibrium equation and the condition ofcontact, are obtained:

$$\left.\begin{aligned}P_i^{(v)}+P_i^{(s)}&=0\\R_i^{(v)}+r_i^{(v)}&=R_i^{(s)}+r_i^{(s)}\end{aligned}\right\}\text{(contact)} \tag{9}$$

$$\left.\begin{aligned}P_i^{(v)}=P_i^{(s)}&=0\\R_i^{(v)}+r_i^{(v)}&>R_i^{(s)}+r_i^{(s)}\end{aligned}\right\}\text{(non}-\text{contact)} \tag{10}$$

where $R_i(s)$ and $R_i(v)$ denote the initial radii of the stent and the blood vessel, respectively, at i.

Then, by replacing $\{P(v)\}$ and $-\{P(s)\}$ with $\{P\}$, the equation of contact force is obtained as follows:

$$\left[C^{(v)}+C^{(s)}\right]\{P\}=\left\{R^{(s)}-R^{(v)}\right\} \tag{11}$$

Equation (11) is solved for $\{P\}$ to obtain the distribution of the contact force between the stent and the blood vessel wall. Since the rigidity of the strut section of the stent is considerably much lower than that of the wire section, $r_{ji}(s)$ is negligible, except for the case in which i and j are located in the wire section of the stent. Therefore, in the present study, the influence matrix $[C(s)]$ is composed by diagonally placing the partial matrix for the wire section, and the contact force on the strut section of the stent is not evaluated. The Gaussian elimination method is used to solve the simultaneous equations.

5.2 Calculation of distribution of straightening force on vascular wall

Straightening of the blood vessel occurs when a straight stent is inserted into a curved vessel. This insertion causes a straightening force to act on the vascular wall. The straightening effect should be evaluated while considering the interaction process of the expansion of the stent with the curved blood vessel. However, solving this problem is extremely complicated. Therefore, the straightening of the blood vessel is assumed to be independent from the expansion of the vessels by the stent, and it is simplified as shown in Fig. 12. The stent is approximated by a beam with the flexural rigidity obtained in the previous section, and the stent is modeled as a laminated beam by combining with a curved beam, which is used a model of the blood vessel.

The beam models are divided into n intervals. When a unit force is applied to point i of model m (m = s, v, which correspond to the stent and the vessel, respectively), the deflection at point j is denoted as $d_{ji}(m)$. The n-order influence matrix $[D(m)]$ is defined in terms of $d_{ji}(m)$ as follows:

$$\left[D^{(m)}\right]=\left[\left\{d_1^{(m)}\right\},\left\{d_2^{(m)}\right\},\ldots,\left\{d_n^{(m)}\right\}\right] \tag{12}$$

where

$$\left\{d_i^{(m)}\right\}=\left(d_{1i}^{(m)},d_{2i}^{(m)},\ldots,d_{ni}^{(m)}\right)^{\mathrm{T}} \tag{13}$$

The force $\{F(m)\}$ distributed along the beam is related to the deflection $\{d(m)\}$ so as to satisfy the following equation:

$$\left[D^{(m)}\right]\left\{F^{(m)}\right\}=\left\{\delta^{(m)}\right\} \tag{14}$$

The equilibrium equation and contact condition are given by the following equations:

$$F_i^{(s)}+F_i^{(v)}=0 \tag{15}$$

$$\delta_i^{(s)}=\delta_i^{(i)}+\delta_i^{(v)} \tag{16}$$

where $d_i(i)$ denotes the initial deflection of the blood vessel at point i, and $d_i(s)$ and $d_i(v)$ are the deflections of the stent and the blood vessel, respectively. Replacing $-\{F(v)\}$ and $\{F(s)\}$ with $\{F\}$, the equation of straightening force is obtained as follows:

$$\left[D^{(v)} + D^{(s)}\right]\{F\} = \left\{\delta^{(i)}\right\} \tag{17}$$

5.3 Limitations of these calculation methods

The methods will be useful for improving stent shape in order to reduce the peak force acting on the vascular wall. Although these methods are useful to calculate a contact force and a straightening force on a vascular wall, there are the following limitations:

1. The isotropic material property was assumed for the artery.
2. The cross-sectional distortions and thickness changes of the stent and blood vessel were ignored.
3. The friction on the contact surface between the stent and blood vessel was ignored.
4. It was consider that the straightening of the blood vessel and the expansion of the vessels by the stent were independent each other.
5. The stent and blood vessel were simplified by using axisymmetrical models in computation of the contact force and using the laminated beam in calculation of the straightening force, respectively.

6. DESIGN METHOD OF STENT SUITABLE FOR DIVERSE CLINICAL MANIFESTATION

The stent must have the radial stiffness sufficient to expand the stenotic part in the blood vessel outward. Simultaneously, it must be sufficiently flexible to conform to the vascular wall. Neither the symptom nor the blood vessel shape is always in the same state. Therefore, it is more important to design stents suited to each unique symptom of every patient. (Colombo et al., 2002) made evaluations of stent 'deliverability,' 'scaffolding,' 'accurate positioning,' and so on for the average lesion of the coronary artery. They proposed a guideline for use in determining a suitable stent. This guideline was based on clinical trials. Therefore, it is governed by the doctor's sense. One more noteworthy point is that the best- suited stent cannot be actually available for every specific symptom because a stent must be chosen only from among commercially available stents.

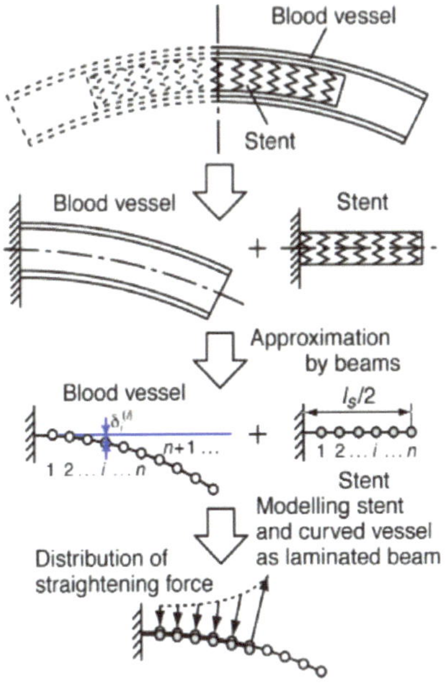

Fig. 12. Model for calculation of the straightening force on vascular wall. The stent andblood vessel are simply modeled as a laminated beam.

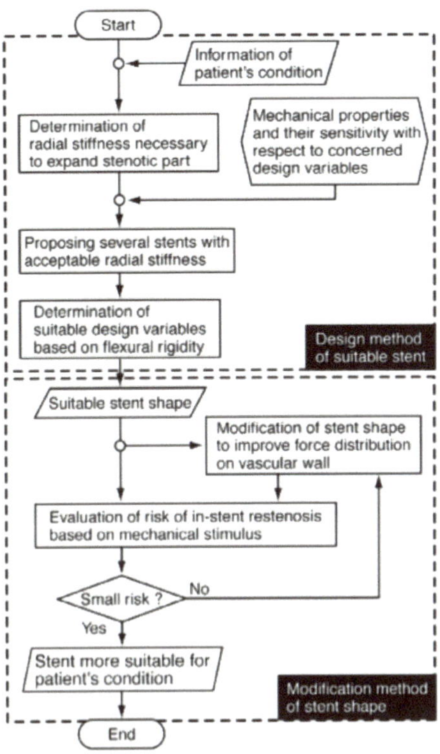

Fig. 13. Flow of designing stent suitable for diverse clinical manifestation. Designing a stent suitable for diverse clinical manifestation consists of two steps.

In the first step, it is possible to design a stent necessary to expand the stenotic part of a blood vessel. In the second step, the designed stent is modified to suit the patient's symptom better.

6.1 Design method of stent having suitable mechanical properties to expand stenotic artery

6.1.1 Mechanical properties of artery model

When considering expanding the stenotic part in the blood vessel as presented in Fig. 14, it is important to know the vascular mechanical properties. Arteries in a living body are always pressured inside. Simultaneously, they are tensed with strain produced along its axial direction by several tens of percent. In this sense, it can be regarded as a thick cylinder being in a multiaxial stress state. It is necessary to assume such a multiaxial stress test to measure mechanical properties of the blood vessel. This means that it is extremely difficult to obtain mechanical properties of the blood vessel because the mechanical properties are associated with various levels and types of blood vessels.

In this study, the pressure strain elastic modulus propounded by (Peterson et al., 1960), which is easy to manipulate, is used. The pressure strain elastic modulus $E_{p,v}$ of a blood vessel can be expressed by the following equation

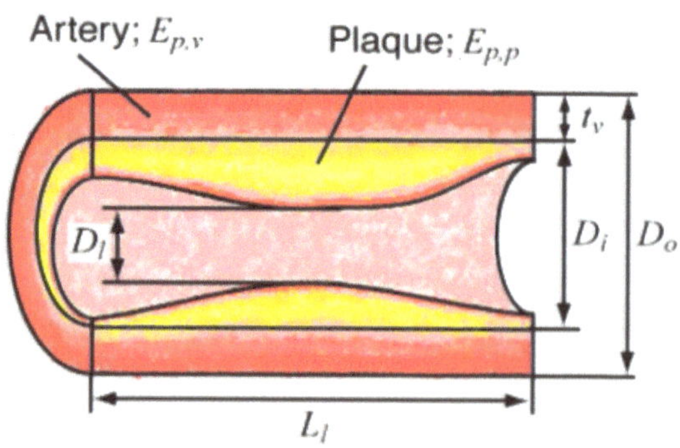

Fig. 14. Dimension of blood vessel with stenosis

$$E_{p,v} = \frac{\Delta p}{\Delta D_o / D_o} \quad (18)$$

where D_o is the blood vessel's outer diameter, and ΔD_o and Δp are the increase in the blood vessel diameter and the increase in the internal pressure respectively. The pressure strain elastic modulus has been used widely in clinical studies. Therefore, many reports on that subject are available. Table 3 presents value of the pressure strain elastic modulus for each type of human blood vessel, extracted from references (Hayashi et al., 1980; Hayashi, 2005; Stratouly et al., 1987; Gow & Hadfield, 1979), with modification.

Table 3. Pressure strain elastic modulus for each type of human artery

Artery	Ep,v	Reporter
Arteria pulmonalis	0.016	Greenfield and Griggs (1963)
Ascending aorta	0.076	Patel et al. (1964)
Basilar artery	0.186 (β=14)	Hayashi et al. (1980)
Common carotid artery	0.049	Arndt et al. (1969)
Common iliac artery	0.120	Stratouly et al. (1987)
Coronary artery	0.602	Gow and Hadfield (1979)
Femoral artery	0.433	Patel et al. (1964)
Thoracic aorta	0.126	Luchsinger et al. (1962)

6.1.2 Determination of radial stiffness necessary to expand stenotic part in artery

To determine the radial stiffness of the stent necessary to expand the stenotic part in the blood vessel, the requirements are, in addition to the pressure strain elastic modulus $E_{p,v}$ of the blood vessel, shown as follows. The outer and inner diameters, D_o and D_i, of the blood vessel in the normal state, the least diameter D_l produced by the stenotic part, and the length of the stenotic part L_l are required. The pressure strain elastic modulus of the plaque $E_{p,p}$ is required. Also required is the inner diameter after the treatment is made, D_t, which is an indicator to use as the target setting for the percentage by which to improve the blood flow level there. Given all the values listed above, the calculations for the necessary radial stiffness are made as follows.

$$p^* = \left(E_{p,v} + E_{p,p}\right)\frac{D_t - D_l}{D_o} = E_{p,vl}\frac{D_t - D_l}{D_o}$$

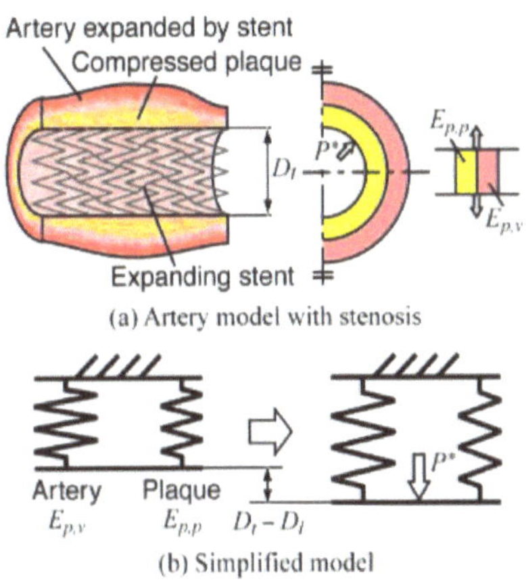

(a) Artery model with stenosis

(b) Simplified model

Fig. 15. Simplified modeling of expansion of stenosis in artery by insertion of stent

By knowing that the increase in the blood vessel diameter is obtainable from circumferential strain of a cylinder, the blood vessel with stenosis can be modeled by simply using springs connected in parallel, as presented in Fig. 15. Thus, the internal pressure in blood vessel p^*, which is necessary to expand the stenotic part, can be obtained by

$$p^* = \left(E_{p,v} + E_{p,p}\right)\frac{D_t - D_1}{D_o} = E_{p,vl}\frac{D_t - D_1}{D_o} \tag{19}$$

When measuring the pressure strain elastic modulus of the plaque $E_{p,p}$ is difficult, it is possible to replace the sum of the elastic moduli, $E_{p,v}$ + $E_{p,p}$, by the pressure strain elastic modulus of the diseased blood vessel $E_{p,vl}$, which can be easily measured.

As a matter of fact, the diameter of the stent, which is inserted into the stenotic part, is greater than the target vascular diameter after treatment. In conclusion, from the stent diameter d_s and the target vascular diameter D_t obtained after treatment, the radius reduction Δr in stent after inserting can be calculated by means of the following equation:

$$\Delta r = \frac{d_s - D_t}{2} \tag{20}$$

By substituting equations (19) and (20) into $K_p = p/\Delta r$, which was defined by (Yoshino et al., 2008) to obtain the radial stiffness, the radial stiffness K_p necessary to expand the stenotic part can be obtained from

$$K_p^* = 2\left(E_{p,v} + E_{p,p}\right)\frac{D_t - D_1}{D_o\left(d_s - D_t\right)} = 2E_{p,vl}\frac{D_t - D_1}{D_o\left(d_s - D_t\right)} \tag{21}$$

The obtained K^* is the least required stiffness to expand the stenotic part in the blood vessel. Therefore, when designing a stent, the chosen stiffness should be greater than this K^* value.

Now consider the case where a stent with a diameter of 6 mm is inserted into a coronary artery. The symptoms shown in Table 4 are examples based on references (Gow & Hadfield, 1979; Le Floc'h et al., 2009). Based on this data, the radial stiffness K^* necessary to expand the stenotic part in the blood vessel is calculated by equation (21) to be K_p^* = 366.7 MPa/m.

The required radial stiffness K_p^* is plotted as a broken line on the radial stiffness map shown in Fig. 16. From this line, it is possible to determine the design variables for a stent with sufficient radial stiffness to expand the stenotic part of the blood vessel. Based on these, the designer can proceed to work with some design propositions.

Table 4. Assumed patient symptom information

Coronary artery	Parameter s
Outer diameter, D_o (mm)	4.90
Inner diameter, D_i (mm)	4.06
Least diameter of lesion, D_l (mm)	2.5
Length of lesion, L_l (mm)	10
Total flexion angle (deg.) (Flexion angle (deg.))	90 (45)
Rate of stenosis by ECST method (%)	38.4
Pressure strain elastic modulus of artery, $E_{p,v}$ (MPa)	0.602
Pressure strain elastic modulus of diseased artery, $E_{p,vl}$ (MPa)	0.628
Inner diameter after treatment, D_t (mm)	4.56

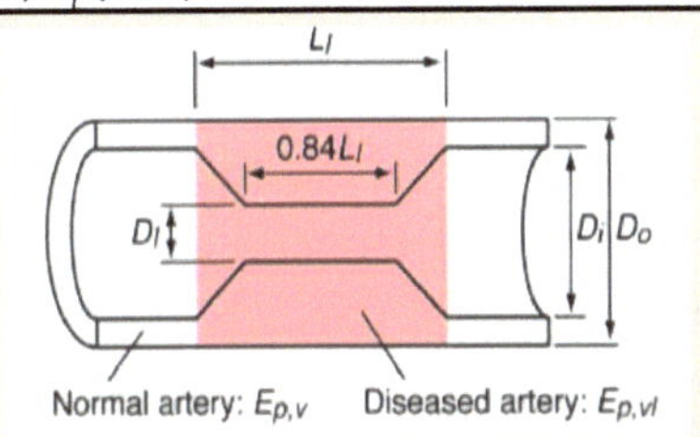

Schematic view of assumed symptom

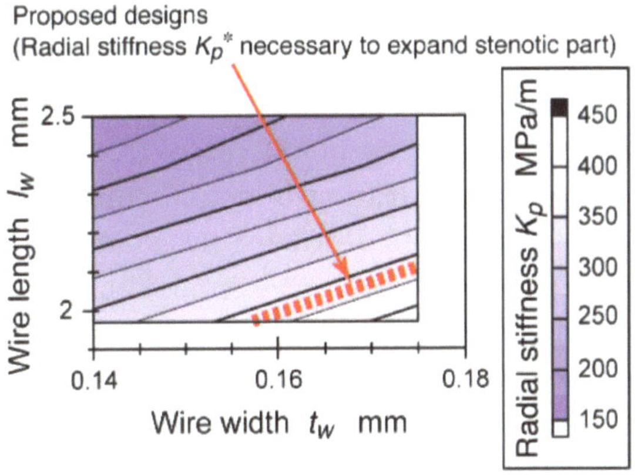

Fig. 16. Proposed designs having the radial stiffness necessary to expand the stenotic part of a blood vessel. By assuming that a stent is inserted into a coronary artery, a value of 0.602 MPa was used for the pressure strain elastic modulus *Ep,v* of the normal part of the coronary artery, and a value of 0.628 MPa for the pressure strain elastic modulus *Ep,vl* of the diseased artery. For the coronary artery model, *Do* = 4.90 mm, *Di* = 4.06 mm, and *Dl* = 2.5 mm were assumed. The corresponding part of the map is magnified and displayed.

6.1.3 Range of selectable flexural rigidity and the dilemma of selecting the design

After inserting an originally straight stent into a curved blood vessel and leaving it there, the stent generally conforms to the blood vessel shape. Nevertheless, because the flexural rigidity of the stent is greater than that of the blood vessel, the blood vessel tends to become straighter. This phenomenon of straightening of the blood vessel was previously described in detail, and a method to calculate the force resulting in the straightening of the blood vessel was described in previous section. It is apparent that this force depends on the flexural rigidity of the stent, and that greater flexural rigidity results in a larger force. A force too large can damage the vascular wall. Consequently, it is important to choose the most appropriate flexural rigidity of the stent.

By plotting the proposed designs obtained from the required radial stiffness K^* on a flexural rigidity map, the broken line shown in Fig. 17(a) is obtained. For one radial stiffness value, multiple flexural rigidity values can be selected as long as they are within the range given in the figure. None of the limiting values for the flexural rigidity, which might prevent damage on the vascular wall or neointimal thickening, are yet quantitatively available. Therefore, the flexural rigidity should be made smaller to decrease the force acting on the vascular wall. In other words, the designer should select the smallest flexural rigidity from the range of selectable values shown in Fig. 17(a).

Figure 17(b) shows the stent shape designed in consideration of the rules described above. The designed stent is referred to as SDCO and has a wire length l^* of 2.01 mm, and a wire width t^* of 0.16 mm. In addition, the length of the designed stent is determined so that it occupies the blood vessel from the stenotic part to the normal part at each end in consideration of actual clinical use. The length of the SDCO is 22.0 mm (the stent consists of 8 wire sections and 7 strut sections). The flexural rigidity of the SDCO is K^* = 25.1×10-6 Nm2. Similarly, the shear rigidity of the SDCO, based on the map of the rigidity, is K_s^* = 1.33 N.

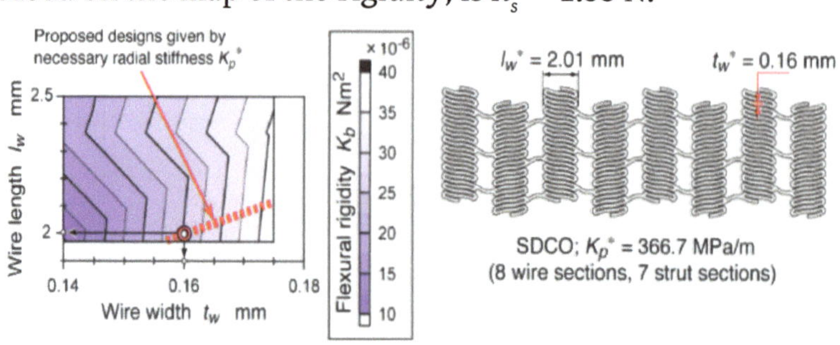

(a) (b)

Fig. 17. Selection of the proposed design from the viewpoint of flexural rigidity. (a) The selectable range of flexural rigidities is indicated by the broken line. The

corresponding part of the map is magnified and displayed as in Fig. 16. (b) A stent 6 mm in diameter is designed to suit the assumed symptom of the coronary artery.

6.2 Evaluation of the risk of instent restenosis based on the mechanical stimulus

Figure 18 shows the distributions of the contact force and the straightening force when the SDCO is inserted into the coronary artery. These distributions were calculated by the previously described method, which was proposed by (Yoshino et al., 2011).

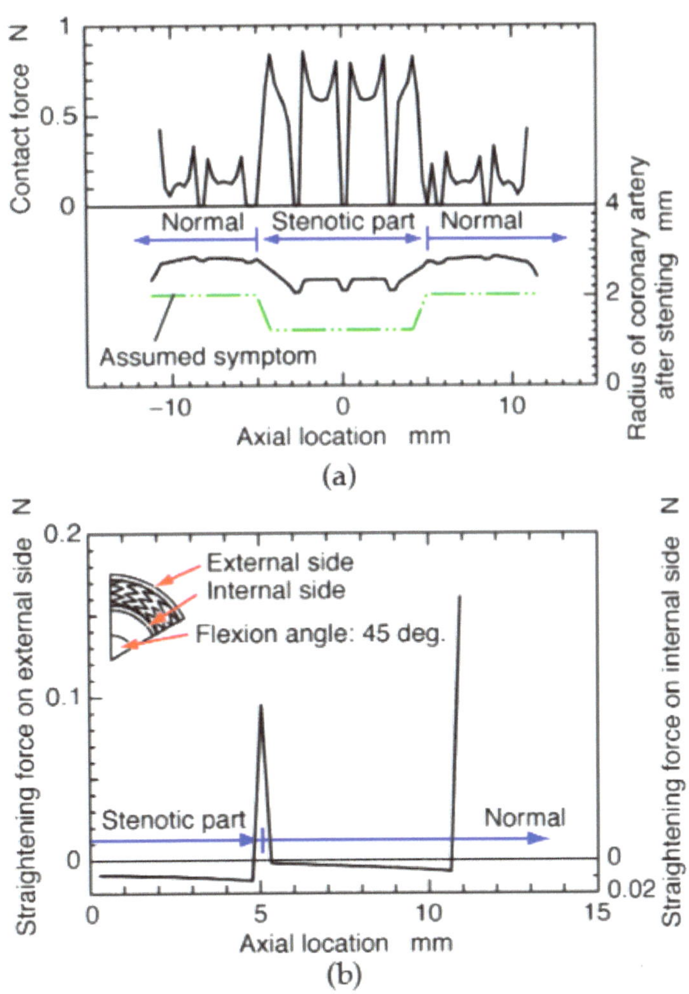

Fig. 18. Computational results of forces which are exerted on the coronary artery wall by insertion of the SDCO. (a) The distribution of the contact force between the SDCO and the coronary artery (upside), and the radius of the coronary artery

after stenting (downside). The dot-dashed line indicates the initial shape of the artery wall with the assumed symptom. (b) The distribution of the straightening force on the coronary artery wall by insertion of the SDCO. The right side of the distribution is displayed from a consideration of the geometrical symmetry.

First, let us focus attention on the force on the vascular wall. Although the contact force is large at the stenotic part, the generation of this large force is unavoidable for expansion of the stenotic part. The contact force is also concentrated at both ends of the stent. The straightening force on the vascular wall is concentrated at the end of the stenotic part in addition to both ends of the stent.

It has been reported that the hyperplasia of smooth muscle cells and the neointimal proliferation occur at the stented part of an artery (Grewe et al., 2000; Clark et al., 2006). It is assumed that the hyperplasia and proliferation are caused by the mechanical stimulus acting on the vascular wall due to insertion of the stent. (Schweiger et al., 2006) reported that in-stent restenosis occurred primarily at both ends of the stent during the period one to three months after stenting. In the study of (Lal et al., 2007), the majority of patients in the target patient population showed in-stent restenosis at the stent ends. (Yazdani and Berry, 2009) cultured a stented native porcine carotid artery under physiologic pulsatile flow and pressure conditions for a week. They confirmed that the proliferation of smooth muscle cells occurred significantly at both ends of the stent. By summarizing these reports and evaluation results of the forces on the vascular wall, it is concluded that the force concentration provokes neointimal thickening due to the hyperplasia of smooth muscle cells, i.e., in-stent restenosis.

Next, the expansion of the vascular wall is examined. The normal part of the artery is expanded to become much larger than the target diameter. This excessive expansion causes expansion of the entire stented part. As a result, stagnation and vortices of blood flow are induced, further increasing the potential for stent thrombosis.

The designed stent has mechanical properties sufficient to expand the stenotic part of theartery, but is not suitable for the normal part of the artery. Thus, it requires modification.

6.3 Effective method to modify the stent shape in consideration of the risk of in-stent restenosis

6.3.1 Design objective for the shape modification

In the previous section, it was shown that although a stent suitable for the assumed symptoms could be designed using the proposed design method, further modifications were still required. Two kinds of design objectives are set up for shape modification.

Objective for the mechanical stimulus: as shown in Fig. 18, the contact force on the normal part of the artery is concentrated at both ends of the stent. This force concentration provokes neointimal thickening from the hyperplasia of smooth muscle cells. Therefore, the contact force must be reduced at both ends of the stent. However, this force should be larger than a certain limit so that stent functions on the lesion.

Objective for the blood flow: as stated above, the normal part of the artery is expanded to become much larger than the target diameter. This expansion state of the artery causes stagnation and vortices of blood flow. As a result, stent thrombosis may be induced. It is very important to reduce stagnation and vortex creation to decrease the risk of stent thrombosis. Therefore, the vascular wall should be expanded flatly by the insertion of the stent. The flat expansion of the artery can prevent the generation of stagnation and vortices in the stented artery.

6.3.2 Modification method of the stent shape to suit the clinical manifestation

The stent designed in the previous section has a uniform radial stiffness K^* along its axial direction. The radial stiffness K^* causes excessive expansion at the normal part of the artery because K^* was calculated for expansion of the stenotic part, which is generally stiffer than the normal part. The objective for the mechanical stimulus can be attained by decreasing the radial stiffness to a value corresponding to that of a normal artery. At the same time, the objective for blood flow must be attained. Therefore, the stent should have a radial stiffness K^{**} to expand the vascular wall at normal part only for the stent thickness t. This can achieve the flat expansion of the artery. By changing the stent radial stiffness at the normal part from Kp^* to Kp^{**}, the contact force can also decrease. In this study, it was decided to adopt the radial stiffness K^{**} as a compromise between the two design objectives.

The methods using the influence matrix described in Section 5 are used for the shape modification. Figure 19 shows the concept for modification of the stent shape. The distributed contact force P_i^* on the stenotic part can be calculated based on the radial stiffness K^*, as follows:

$$P_i^* = \frac{\pi D_t K_p^* \Delta r_t l_w^*}{n_{CP}} \tag{22}$$

where D_t is the inner diameter of the blood vessel after treatment, Δr_t is the increase in the vascular radius of the target, l^* is the wire length of the stent designed based on the radial stiffness Kp^*. Also, nCp represents the number of calculation points on the wire section.

For simplicity, it was assumed that a uniform contact force was exerted on the vascular wall.

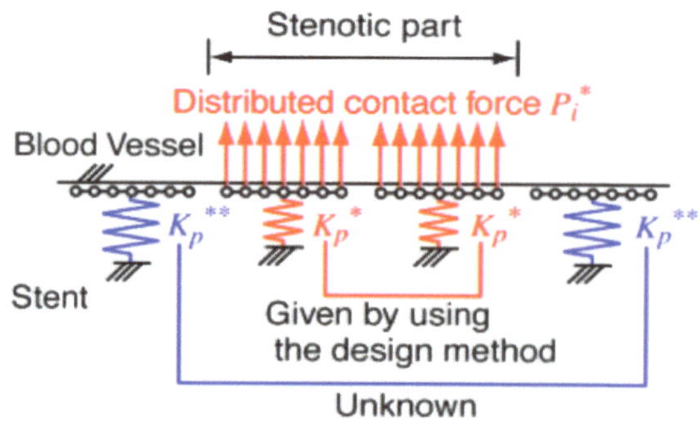

Fig. 19. Concept for modification of the stent shape. The distributed contact force is obtained from the radial stiffness of the stent on the stenotic part by using the design method. The force that should be applied on the normal artery is calculated by using the obtained distributed contact force on the stenotic part.

As described in Section 5, the radial displacement of the vascular wall $\{r(v)\} = (r_1(v), r_2(v), ..., r_n(v))T$ due to the unknown contact force $\{P\} = (P_1, P_2, ..., P_n)T$ is given as follows:

$$\left[C^{(v)}\right]\{P\} = \left\{r^{(v)}\right\} \tag{23}$$

where $[C(v)]$ is the influence matrix of the blood vessel, defined by the radial displacement of the vascular wall due to the unit radial force. The calculated contact force P_i^* is substituted into equation (23), and the influence matrix $[C(v)]$ is downsized to the matrix $[C_{normal}(v)]$ of the normal blood vessel.

$$\left[C^{(v)}_{normal}\right]\left\{P^{**}\right\} = \left\{r^{(v)}_{normal}\right\} + \left[C^{(v)}_{stenosis}\right]\left\{P^{*}\right\} \tag{24}$$

$[C_{stenosis}(v)]$ is the influence matrix of the stenotic part, and $\{r_{normal}(v)\}$ is the radial displacement of the normal part of the blood vessel wall. Equation (24) is solved for $\{P^{**}\}$ to obtain the distributed force P_i^{**} that can expand the normal part to the stent thickness t.

Here, it is assumed that the wire length after modification is l^{**} and the calculation points from k to l are included in the modified wire section. The required radial stiffness K^{**} is defined as follows:

$$K_p^{**} = \frac{\sum_{i=k}^{l} P_i^{**}}{\pi D_t l_w^{**} t} \tag{25}$$

The K^{**} value calculated by equation (25) is plotted on the radial stiffness map. As a result, the wire width t_w^{**} after modification is determined from the l_w^{**} value and the curve of K^{**} on the map. Therefore, the designed stent has to be modified to the shape of l_w^{**} and t_w^{**} at the normal part of the blood vessel. Considering that an increase in the stent length is undesirable, the designer should keep the wire length l_w^{**} equal to l^{**} after the modification. However, it is possible that the wire width t^{**} after the modification cannot be obtained because of the mismatch between the assumed l_w^{**} value and the curve of K^{**}. In this case, the designer can increase the wire length l_w^{**}, perform the same procedures, and determine the l^{**} and t^{**} values.

The required radial stiffness $Kp1^{**}$ at both ends of the SDCO is 50.3 MPa/m. The radial stiffness $Kp2^{**}$ at the normal part of the artery except for both stent ends is also calculated as 155.2 MPa/m. From the dot-dashed curves of K^{**} and K^{**} shown in Fig. 20(a), it is determined that l^{**} is 2.52 mm, t^{**} is 0.087 mm, l^{**} is 2.01 mm, and t^{**} is 0.094 mm. Figure 20(b) shows the modified shape of the SDCO.

The flexural rigidity $Kb1^{**}$ at both ends of the SDCO is 5.75×10-6 Nm2, and the shear rigidity $Ks1^{**}$ is 0.20 N. At the normal part of the artery except for both stent ends, the flexural rigidity $Kb2^{**}$ of the SDCO is 6.00×10-6 Nm2, and the shear rigidity $Ks2^{**}$ is 0.42 N.

6.3.3 Confirmation of the effect of the shape modification

Figure 21 shows a comparison of the force on the vascular wall by insertion of the stent before and after the modifications. After the modifications of the SDCO, an approximately 80 % reduction in the concentrated contact force was attained (Fig. 21(a)). Furthermore. the concentrated straightening force at the stent ends after modification was reduced to approximately 35 % of that before modification of the SDCO (Fig. 21(b)).

On the other hand, it is recognized that straightening force increases at the stenotic-healthy tissue interface (the axial location is 5 mm in Fig. 21(b)). The concentration of straightening force also occurs at the axial location of 8 mm. After modification of the stent shape, the flexural and shear rigidities of the stent vary with the axial location. The bending state of the stent changes at the changing point of the rigidities, which corresponds to the turn of the stent shape. Therefore, straightening force increases due to changing of the stent bending state based on rigidities changing. Although the risk of the vessel rupture slightly increases at the stenotic-healthy tissue interface, it is achieved that the straightening force is significantly reduced at both ends of the stent, where the hyperplasia of smooth muscle cells is frequently reported.

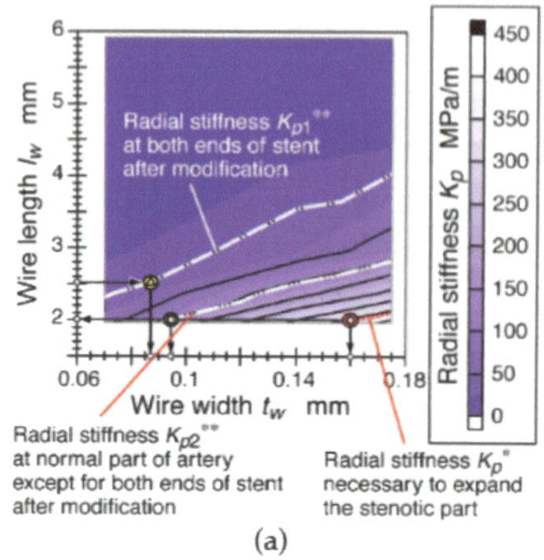

(a)

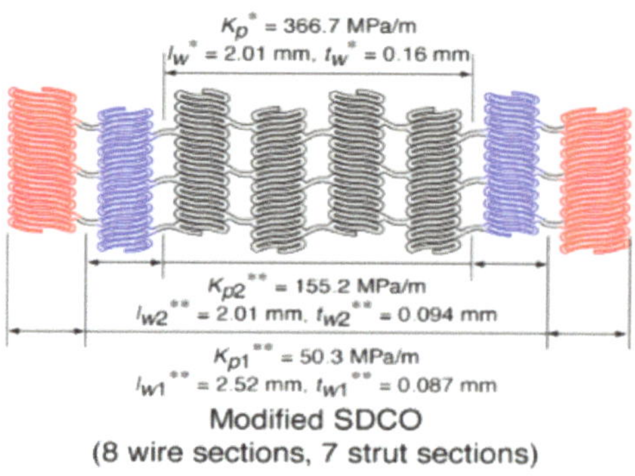

(b)

Fig. 20. The SDCO stent 6 mm in diameter is modified to suit the assumed symptom of the coronary artery. (a) Design variables after modification are determined by using the map of the radial stiffness. (b) The modified SDCO has a nonuniform shape along its axial direction.

The redesigned stent can extend the vascular wall in a more planar fashion. Consequently, altering the intended stent by the suggested method can alleviate force concentration at both extremities of the stent by achieving uniform expansion of the vascular wall.

7. CONCISE SUMMARY

This chapter delineates the assessment of mechanical qualities of a stent and the design/selection of a stent appropriate for various clinical manifestations through the effective utilization of the evaluated data. In summary, it can be proven that two objectives were accomplished: the design and selection of a stent appropriate for various clinical manifestations. The stent, built using the method outlined in this chapter, exhibits nonuniform forms along its axial direction. This configuration is fundamentally innovative in contrast to the traditional stent design, which maintains uniformity along its axial axis. Consequently, the stent configuration developed by this process is regarded as a new-generation variant.

REFERENCES

1. Babapulle, M. N.; Joseph, L., Bélisle, P., Brophy, J. M., & Eisenberg, M. J. (2004). A hierarchical Bayesian meta-analysis of randomised clinical trials of drug-eluting stents. *Lancet*, Vol. 364, No. 9434, (August 2004) 583-591, ISSN 0140-6736 (print version).
2. Carnelli, D.; Pennati, G., Villa, T., Baglioni, L., Reimers, B., & Migliavacca, F. (2010). Mechanical properties of open-cell, self-expandable shape memory alloy carotid stents. *Artificial Organs*, Vol. 35, No. 1, (January 2011) 74-80, ISSN 0160-564X (print version).
3. Clark, D. J.; Lessio, S., O'Donoghue, M., Tsalamandris, C., Schainfeld, R., & Rosenfield, K. (2006). Mechanisms and predictors of carotid artery stent restenosis--A serial intravascular ultrasound study. *Journal of the American College of Cardiology*, Vol. 47, No. 12, (June 2006) 2390-2396, ISSN 0735-1097.
4. Colombo, A.; Stanlovic, G., & Moses, J. W. (2002). Selection of coronary stents. *Journal of the American College of Cardiology*, Vol. 40, No. 6, (September 2002) 1021-1033, ISSN 0735-1097.
5. Duda, S. H.; Wiskirchen, J. Tepe, G., Bitzer, M., Kaulich, T. W., Stoeckel, D., & Claussen, C. D. (2000). Physical properties of endovascular stents: an experimental comparison. *Journal of Vascular and Interventional Radiology*, Vol. 11, No. 5, (May 2000) 645-654, ISSN 1051-0443.
6. Gow, B. S. & Hadfield C. D. (1979). The elasticity of canine and human coronary arteries with reference to postmortem changes. *Circulation Research*, Vol. 45, No. 5, (November 1979) 588-594, ISSN 0009-7330.
7. Grewe, P. H.; Deneke, T., Machraoui, A., Barmeyer, J., & Müller, K.-M. (2000). Acute and chronic tissue response to coronary stent implantation: pathologic findings in human specimen. *Journal of the American College of Cardiology*, Vol. 35, No. 1, (January 2000) 157-163, ISSN 0735-1097.
8. Hayashi, K. (2005). *Biomechanics*, Corona Publishing, Tokyo, Japan. (in Japanese)
9. Hayashi, K.; Handa, H., Nagasawa, S., Okumura, A., & Moritake, K. (1980). Stiffness and elastic behavior of human intracranial and extracranial arteries. *Journal of Biomechanics*, Vol. 13, No. 2, 175-184, ISSN 0021-9290.

10. Kastrati, A.; Mehilli, J., Pache, J., Kaiser, C., Valgimigli, M., Kelbaek, H., Menichelli, M., Sabaté, M., Suttorp, M. J., Baumgart, D., Seyfarth, M., Pfisterer, M. E., & Schömig, A. (2007). Analysis of 14 trials comparing sirolimus-eluting stents with bare-metal stents. *The New England Journal of Medicine*, Vol. 356, No.10, (March 2007) 1030-1039, ISSN 0028-4793 (print version).

11. Lagerqvist, B.; James, S. K., Stenestrand, U., Lindbäck, J., Nilsson, T., & Wallentin, L. (2007). Long-term outcomes with drug-eluting stents versus bare-metal stents in Sweden. *The New England Journal of Medicine*, Vol. 356, No. 10, (March 2007) 1009-1019, ISSN 0028-4793 (print version).

12. Lal, B. K.; Kaperonis, E. A., Cuadra, S., Kapadia, I., & Hobson, R. W. 2nd. (2007). Patterns of in-stent restenosis after carotid artery stenting: classification and implications for long-term outcome. *Journal of Vascular Surgery*, Vol. 46, No. 5, (November 2007) 833-840, ISSN 0741-5214.

13. Le Floc'h, S.; Ohayon, J., Tracqui, P., Finet, G., Gharib, A. M., Maurice, R. L., Cloutier, G., & Pettigrew, R. I. (2009). Vulnerable atherosclerotic plaque elasticity reconstruction based on a segmentation-driven optimization procedure using strain measurements: theoretical framework. IEEE transactions on medical imaging. Vol. 28, No. 7, (July 2009) 1126-1137, ISSN 0278-0062.

14. Mori, K. & Saito, T. (2005). Effects of stent structure on stent flexibility measurements. *Annals of Biomedical Engineering*, Vol. 33, No. 6, (June 2005) 733-742, ISSN 0090-6964 (print version).

15. Morice, M.-C.; Serruys, P. W., Sousa, J. E., Fajadet, J., Hayashi, E. B., Perin, M., Colombo, A., Schuler, G., Barragan, P., Guagliumi, G., Molnar, F., & Falotico, R. (2002) A randomized comparison of a sirolimus-eluting stent with a standard stent for coronary revascularization. *The New England Journal of Medicine*, Vol. 346, No. 23, (June 2002) 1773-1780, ISSN 0028-4793 (print version).

16. Nordmann, A. J.; Briel, M., & Bucher, H. C. (2006). Mortality in randomized controlled trials comparing drug eluting vs. bare metal stents in coronary artery disease: a meta- analysis. *European Heart Journal*, Vol. 27, No. 23, (December 2006) 2784-2814, ISSN 0195-668X (print version).

17. Peterson, L. H.; Jensen, R. E., & Parnell, J. (1960). Mechanical properties of arteries in vivo. *Circulation Research*, Vol. 8, No. 3, (May 1960) 622-639, ISSN 0009-7330.

18. Schweiger, M. J.; Ansari, E., Giugliano, G. R., Mathew, J., Islam, A., Morrison, J., & Cook, J. R. (2006). Morphology and location of restenosis following bare metal coronary stenting. *The Journal of Invasive Cardiology*, Vol. 18, No. 4, (April 2006) 165-168, ISSN 1042-3931.

19. Stratouly, L. I.; Cardullo, P. A., Anderson, F. A. Jr., Durgin, W. W., & Wheeler, H. B. (1987). The use of ultrasound imaging in the in-vivo determination of normal human arterial compliance. *Proceedings of the 13th Annual Northeast Bioengineering Conference*, pp.435-437, Philadelphia, PA, U.S., March 1987, Defense Technical Information Center, Fort Belvoir.

20. Tamakawa, N.; Sakai, H., & Nishimura, Y. (2008). Prediction of restenosis progression after carotid artery stenting using virtual histology IVUS. *Journal of Neuroendovascular Therapy*, Vol. 2, No. 3, (December 2008) 193-200, ISSN 1882-4072.

21. Yazdani, S. K. & Berry, J. L. (2009). Development of an in vitro system to assess stent- induced smooth muscle cell proliferation: a feasibility study. *Journal of Vascular and Interventional Radiology*, Vol. 20, No. 1, (January 2009) 101-106, ISSN 1535-7732.

22. Yoshino, D. & Inoue, K. (2010). Design method of self-expanding stents suitable for the patient's condition. *Proceedings of the Institution of Mechanical Engineers, Part H: Journal of Engineering in Medicine*, Vol. 224, No. 9, (September 2010) 1019-1038, ISSN 0954-4119 (print version).

23. Yoshino, D.; Inoue, K., & Narita, Y. (2008) Mechanical properties of self-expandable stents: a key to product design of suitable stents. *Proceedings of the 7th International Symposium on Tools and Methods of Competitive Engineering*, Vol. 1, pp. 659-672, ISBN 978-90-5155-044-3, Izmir, Turkey, April 2008.

24. Yoshino, D.; M. Sato, & K. Inoue. (2011). Estimation of force on vascular wall caused by insertion of self-expanding stents. *Proceedings of the Institution of Mechanical Engineers, Part H: Journal of Engineering in Medicine*, Vol. 224, No. 8, (August 2011) 831-842, ISSN 0954-4119 (print version).

9
Chapter
SPIN AND SPIN RECOVERY IN AIRCRAFT

1. INTRODUCTION: AN OVERVIEW

Spin is a highly intricate maneuver of an airplane. It is, indeed, a curvilinear unstable flying regime, characterized by the aircraft's rotation accompanied by concurrent linear movements along all three axes, thus exhibiting six degrees of freedom. Consequently, there are no fully developed and precise analytical methods for this type of issue.

2. TYPES OF SPIN

Unwanted complex movements of aircraft are shown in Fig.1. In the study of these regimes, one should pay attention to the conditions that lead to their occurrence. Attention should be made to the behavior of aircraft and to determination of the most optimal way of recovering the aircraft from these regimes. Depending on the position of the pilot during a spin, the spin can be divided into upright spin and inverted spin. During a upright spin, the pilot is in position head up, whilst in an inverted spin his position is head down.

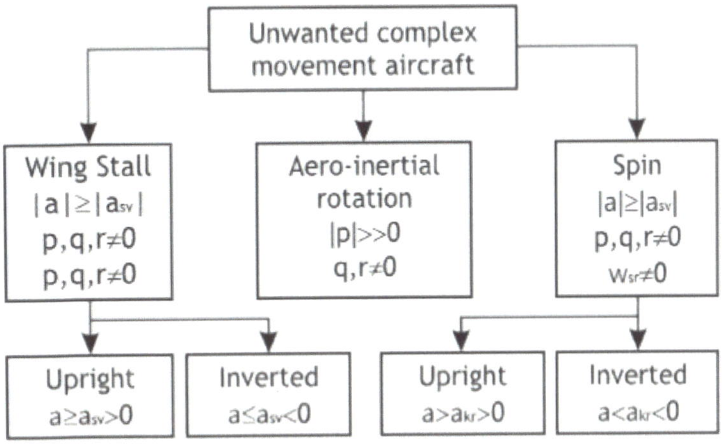

Fig. 1. Unwanted rotations of aircraft

The upright spin is carried out at positive supercritical attack angles, and the inverted spin at negative supercritical attack angles. According to the slope angle of the aircraft longitudinal axis against the horizon, spin can be steep, oblique and flat spin (Fig.2). During a steep spin, the absolute value of the aircraft slope angle is greater than 50 degrees, i.e. angle $|\upsilon| > 50°$, during an oblique spin $30° \le |\upsilon| \le 50°$, and during a flat spin $|\upsilon| < 30°$. According to direction of aircraft rotation, spin can be divided into a left and right spin. In a left spin (upright and inverted), the aircraft is rotating leftward, and during a right spin (upright and inverted) rightward. If the aircraft is observed from above, in a right upright and in a left inverted spin, the center of gravity of the aircraft will move in a clockwise direction, and vice versa while at a left upright and a right inverted spin.

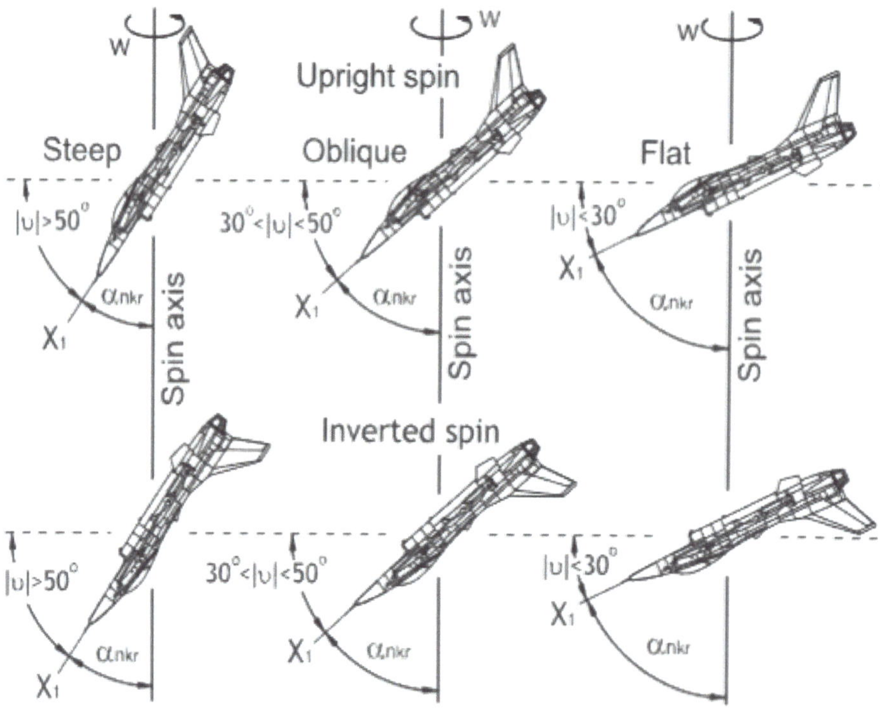

Fig. 2. The positions of aircraft in the vertical plane at entry and during the upright and inverted spin

The axis of spin is a an axis of a spiral by which the center of gravity is moving during a spin, and the spin radius is the radius of the horizontal projection of that spiral.

There exists inward and outward sideslip of the aircraft during a spin. The inward sideslip is when the air stream encounters the aircraft from the side of the inner wing, the wing in which direction the aircraft is rotating during a spin. The outward sideslip occurs when the air stream encounters the aircraft from the side of the outer wing.

Modern supersonic aircrafts (unlike older supersonic and even more the subsonic aircrafts) are characterized by a greater diversity of spin. This can be explained by the influence of constructive-aerodynamic properties of such aircrafts. Even for the same supersonic aircraft, the characteristics of upright and inverted spin can substantially differ depending on the initial angles of initiation (height, centering, etc.), regime length, position of rudder and ailerons during spin, etc. As a rule, these aircrafts have a distinct unevenness of motion and great fluctuations during spin.

The pilot has to study and reliably recognize the characteristic features of every spin type. Thus, he can quickly and accurately determine the type of spin and properly recover the aircraft from such a dangerous and complex motion. In modern aircrafts, spin can be divided into several types similar by characteristic features. The magnitude and character of changes in the angular velocities and load during a spin are taken into account as characteristic features. These determine the conditions for recovery from this regime, i.e. the value and sequence of rudder deflection. The given type includes all regimes of spin where in order to recover the aircraft the same control has to be applied. According to this principle, spin regime classifications are given on Figs. 3 and 4. In compliance with Figures 3 and 4, upright spin has four, and the inverted spin has three regimes for recovery. On these images, letters "N" and "L", alongside numerated ways of recovery, denote the upright and inverted spin, respectfully.

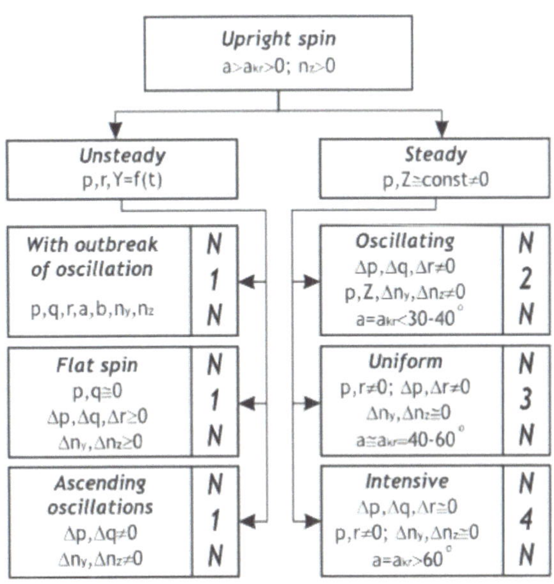

Fig. 3. Types of upright spin in modern aircrafts

Four basic methods can be applied for recovering from a upright spin (letter "N" denotes that the method refers to a upright spin):

- Method N1N - recovery from spin with simultaneous positioning of elevator and rudder in neutral position at neutral position of ailerons;

- Method N2N - recovery from spin by full rudder deflection opposite to spin with a delayed setting (2 - 4 sec) of the elevator in neutral position at neutral position of ailerons;
- Method N3N - recovery from spin by rudder deflection, and a delayed (3 to 6 sec.) full elevator deflection opposite to spin at neutral aileron position.
- Method N4N - recovery from spin as by method N3N but with simultaneous rudder and aileron deflection, possibly by full for recovery (for supersonic aircrafts aileron deflection is the appropriate deflection in the side of spin)

Recovery of modern aircrafts from inverted spin is carried out by three basic methods (letter "L" denotes that the method applies to recovery from inverted spin):

- Method N1L - recovery from spin by simultaneous elevator and rudder positioning in neutral with neutral aileron position;
- Method N2L - recovery from spin by rudder deflection totally opposite to spin with delayed (2 to 4 sec.) elevator deflection into neutral position at neutral aileron position;
- Method N3L - recovery from spin by rudder deflection and an elevator deflection (delayed 2 to 4 sec.) totally opposite to spin at neutral aileron position.

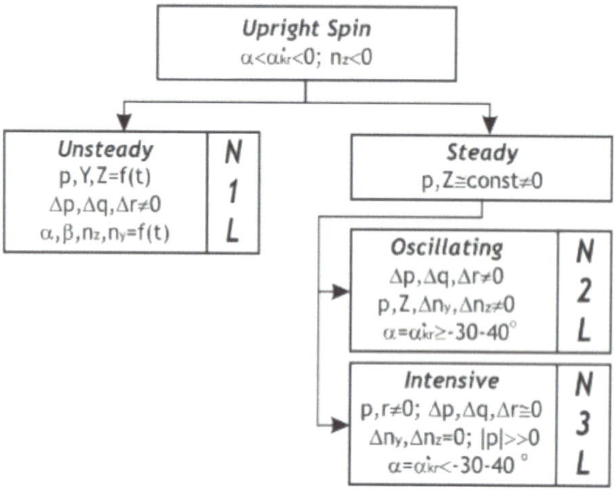

Fig. 4. Inverted spin types in modern aircrafts

3. STEADY SPIN

A steady spin is a spin in which the aircraft does not change direction of rotation neither in roll or yaw (parameters p and r are unchanged sign). Rotation is very intense, and the average values of angular velocities do not change. Upright and inverted steady spins can be oscillating and uniform, and the upright can also be intensive.

The oscillating steady spin is characterized by very large changes in the amplitude of the rolling angular velocity and angular velocity slight changes yaw and pitch. It is usually a steep spin with average angles of attack of 30° or 40° . Changes of basic parameters during a left upright steady oscillating spin are shown on figure 5. Figure reflects an aircraft induced into spin at an altitude of 12.5 *km* at a velocity of 250 <u>*km*</u> . After a time interval t = 5 s a deflection of commands is carried out according to spin. Pilot has completely pulled the yoke to himself which has caused a negative rudder deflection (δhk = -20°).

A positive deflection of the rudder is carried out δvk = +20°, whilst ailerons are in neutral position. After a time interval t = 6 *s* parameters become stable so the average angular velocity of yawing is r = -5 <u>*rad*</u> and of rolling p = -0.75 <u>*rad*</u> , the average normal load coefficient is nz = 1.25. However, after a time interval t = 8 s, the aircraft starts to oscillate and the following values are obtained: nz = 1.4, p = 3.6 <u>*rad*</u> and r = 0.6 <u>*rad*</u> . During that period the velocity on path V has oscillated between stabilized angular velocities of yawing and rolling, and an increase in velocity, which caused an increase in the normal load coefficient nz, so that during the interval t = (20 ÷ 25) *s* the normal load coefficient was nz = 2 ÷ 3.

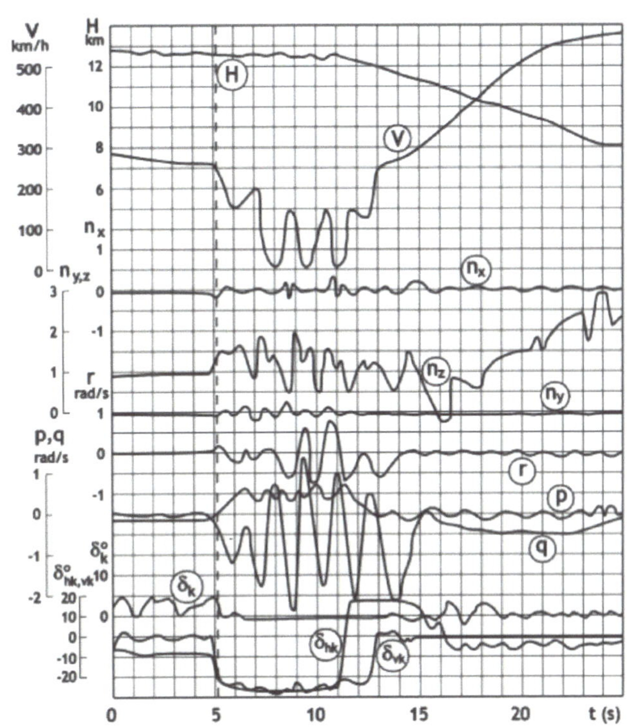

Fig. 5. Left upright steady oscillating spin

Unlike the oscillatory spin, the steady uniform spin is characterized by small amplitude oscillations of the aircraft, as well as intensive rotation of invariant direction. Figure 6 shows an example of a right-hand steady uniform spin. It can be seen that the aircraft was induced into spin at an altitude of H = 10.500 m and a velocity of V = 320 $\underline{km}$. After an interval t = 6 s, commanding surfaces were deflected. First, a negative aileron deflection (δk = 10°) can be observed resulting in a negative rolling moment, i.e. to lowering the right and lifting the left wing. Then the direction control was deflected in a negative direction (δvk = 20°) which lead to a yaw to right. In addition, the yoke was drawn onto the pilot and thus a negative deflection (δhk = -20°) was carried out on the elevator. At all time during spin, i.e. during interval t = (5 ÷ 40) s, parameters (angular velocity of rolling p, angular velocity of yaw r, and velocity on path V) oscillated very little around the following values: psr = -0.5 $\underline{rad}$, rsr = -0.5 $\underline{rad}$ i Vsr = 250 $\underline{km}$. In addition, there was a noteworthy smaller loss of altitude of ΔH = 2.500 m, than in the oscillating spin.

4. UNSTEADY SPIN

In some aircrafts, spin can periodically change the direction of rotation relative to normal and longitudinal axis. In some aircrafts, spin might even stop to oscillate, i.e. it will continue with certain oscillations. Spin at which the aircraft periodically changes direction of rotation regarding the normal and longitudinal axis, or even stops oscillating, is called unsteady.

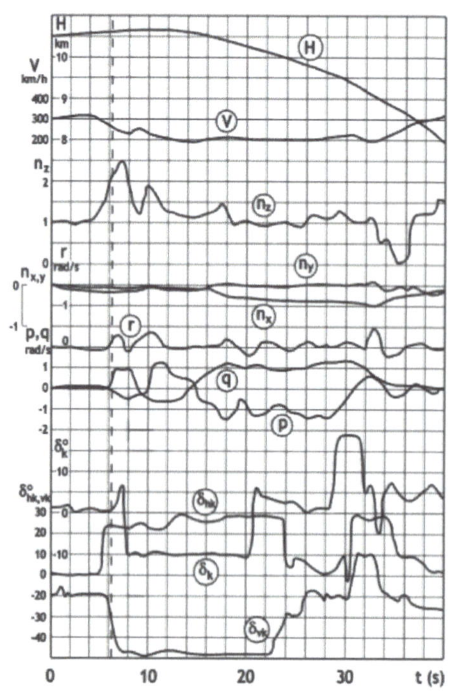

Fig. 6. Right-hand steady uniform spin

During an unsteady spin, there is a non-uniform rotation with large amplitudes of aircraft parameter change. During the regime, usually there is a tendency towards an arbitrary transition of the aircraft from a spin of one direction into a spin of another direction, or from upright into inverted spin.

Unsteady spin in modern aircrafts has three forms: spin continuing with outbreak oscillations (diagram shown in Figure 7), spin continuing as a shape of a falling leaf on a spiral path (diagram shown in Figure 8), and spin during which oscillations of the aircraft are ascending (diagram shown in Figure 9).

4.1 Spin continuing with outbreak oscillations

Diagram of parameter change for this shape of spin is shown in Figure 7. From this diagram, it is obvious that this is a left upright spin. In addition, it can be noticed that there is a periodic change of parameters, especially the angular velocity of roll *p*, angular velocity of yaw *r*, and the coefficient of normal load *nz*. Every cycle of outbreak (change of angular velocities in roll and yaw) stands for about 15 *s*. Between these cycles the aircraft has stopped rotating around it's normal axis $r = 0$. Period of oscillations in every cycle was $t = 2.5$ *s*. Velocity on path (*V*), during spin was approximately constant $V = 280 \frac{km}{h}$, whilst altitude loss during an interval of $t = 90$ *s* was $\Delta H = 8.000$ *m*.

4.2 Spin continuing in the shape of a falling leaf on a spiral path - flat spin

The parameter change in this shape of spin is shown in Fig. 8. During this spin there were periodic changes in magnitude and direction of angular velocities (*p*, *r*), and with harsh changes of the aircraft position in space.

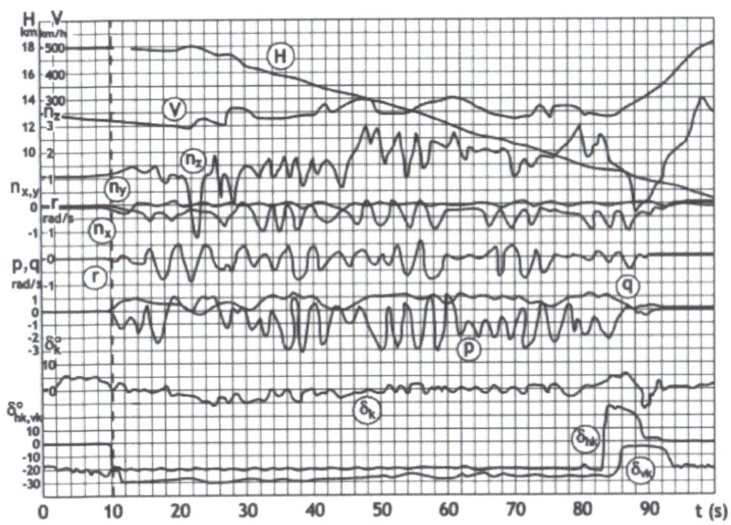

Fig. 7. Spin with outbreak oscillations

The aircraft was leaning from one wing to the other, with a nose deflection to the left and right, so that it resembles the movement of falling leaves. During this regime, the aircraft's center of gravity shifted according to a spiral path. The diagram shows that the normal load coefficient *nz* has a significant oscillation. During a period of $t = 40\ s$ loss in altitude was $\Delta H = 4.000\ m$.

4.3 Spin continuing with ascending (intensifying) oscillations

In this form of spin there is a significant increase of oscillations in pitch that lead to an increase in oscillations of the normal load coefficient nz, that began in $t = 10\ s$. At time $t = 55\ s$ there is a rise in the normal load coefficient $nz = 6$. Also, present are oscillations with a high frequency in angular velocity of rolling p, and in the oscillation of the vertical rudder in the interval $t = (40 \div 60)\ s$. Figure 9 shows the diagram of parameter change for this shape of spin. It is important to note that on these diagrams the velocity on the path V and altitude H are not trustworthy because of the inaccuracy of the measuring device (Pitot tube). This came as a result of the flight regime at high angles of attack where, because of the separated flow, the flow parameters were significantly altered.

5. PHASES OF SPIN

A typical spin can be divided into two phases shown in Fig.10, and they are: spin entry, incipient spin, steady spin and recovery from spin.

5.1 The spin entry phase

The spin entry phase begins with the aircraft being at attack angles higher than the critical angle of attack.

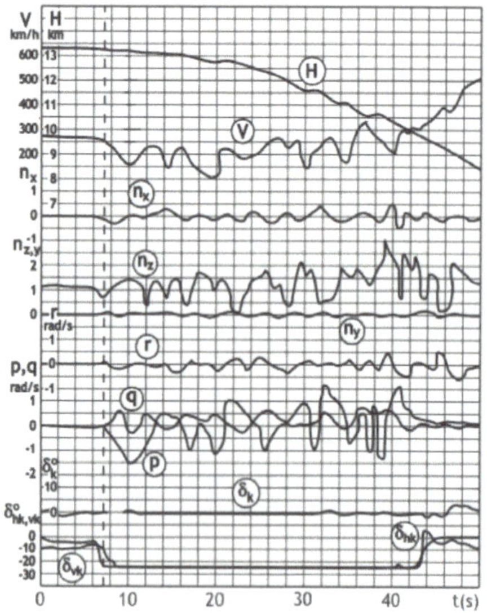

Fig. 8. Spin with outbreak oscillations

This is the condition known as wing stall. Influenced by many factors (geometrical or aircraft aerodynamic asymmetry, rudder or aileron deflection, etc.) or disturbances (vertical wind stroke), the flow around the aircraft is asymmetrical. Due to this asymmetrical flow, there are aerodynamic rolling and/or yawing moments and angular rolling and/or yawing velocities. This means that the aircraft holds an uncontrolled rotation with respect to all three abiding axes, i.e., an autorotation arises. During this rotation, the angle of attack can periodically be less than the critical angle of attack, i.e. the nose of the aircraft can periodically ascend and/or descend. At this stage, it is not possible to determine what type of spin will develop.

5.2 Incipient spin phase

The aircraft motion during the incipient spin phase is unsteady. Forces and moments acting on the aircraft are not in equilibrium. There are also, linear and angular accelerations. The aircraft rotates around the inclined axis, which is changing direction from horizontal to vertical. At this stage, the spin type can be determined.

5.3 Steady spin phase

During this phase, the aerodynamic and inertial forces and moments come into equilibrium. The aircraft is rotating downward around the vertical axis. The motion is steady, i.e. all motion parameters (attack angle, angular velocity, altitude loss per turn, time interval per turn, etc.) are constant.

5.4 Spin recovery phase

This stage begins by deflecting aircraft commanding surfaces into position for spin recovery. Autorotation stops and the aircraft enters dive as to increase flight speed. When the aircraft "accumulates" sufficient speed reserve, the pilot starts to pull out the aircraft from dive. This phase ends with the aircraft transitioning into horizontal flight.

6. METHODS FOR UPRIGHT SPIN RECOVERY

As stated before, spin can only occur at overcritical attack angles. Due to this, for the aircraft to recover from spin it is necessary to decrease the attack angle, convert the aircraft to below critical attack angles, at which autorotation stops. This comprises the basic task for spin recovery.

For some time after the discovery of the physical image of spin, propositions were made to decrease the attack angle with an appropriate deflection of the elevator. This method proved to be efficient only for some cases, when aerodynamic pitching moments, produced by elevator deflection, were greater than the inertial pitching moments by absolute value.

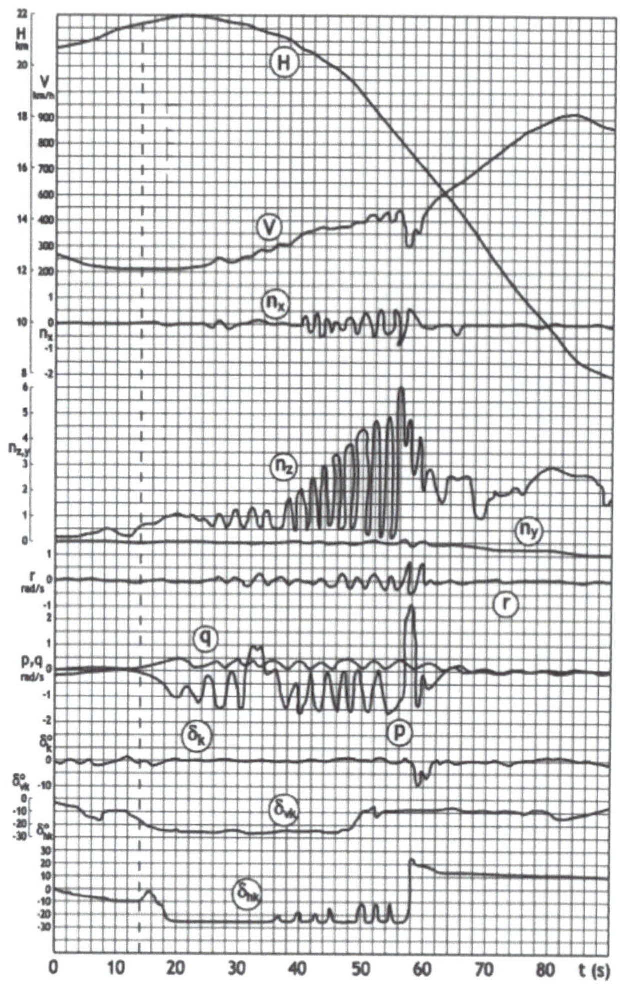

Fig. 9. Spin with ascending oscillations

For some time after the discovery of the physical image of spin, propositions were made to decrease the attack angle with an appropriate deflection of the elevator. This method proved to be efficient only for some cases, when aerodynamic pitching moments, produced by elevator deflection, were greater than the inertial pitching moments by absolute value.

However, when using such a control method, in most cases the aircraft did not recover from spin, even at full deflection of elevators. The spin became only steeper, but it did not stop, especially for aircrafts with rear centering.

In development of aircrafts their mass increased, which meant an increase in inertial moments, specially the inertial pitching moment. Therefore, in order to recover from spin it was necessary to increase the aerodynamic pitching moments, as well. All the same, it became clear that merely elevators could not achieve sufficient aerodynamic pitching moments for the purpose. Investigations have shown that in order to facilitate (sometimes, even to secure) spin recovery, previously the angular rate of rotation should be reduced (inertial moments decreased) for which an inward sideslip should be introduced, i.e. an aerodynamic yawing moment. This was achieved by deflecting the rudder opposite to spin.

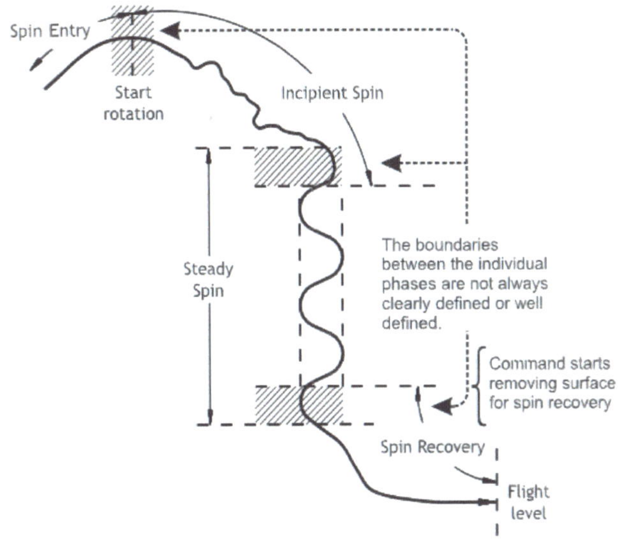

Fig. 10. Phases of a typical spin

As a result, the first scientifically based method for spin recovery was developed, by which a deflection was made opposite to spin, first by the rudder, then, after a delay (required because the inward sideslip, created by rudder deflection, could reduce the angular rate of autorotation), by the elevator. This was a, so-called standard method for spin recovery. However, it turned out that one standard method was not sufficient for modern aircrafts, characterized by vast diversity in spin regime.

Creating aerodynamic yawing moments, which in turn cause inward sideslip, is a powerful means for stopping, or at least greatly reducing the autorotation. Characteristics of aircraft recovery depend on the capability to achieve the best ratio between aerodynamic pitching and yawing moments, and between aerodynamic and appropriate inertial moments.

Four basic methods of spin recovery for modern aircrafts are as follows (letter "N" denotes method related to upright spin):

- Method N1N - spin recovery by simultaneous rudder and elevator positioning in neutral, with ailerons in neutral position;
- Method N2N - spin recovery by a full rudder deflection opposite to spin and a sequential elevator positioning in neutral(delayed by 2 - 4 sec), with ailerons in neutral position;

- Method N3N - spin recovery by rudder deflection, and sequential elevator deflection after 3 - 6 sec, completely opposite to spin, with ailerons in neutral position; and
- Method N4N - spin recovery by N3N method, but with simultaneous deflection of rudder and ailerons possibly at full for recovery (for supersonic aircrafts aileron deflection for recovery corresponds to a deflection to the side of spin.)

Deflection of controls and commanding surfaces during recovery from a left upright spin by the four basic methods is shown in Fig.11. Conditional labeling of yoke and pedal deflection that were adopted are shown in Fig.12.

The stated methods of spin recovery are structured by their efficiency or "force" increase, i.e. by the increase in aerodynamic moments created with rudder and elevator deflections for spin recovery. Therefore, the "weakest" method will be N1N and the "strongest" N4N.

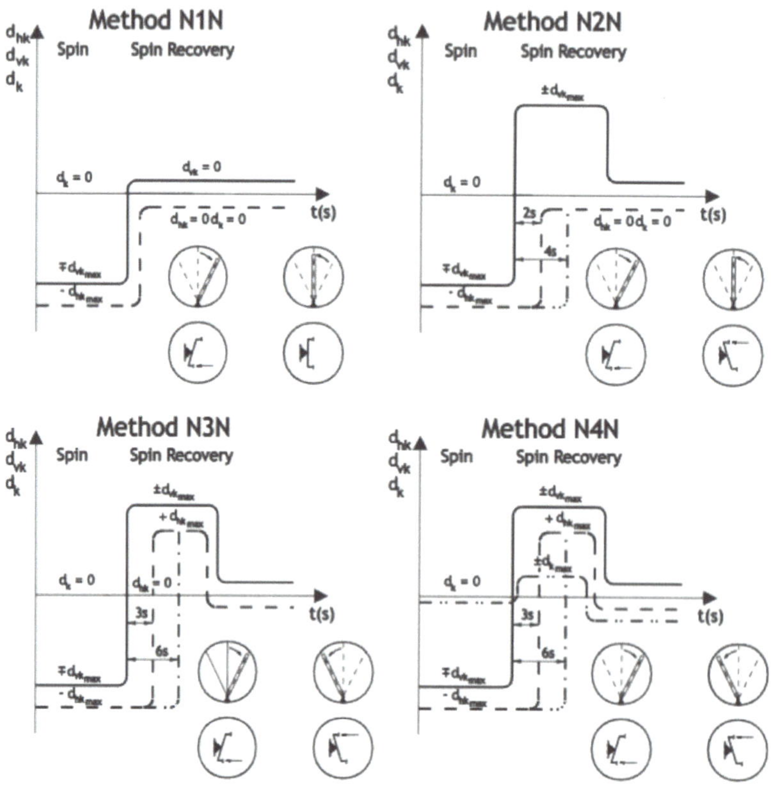

Fig. 11. Spin recovery methods for modern aircrafts from left upright spin (conditional notation of pedal and yoke position is shown in Fig.12)

Method N1N is recommended for aircraft recovery from upright unstable spin, method N2N for recovery from upright stable wavering spin, method N3N for recovery from upright stable uniform spin, and method N4N for recovery from upright stable intensive spin. These methods, as a rule, allow faster (with minimum time and altitude loss during recovery) and safer recovery of modern aircrafts from all possible upright spin regimes.

When the aircraft falls into spin, usually it is necessary to position rudder completely on the side of spin so that their efficiency will be best while deflecting for recovery. In this case, first, a maximum rudder deflection (greatest stroke) is obtained, and second, a dynamic ("shock") effect is used during an abrupt deflection of the rudder from one end position to the other. This kind of method applies only when recovering aircraft from sufficiently stable spin regimes. However, since the pilot cannot know beforehand the type of spin that will be created (stable, unstable,...), as a rule, he has to completely position the rudders onto the side of spin.

The discussed methods for modern aircrafts' upright spin recovery (as well as method for inverted spin recovery), were developed by special investigations during flight. The possibility of applying four methods instead of a standard one (method N3N) substantially increases the assurance of spin recovery, which means a safe flight. On the other hand, this requires additional attention of the pilot, because now he has to choose among the four required methods to recover from spin and to remember the sequence of rudder activities for every method.

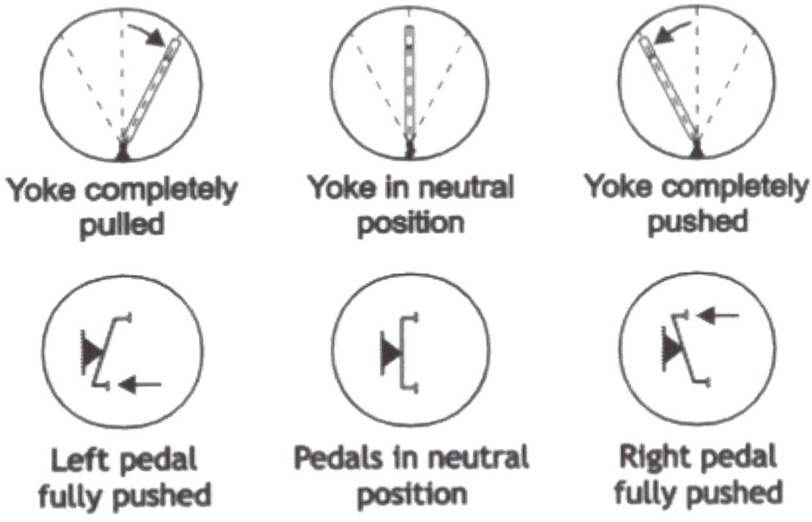

Fig. 12. Conditional annotation of deflections for yoke and pedals

However, in opinion of many pilots of high and medium proficiency, this problem does not cause extra caution, as it may seem at first glance. First, these spin recovery methods differ from one another only by "force", i.e. by the magnitude of required deflections and the interval of delay between them. Second, same rudders are used for spin recovery (only spin recovery by method N4N requires additional deflection of ailerons), and the direction of controls deflections is the same.

For subsonic aircrafts, just the use of "strong" method was generally appropriate, since there was danger of a "lack" of rudder for spin recovery. Hence, the required efficiency of subsonic aircraft rudders was conditioned, mostly by the need to ensure spin recovery, i.e. the ability to create sufficient aerodynamic moments to stop autorotation.

On the other hand, supersonic airplanes have the risk of excessive rudder "delivery" during recovery, i.e. creating excessively large aerodynamic moments. The possibility of exhibiting this danger has to do with that that the required efficiency of the supersonic airplane rudder is chosen starting from the requirement to provide maneuverability in flight at high Mach numbers, contrary to subsonic aircrafts. Hence, as a rule, the efficiency is more than sufficient for a spin recovery. Excessive rudder deflection can significantly impair characteristics of recovery (for example, unnecessarily increase the steepness of dive after spin, and consequently loose altitude for spin recovery, etc.), or lead to impossible spin recovery (the plane changes from upright to inverted spin, from left to right spin, etc.). This explains the necessity to apply the "weak" methods (N1N and N2N) for spin recovery.

However, only "strong" methods should be used for recovery of supersonic airplanes from stable intensive and uniform spins.

Therefore, for recovery of modern supersonic aircrafts from upright spin it is necessary to use "weak", as well as "strong" methods. It should be in mind at all times that, "strong" methods in any case are not above nor can they replace the "weak" methods. Each method has its area of application.

When choosing a recovery method the pilot must be guided only by the character of spin at the time he decided to recover from spin. Other data (for example, flight altitude) can be used only as supplementary for refining and hastening the determination of regime characteristics.

In all cases, it is necessary to deflect the rudders, if possible, more abrupt. Slow, sluggish deflecting will worsen the characteristics of spin recovery, and, sometimes, make it impossible to recover.

If recovery of aircraft is particularly difficult, the pilot must carefully choose the moment to start recovering, i.e. moment when to deflect the first rudder for recovery. The best moment for the recovery start is considered to be the moment when the plane stops yawing, the nose starts declining (in a inverted spin - nose starts inclining) and similar.

It is better for the pilot to count seconds instead of turns when deflecting rudder and elevator. Practice has shown that even small changes in the regime (accelerating or decelerating the aircraft) make it difficult and often impossible to properly count turns, especially at the notable positions of the aircraft in space (during periodic switching to inverted, etc.). It is not advantageous to count turns of the aircraft during spin, not only for recovery but during the regime, as well. There are several reasons for this. First, in an unstable spin regime (for example, in a flat spin where the aircraft's motion acts like a "falling leaf") the term "turn" is meaningless. Second, for the pilot, in order to evaluate the situation, it is of more importance that he knows the time of the acting regime and the altitude loss during that time. In addition, it is always easier to count seconds of duration (also much safer), and often, sometimes, it could be the only possible way to determine the mentioned interval and delay of aircraft recovery from spin. The interval and delay are usually determined count seconds aloud.

The control of altitude change during a spin is one of the most important conditions of a safe flight, especially entering spin at low altitudes.

Therefore, the pilot must follow the altimeter reading during a spin. The altimeter shows absolute values of altitude with great errors. This is due to a greater change in the flow over the Pitot-tube at higher attack angles and yaw angles, and at high rotation rates during spin. Even so, this instrument allows a proper determination of the altitude loss (altitude difference).

Flight practice shows that the most often applied methods for spin recovery for supersonic aircraft are N1N and N2N. If the first attempt to recover from a spin fails (for example, with the method N1N), i.e. the self-rotation did not stop, the pilot has to reset flight commands into position for spin, and, after 2 - 4 sec repeat recovery but with a "stronger" method (N2N). Obviously, in the first attempt the pilot applied a "weak" method due to an improper determination of the character of the spin.

The choice of the required method for spin recovery greatly depends on the weight composition of the aircraft - mass distribution and centering.

A known fact is that subsonic aircrafts, with rectangular wings, characteristically have a large weight distribution along the wingspan, which in turn adds up to generating significant inertial rolling and yawing moments during spin. On the other hand, modern supersonic aircrafts have a characteristically large mass distribution along the fuselage axis, which contributes to generation of significant inertial pitching and yawing moments.

The sequence of actions with the rudders for spin recovery depends on the nature of the interaction of inertial rolling, yawing, and pitching moments, with the inertial moments created by rudder deflection during autorotation.

7. METHODS FOR RECOVERY FROM INVERTED SPIN

To recover modern aircrafts from an inverted spin there are three basic methods (letter “L” denotes the method for recovery from an inverted spin):

- Method N1L - spin recovery by simultaneous positioning elevator and rudder into neutral, with ailerons in neutral position;
- Method N2L - spin recovery by deflecting the rudder fully opposite, followed by a delayed (2 - 4 sec) elevator positioning into neutral, with ailerons in neutral position;
- Method N3L - spin recovery by fully deflecting the both rudder and elevator (delayed for 2 - 4 sec) opposite to spin direction, with ailerons in neutral position.

The N1L method is recommended for recovery from an unstable inverted spin, the N2L method from a stable wavering spin, and method N3L from an inverted stable uniform spin.

Deflection of controls and commanding surfaces during inverted spin recovery according to the three basic methods is shown in Fig.13, with the adopted notation shown in Fig.12.

For supersonic aircrafts, most often in use is the method N2L, because these aircrafts are most characterized by an unstable wavering inverted spin. In general, supersonic airplanes rarely fall into inverted spin; more often, it is a upright spin.

As a rule, airplanes of usual geometry more easily recover from an inverted spin than from a upright spin. This is explained by the fact that during this regime the autorotation is weaker, the rudder is more efficient (practically it is out of the wing and stabilizer flow), the efficient arrow angle of the vertical surfaces is decreased, and the average absolute values of attack angles are reduced.

However, despite of everything already said, for the pilot the inverted spin is always more difficult than the upright spin. This is conditioned by the unusual position of the pilot: hanging on restraint harnesses, head down, and a negative load ($nz < 0$) tends to detach him from the seat. In such conditions, the pilot could drop the yoke and release the pedals (he could lose control of the aircraft, especially if he is not firmly seated).

Sometimes, during aircraft wavering the pilot has great difficulty to visually determine type of spin - upright or inverted. This is expressed if the aircrafts longitudinal axis is close to the vertical axis - the plane will “swirl” at low, by absolute value, negative overcritical attack angles, which is typical for a stable inverted spin. In this case, in order to stop autorotation the rudders have to be put in neutral position.

In the absence of or inability to use visual landmarks (also, to have control over a classical spin), the pilot can easily determine the type of spin by sensation: if the seat is pressuring the pilot - it is a upright spin, if the pilot is detaching from his seat, i.e. hanging on restraint harnesses - it is an inverted spin). However, if the aircraft exits at high negative attack angles that randomly change during the inverted spin regime, determining spin type by sensation is inapplicable.

During an inverted spin, it is more difficult to determine the direction of rotation (whether the plane is rotating to the right or to the left). In a classical spin, when the position of the nose of the aircraft to the horizon is practically unchanged, the direction of rotation can easily be determined according to the angular speed of rotation. However, when in inverted spin at high angular velocities of rolling, uneven motion of roll and pitch, it is impossible to determine the aircraft's direction of rotation with the mentioned method. The situation becomes more complex, because during inverted spin rolling motion is opposite to rotation. For a pilot seated in front cockpit look forward this means that, for example, in a right-hand spin the direction of rotation will into the left.

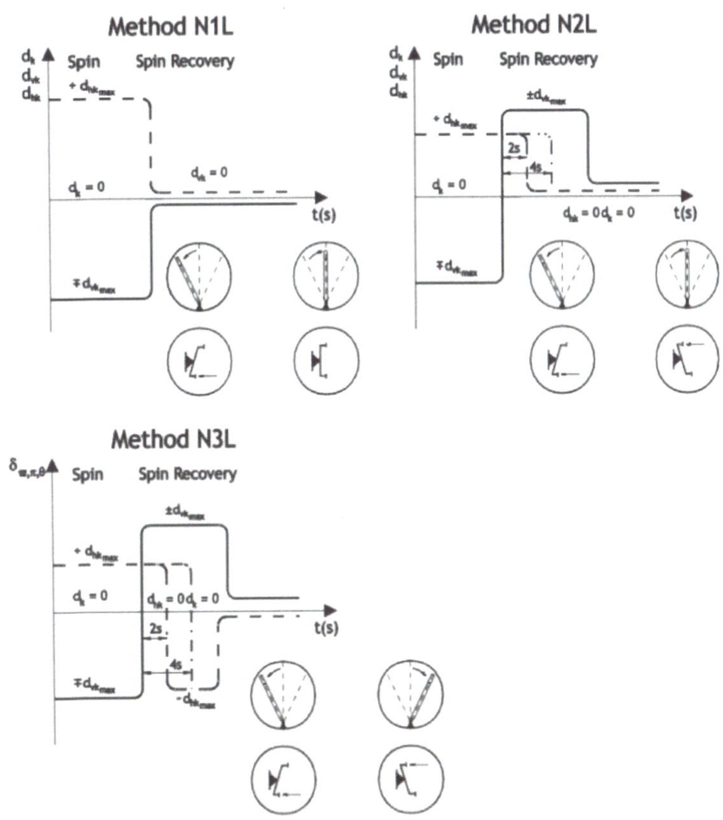

Fig. 13. Three basic methods of recovery from an inverted left spin for modern aircrafts (conditional annotation of yoke and pedal position is same as in Fig.12.)

In a upright spin, the situation is opposite: directions of roll and yaw are the same. As so, for example, in a right upright spin the pilot can see the nose of the aircraft turning to the right and plane leaning to the same side. A more experienced pilot in this matter can distinguish a upright spin from an inverted.

Pilots lacking of sufficient flights for spin recovery training often determine spin direction according to rolling direction, but not yawing direction, because the rolling angular velocity is usually higher than the yawing angular velocity (except for a flat spin). Determination of spin direction in this manner is applied more often (rolling angular velocity increasing). Use of this manner of determining spin during an inverted spin will only disorient an insufficiently trained pilot.

Therefore, when recovering from this regime, it is necessary to have safe means of control to facilitate easy conservation of spatial orientation and assure possibilities for effective and proper actions with wings. Such means could be the yaw indicator (it hand always turns in yaw direction regardless on spin type) and attack angle indicator, and if it is not present - a normal load indicator. Attack angle indicator allows the pilot to reliably determine the spin type (upright or inverted), and the yaw indicator - its direction (left or right).

8. SPIN MODELING

Modeling is the most accurate graph-analytical method for determining aircraft characteristics prior to flight tests. Modeling of flight conditions, with initial data correction, quite properly reflects timely development of aircraft motion and enables more complete conclusions about flight test results.

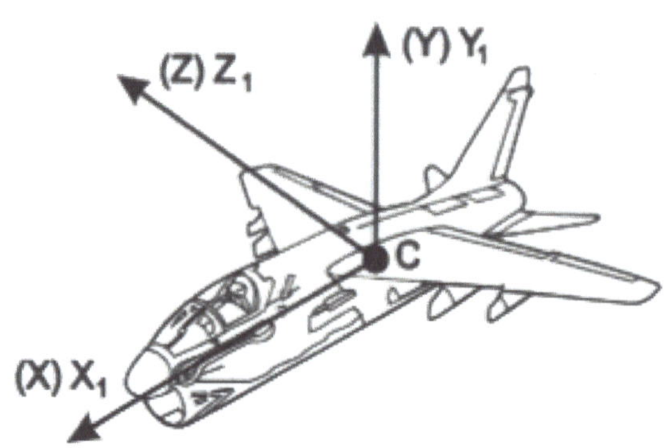

Fig. 14. Attached Coordinate System)

The differential equations of aircraft motion relative to its center of gravity is obtained from the Law of conversation of momentum. This is the moment equation. Projecting these equations to axes of the attached coordinate system (X_1, Y_1, Z_1), shown in Fig.14, whose axes we denote as (X, Y, Z) for simplicity, the following system of differential equations is derived:

$$m\left(\frac{dV_x}{dt}+q\,V_z-r\,V_y\right)=R_x+G_x$$
$$m\left(\frac{dV_y}{dt}+r\,V_x-p\,V_z\right)=R_y+G_y \qquad (1)$$
$$m\left(\frac{dV_z}{dt}+p\,V_y-q\,V_x\right)=R_z+G_z$$

$$I_x\frac{dp}{dt}+(I_z-I_y)\,q\,r+I_{xy}\left(p\,r-\frac{dp}{dt}\right)=\mathcal{M}_x$$
$$I_y\frac{dq}{dt}+(I_x-I_z)\,p\,r+I_{xy}\left(q\,r-\frac{dp}{dt}\right)=\mathcal{M}_y \qquad (2)$$
$$I_z\frac{dr}{dt}+(I_y-I_x)\,p\,q+I_{xy}\left(p^2-q^2\right)=\mathcal{M}_z$$

where:

- *Vx* , *Vy*, *Vz* correspond to projections of the velocity of aircraft center of gravity with respect to the axes of the adopted coordinate system;
- *p*, *q*, and *r* represent projections of the aircraft angular velocities to the axes of the adopted coordinate system;
- *Ix*, *Iy*, *Iz* - aircraft axial inertia moments with respect to axes of the adopted coordinate system;
- *Ixy* - aircraft centrifugal inertia moment;
- *Rx* , *Ry*, *Rz* - projections of aerodynamic forces acting on the aircraft with respect to axes of the adopted coordinate system;
- *Gx*, *Gy*, *Gz* - projection of aircraft weight with respect to axes of the adopted coordinate system;
- *Mx*, *My*, *Mz* - projection of resulting moments from external forces acting on aircraft, with respect to axes of the adopted coordinate system.

8.1 Motion equations for modeling

In order to simplify the task during tests, following assumptions are made: the angle of sideslip is small so that $sin\beta$ β and $cos\beta$ 1, effects of Mach and Reynolds number are ignored, and motion is investigated with engines turn off. This study uses known kinematic relations that are, with the adopted simplifications, equal to:

$$\theta^{\cdot} = r\sin\gamma + q\cos\gamma$$
$$\gamma^{\cdot} = p + (q\sin\gamma - r\cos\gamma)\tan\theta \qquad (3)$$
$$H^{\cdot} = V\cos\alpha\sin\theta - V\sin\alpha\cos\gamma\cos\theta - V\beta\sin\gamma\cos\theta$$

where:

θ - pitch angle, angle between *X* axis and horizontal plane,

γ - angle of transverse inclination, angle between *Y* axis and vertical plane.

If it is assumed that velocity and altitude are unchanged, i.e. $V = const$ and $H = const$, and if the right-hand sides of Eq.(3) are approximated by a Taylor polynomial, the following simplified system of equations is obtained, a so-called **System I**:

$$
\begin{aligned}
\dot{\alpha} &= a_{11}\frac{1}{\cos\alpha} + a_{12}\frac{\beta p}{\cos\alpha} + a_{13}\, q + a_{14}\frac{\cos\theta\cos\gamma}{\cos\alpha} + a_{15} \\
\dot{\beta} &= a_{21}\,\beta + a_{22}\, r\cos\alpha + a_{23}\, p\sin\alpha + a_{24}\cos\theta\sin\gamma + a_{25} \\
\dot{p} &= a_{31}\, r q + a_{32}\,\beta + a_{33}\, p + a_{34}\, r + a_{35} + a_{36} \\
\dot{r} &= a_{41}\, p q + a_{42}\,\beta + a_{43}\, p + a_{44}\, r + a_{45}\, q + a_{46} + a_{47} \\
\dot{q} &= a_{51}\, p r + a_{52}\,|\,\beta\,| + a_{53}\, r + a_{54}\, q + a_{55} + a_{56}\,\dot{\alpha} + a_{57} \\
\dot{\theta} &= a_{61}\, r\sin\gamma + a_{62}\, q\cos\gamma \\
\dot{\gamma} &= a_{71}\, p + a_{72}\, r\cos\gamma\tan\theta + a_{73}\, q\sin\gamma\tan\theta
\end{aligned}
\tag{4}
$$

The coefficients involved in the system of equations (4) are defined by expressions:

$$
\begin{aligned}
&a_{11} = -\frac{S\rho V}{2m}C_z(\alpha) \qquad a_{12} = -1 \qquad a_{13} = 1 \qquad a_{14} = \frac{g}{V} \\
&a_{15} = -\frac{S\rho V}{2m}C_{z\delta_{hk}}(\alpha)\,\Delta\delta_{hk}(t) \qquad a_{21} = \frac{S\rho V}{2m}C_{y_\beta}(\alpha) \\
&a_{22} = 1 \qquad a_{23} = 1 \qquad a_{24} = \frac{g}{V} \qquad a_{25} = \frac{S\rho V}{2m}C_{y\delta_{vk}}(\alpha)\,\Delta\delta_{vk}(t) \\
&a_{31} = \frac{I_y - I_z}{I_x} \qquad a_{32} = \frac{S\,l\,\rho V}{2\,I_x}C_{l_\beta}(\alpha) \qquad a_{33} = \frac{S\,l\,\rho V^2}{2\,I_x}C_{l_p}(\alpha) \\
&a_{34} = \frac{S\,l\,\rho V^2}{2\,I_x}C_{l_r}(\alpha) \qquad a_{35} = \frac{S\,l\,\rho V^2}{2\,I_x}C_{l_{\delta_{vk}}}(\alpha)\,\Delta\delta_{vk}(t) \\
&a_{36} = \frac{S\,l\,\rho V^2}{2\,I_x}C_{l_{\delta_k}}(\alpha)\,\Delta\delta_k(t) \qquad a_{41} = \frac{I_z - I_x}{I_y} \\
&a_{42} = \frac{S\,l\,\rho V^2}{2\,I_y}C_{n_\beta}(\alpha) \qquad a_{43} = \frac{S\,l\,\rho V^2}{2\,I_y}C_{n_p}(\alpha) \\
&a_{44} = \frac{S\,l\,\rho V^2}{2\,I_y}C_{n_r}(\alpha) \qquad a_{45} = -\frac{I_p\,\omega_p}{I_y} \\
&a_{46} = \frac{S\,l\,\rho V^2}{2\,I_y}C_{n\delta_{vk}}(\alpha)\,\Delta\delta_{vk}(t) \qquad a_{47} = \frac{S\,l\,\rho V^2}{2\,I_y}C_{n_{\delta_k}}(\alpha)\,\Delta\delta_k(t) \\
&a_{51} = \frac{I_x - I_y}{I_z} \qquad a_{52} = \frac{S\,b\,\rho V^2}{2\,I_z}C_{m_\beta}(\alpha) \qquad a_{53} = \frac{I_p\,\omega_p}{I_z} \\
&a_{54} = \frac{S\,b\,\rho V^2}{2\,I_z}C_{m_q}(\alpha) \qquad a_{55} = \frac{S\,b\,\rho V^2}{2\,I_z}C_m(\alpha) \\
&a_{56} = \frac{S\,b\,\rho V^2}{2\,I_z}C_{m_{\dot{\alpha}}}(\alpha) \qquad a_{57} = \frac{S\,b\,\rho V^2}{2\,I_z}C_{m\delta_{hk}}(\alpha)\,\Delta\delta_{hk}(t) \\
&a_{61} = 1 \qquad a_{62} = 1 \qquad a_{71} = 1 \qquad a_{72} = -1 \qquad a_{73} = 1
\end{aligned}
\tag{5}
$$

If assumed that velocity and altitude are changeable over time, i.e. $V = f(t)$ and $H = f(t)$, and if the right-hand side of Eq.(1) and (3) (projections of aerodynamic forces, weights and moments) is approximated by Taylor polynomial, the following simplified system of equations can be obtained, a so-called **System II**:

$$\dot{\alpha} = b_{11}\frac{\rho V}{\cos\alpha} + b_{12}\frac{p\,\beta}{\cos\alpha} + b_{13}\,q + b_{14}\frac{\cos\theta\,\cos\gamma}{V\,\cos\alpha} + b_{15}\frac{\dot{V}\,\tan\alpha}{V} + b_{16}\frac{\rho V}{\cos\alpha}$$

$$\dot{\beta} = b_{21}\,\rho\,V\,\beta + b_{22}\,r\,\cos\alpha + b_{23}\,p\,\sin\alpha + b_{24}\frac{\cos\theta\,\cos\gamma}{V} + b_{25}\frac{\dot{V}\,\beta}{V} + b_{26}\,\rho\,V$$

$$\dot{p} = b_{31}\,r\,q + b_{32}\,\rho\,V^2\,\beta - b_{33}\,\rho\,V^2\,p + b_{34}\,\rho\,V^2\,r + b_{35}\,\rho\,V^2 + b_{36}\,\rho\,V^2$$

$$\begin{aligned}
\dot{r} &= b_{41}\,p\,q + b_{42}\,\rho\,V^2\,\beta + b_{43}\,\rho\,V^2\,p + b_{44}\,\rho\,V^2\,r + b_{45}\,q + b_{46}\,\rho\,V^2 + b_{47}\,\rho\,V^2\\
\dot{q} &= b_{51}\,p\,r + b_{52}\,\rho\,V^2\,\,|\,\beta\,| + b_{53}\,r + b_{54}\,\rho\,V^2\,q + b_{55}\,\rho\,V^2 + b_{56}\,\rho\,V^2\,\dot{\alpha} + b_{57}\,\rho\,V^2\\
\dot{\theta} &= b_{61}\,r\,\sin\gamma + b_{62}\,q\,\cos\gamma\\
\dot{\gamma} &= b_{71}\,p + b_{72}\,r\,\cos\gamma\,\tan\theta + b_{73}\,q\,\sin\gamma\,\tan\theta\\
\dot{V} &= b_{81}\frac{\rho\,V^2}{\cos\alpha} + b_{82}\,V\,\dot{\alpha}\,\tan\alpha + b_{83}\,V\,q\,\tan\alpha + b_{84}\frac{V\,r\,\beta}{\cos\alpha} + b_{85}\frac{\sin\theta}{\cos\alpha} + b_{86}\frac{\rho\,V^2}{\cos\alpha}\\
\dot{H} &= b_{91}\,V\,\cos\alpha\,\sin\theta + b_{92}\,V\,\sin\alpha\,\cos\theta\,\cos\gamma + b_{93}\,V\,\beta\,\cos\theta\,\sin\gamma
\end{aligned} \tag{6}$$

The coefficients involved in the system of equations (6) are defined by expressions:

$$\begin{aligned}
& b_{11} = -\frac{S}{2m}C_z(\alpha) \qquad b_{12} = -1 \qquad b_{13} = 1 \qquad b_{14} = g\\
& b_{15} = -1 \qquad b_{16} = -\frac{S}{2m}C_{z_{\delta_{hk}}}(\alpha)\,\Delta\delta_{hk}(t) \qquad b_{21} = \frac{S}{2m}C_{y_\beta}(\alpha)\\
& b_{22} = 1 \qquad b_{23} = 1 \qquad b_{24} = g \qquad b_{25} = -1\\
& b_{26} = \frac{S}{2m}C_{y_{\delta_{vk}}}(\alpha)\,\Delta\delta_{vk}(t) \qquad b_{31} = \frac{I_y - I_z}{I_x}\\
& b_{32} = \frac{S\,l}{2\,I_x}C_{l_\beta}(\alpha) \qquad b_{33} = \frac{S\,l}{2\,I_x}C_{l_p}(\alpha) \qquad b_{34} = \frac{S\,l}{2\,I_x}C_{l_r}(\alpha)\\
& b_{35} = \frac{S\,l}{2\,I_x}C_{l_{\delta_{vk}}}(\alpha)\,\Delta\delta_{vk}(t) \qquad b_{36} = \frac{S\,l}{2\,I_x}C_{l_{\delta_k}}(\alpha)\,\Delta\delta_k(t)\\
& b_{41} = \frac{I_z - I_x}{I_y} \qquad b_{42} = \frac{S\,l}{2\,I_y}C_{n_\beta}(\alpha) \qquad b_{43} = \frac{S\,l}{2\,I_y}C_{n_p}(\alpha)\\
& b_{44} = \frac{S\,l}{2\,I_y}C_{n_r}(\alpha) \qquad b_{45} = -\frac{I_p\,\omega_p}{I_y}\\
& b_{46} = \frac{S\,l}{2\,I_y}C_{n_{\delta_{vk}}}(\alpha)\,\Delta\delta_{vk}(t) \qquad b_{47} = \frac{S\,l}{2\,I_y}C_{n_{\delta_k}}(\alpha)\,\Delta\delta_k(t)\\
& b_{51} = \frac{I_x - I_y}{I_z} \qquad b_{52} = \frac{S\,b}{2\,I_z}C_{m_\beta}(\alpha) \qquad b_{53} = \frac{I_p\,\omega_p}{I_z}\\
& b_{54} = \frac{S\,b}{2\,I_z}C_{m_q}(\alpha) \qquad b_{55} = \frac{S\,b}{2\,I_z}C_m(\alpha) \qquad b_{56} = \frac{S\,b}{2\,I_z}C_{m_{\dot{\alpha}}}(\alpha)\\
& b_{57} = \frac{S\,b}{2\,I_z}C_{m_{\delta_{hk}}}(\alpha)\,\Delta\delta_{hk}(t) \qquad b_{61} = 1 \qquad b_{62} = 1\\
& b_{71} = 1 \qquad b_{72} = -1 \qquad b_{73} = 1 \qquad b_{81} = -\frac{S}{2m}C_x(\alpha)\\
& b_{82} = 1 \qquad b_{83} = -1 \qquad b_{84} = -1 \qquad b_{85} = -g\\
& b_{86} = -\frac{S}{2m}C_{x_{\delta_{hk}}}(\alpha)\Delta\delta_{hk}(t) \qquad b_{91} = 1\\
& b_{92} = -1 \qquad b_{93} = -1
\end{aligned} \tag{7}$$

Notation in previous equations are:

- C_z - lift coefficient;
- C_x - drag coefficient;
- C_m - pitching moment coefficient;
- $C_{z_{\delta_{hk}}}$ - derivative of lift coefficient with respect to angle of deflection of the elevator;
- C_{y_β} - derivative of sideslip force with respect to sideslip angle;
- $C_{y_{\delta_{vk}}}$ - derivative of sideslip force with respect to angle of deflection of the rudder;
- C_{l_β} - derivative of rolling moment coefficient with respect to angle of sideslip;
- C_{l_p} - derivative of rolling moment coefficient with respect to rolling angular velocity;
- C_{l_r} - derivative of rolling moment coefficient with respect to yawing angular velocity;
- $C_{l_{\delta_{vk}}}$ - derivative of rolling moment coefficient with respect to rudder angle of deflection;
- $C_{l_{\delta_k}}$ - derivative rolling moment coefficient with respect to aileron angle of deflection;
- C_{n_β} - derivative of yawing moment coefficient with respect to angle of sideslip;
- C_{n_p} - derivative of yawing moment coefficient with respect to rolling angular velocity;
- C_{n_r} - derivative of yawing moment coefficient with respect to yawing angular velocity;
- $C_{n_{\delta_{vk}}}$ - derivative of yawing moment coefficient with respect to rudder angle of deflection;
- $C_{n_{\delta_k}}$ - derivative of yawing moment coefficient with respect to aileron angle of deflection;
- C_{m_β} - derivative of pitching moment coefficient with respect to angle of sideslip;
- C_{m_q} - derivative of pitching moment coefficient with respect to pitching angular velocity;
- $C_{m_{\dot{\alpha}}}$ - derivative of pitching moment coefficient with respect to derivative of angle of attack over time;
- $C_{m_{\delta_{hk}}}$ - derivative of pitching moment coefficient with respect to elevator angle of deflection;
- $C_{x_{\delta_{hk}}}$ - derivative of drag coefficient with respect to elevator angle of deflection;
- I_p - polar moment of inertia of engine rotor;
- ω_p - angular velocity of engine rotor;
- $\Delta\delta_{hk}$ - change in elevator angle of deflection;
- $\Delta\delta_{vk}$ - change in rudder angle of deflection;
- $\Delta\delta_k$ - change in aileron angle of deflection.

These coefficients are entirely determined with the aid of DATCOM1 reference, and in some additional literature2 their values are defined for the category of light aircrafts.

8.2 **Methods of modeling**

It is obvious that most of the coefficients in Exp.(6) are not constant, but vary with time, i.e. angle of attack.

If assumptions that speed and altitude are constant are rejected, a complete system of equations is obtained. In order to perform modeling a computer is used with its memory loaded with the simplified or complete system of equations, which depends on the regime that has to be studied or the accuracy of results. The computer loaded with data necessary to calculate the coefficients and other values in Eq.(6), for a given time t_1, or the given angle of attack α_1. Data is obtained by testing models in wind tunnel or in flight. Then, the data is entered for moment t_2 or attack angle α_2, and so on. Data is entered until the end of the observed time interval, and it is entered point-by-point. The accuracy of obtained results depends on the magnitude of the time change (time difference between two points), i.e. the angle of attack. The more points are entered, the more accurate the results will be. The computer will integrate and associate values for angle of attack, angle of sideslip, pitching angle and angular velocities for the observed time interval, for which data is entered, and results are obtained as shown in Fig.15, i.e. following functions are defined: $\alpha = f(t)$, $\theta = f(t)$, $\beta = f(t)$, $p = f(t)$, $q = f(t)$ i $r = f(t)$. For System II (Eq.7), following the same analogy, the computer will integrate and associate values from the observed time interval, for which data was entered, and results are obtained as shown in Fig.16.

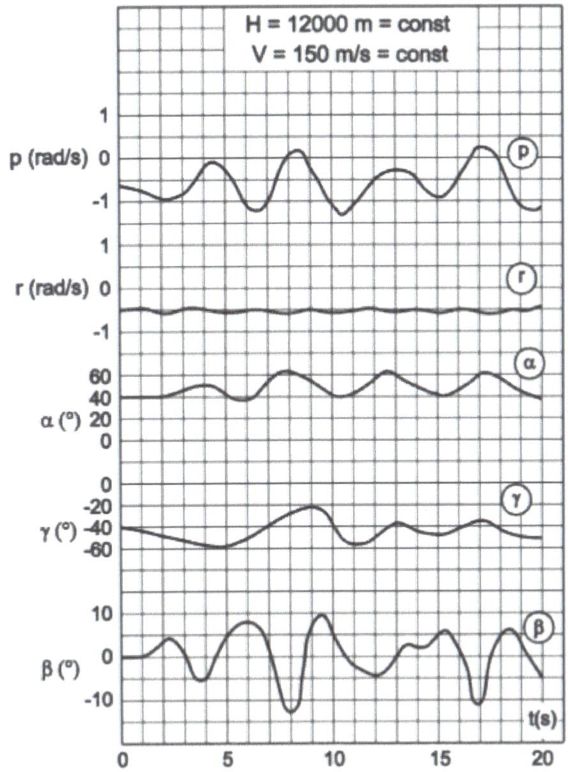

Fig. 15. Diagram for a timely development of a left-hand spin (System I)

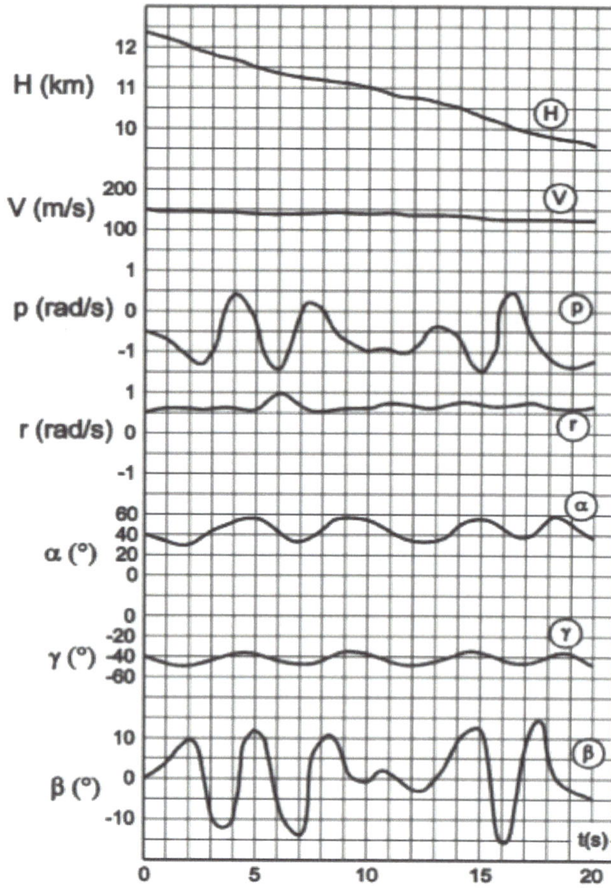

Fig. 16. Diagram for a timely development of a left-hand spin (System II)

8.3 Analysis of results from modeled flat spin (spin ongoing as a falling leaf On a spiral path)

An analysis of relations is made if effects from individual terms in equations of motion need to be determined (Eq.6). To do this, each member will be designated by the letter "A" with appropriate numerical indexes *i*, *j* (*Ai,j*), with "i" being the ordinal number of the equation *i* = (1, *n*), and "j" being the ordinal number of the term in the equation *j* = (1, *m*). For example, the first equation of Eq.6 will look:

$$\alpha^{\cdot} = A11 + A12 + A13 + A14 + A15,$$

where:

$$A_{11} = a_{11}\frac{1}{\cos\alpha},\quad A_{12} = a_{12}\frac{\beta p}{\cos\alpha},\quad A_{13} = a_{13}\,q,$$

$$A_{14} = a_{14}\frac{\cos\theta\,\cos\gamma}{\cos\alpha},\quad A_{15} = a_{15}$$

Ratios of absolute values of terms in Eq.6 are assessed, as of these depend the magnitudes of effects of terms α, β, θ, *p*, *q* and *r*„ and therefore are shown as fractions:

$$\bar{A}_{ij} = \frac{|\ A_{ij}\ |}{\sum_{j=1}^{m} |\ A_{ij}\ |} 100 \qquad [\%]$$

In case of modeling spin with the complete system of equations, terms *Aij* and *A*— *ij* are introduced in the same manner.

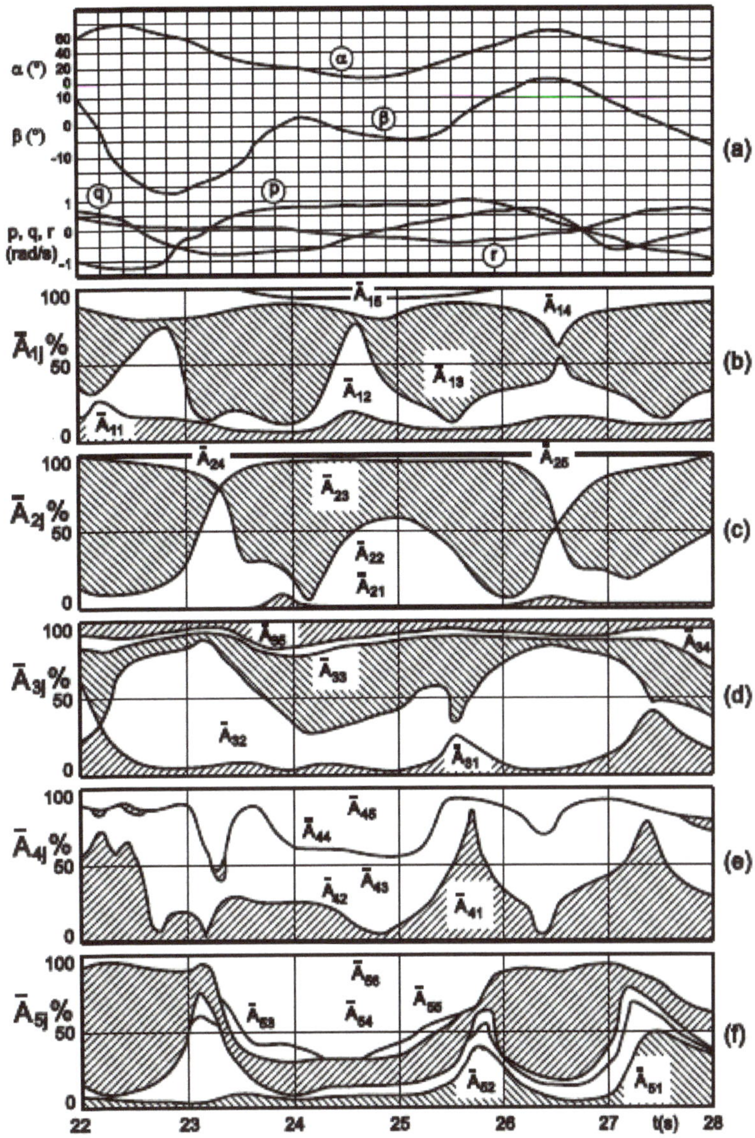

Fig. 17. Diagram of development for a flat spin (spin ongoing as a falling leaf)

Figure 17 shows results obtained for a numerically modeled spin. From diagrams (b), (c), (d), (e), and (f), it can be seen which terms have the most effect on results given in diagram (a). Diagrams (b), (c), (d), (e), and (f), were constituted by showing time on the abscissa, while values of *A— ij* (*i = const*) are "stacked" on the ordinate, one over the other. For example, value *A— i*1 is imposed from 0, value *A— i*2 is imposed from point where *A— i*1 ends, value *A— i*3 is imposed from point where *A— i*2, . . ., ends, value *A— im* is imposed from point where *A— i,m* 1 ends, and ends at 100%. By doing so, these diagrams show which term *A— ij* has the most impact on values of α, β, θ, p, q and r. For example, in diagram (b) it is shown that at moment $t = 23$ s, value of *A—* 13 is higher than of *A—* 11, *A—* 12, *A—* 14 and *A—* 15. This means that *A—* 13 has the most effect on values of function $\alpha = f(t)$ at the given time. When observed what A_{13} is equal to, it can be noticed that the value of angle of attack, at this moment, mostly depends on q, since $a13 = 1$. In this way, every term *Aij* on the right-hand side of Eq.(6) can be analyzed on its impact on the left-hand side of same equations. Analysis of every term *Aij*, brings about a deeper understanding of the physical image of the studied regime. Furthermore, it facilitates the identification of factors that significantly influence such a condition when the aircraft fails to meet essential criteria for spin, allowing for suitable modifications to achieve an aircraft with the requisite technical specifications for spin.

FOOTNOTES

1. D. E. Hoak: *USAF Stability and Control DATCOM*, (N76-73204), Flight Control Devision, Air Force Flight Dynamics Laboratory, Wright-Patterson Air Force Base, Ohio, 1975.
2. D. Cvetkovic: *The adaptive approach to modeling and simulation of spin and spin recovery*, PhD Thesis, Faculty of Mechanical Engineering, Belgrade University, 1997.

REFERENCES

1. R.H. Barnard; D.R. Philpott; A.C. Kermode (2006). *Mechanics of Flight* (11th Edition), Prentice Hall, ISBN: 1405823593
2. Warren F. Phillips (2009). *Mechanics of Flight* (2nd Edition), Wiley, ISBN: 0470539755
3. B. Pamadi (2004). *Performance, Stability, Dynamics, and Control of Airplanes* (2nd Edition), AIAA Education, ISBN: 1563475839
4. Barnes W. McCormick (1994). *Aerodynamics, Aeronautics, and Flight Mechanics*, (2nd Edition), Wiley, ISBN: 0471575062

5. John D. Anderson (2001). *Fundamentals of Aerodynamics* (3rd Edition), McGraw-Hill Science/Engineering/Math, ISBN: 0072373350
6. J. H. Blakelock (1991). *Automatic Control of Aircraft and Missiles*, John Wiley & Sons, Inc., New York, ISBN: 0471506516
7. Bernard Etkin; Lloyd Duff Reid (1995). *Dynamics of Flight: Stability and Control* (3rd Edition), Wiley, ISBN: 0471034185
8. D. E. Hoak (1975). *USAF Stability and Control DATCOM*, N76-73204, Flight Control Division, Air Force Flight Dynamics Laboratory, Wright-Patterson Air Force Base, Ohio
9. D. Raymer (2006). *Aircraft Design: A Conceptual Approach* (4th Edition), AIAA Education Series, ISBN: 1563478293
10. Roger D. Schaufele (2000). *The Elements of Aircraft Preliminary Design*, Aries Pubns, ISBN: 0970198604
11. E. L. Houghton; P. W. Carpenter (2003). *Aerodynamics for Engineering Students* (5th Edition), Butterworth-Heinemann, ISBN: 0750651113

10 Chapter

SURFACE WELDING AS A METHOD FOR RAILWAY MAINTENANCE

1. INTRODUCTION: AN OVERVIEW

The construction of railway systems has historically been a significant catalyst for technical advancement. Beginning in the 1840s, a comprehensive railroad network was established globally. Within a few decades, railways emerged as the primary transportation system, accommodating a consistently rising amount of products and passengers. This swift advancement was accompanied by significant progress in other fields, including steel manufacturing, engine fabrication, civil engineering, and communication (Zerbst et al., 2005). The global railway sector is implementing increased axle loads, elevated vehicle speeds, and greater traffic volumes for the efficient delivery of goods and passengers. The rising needs for high-speed services and increased axle loads at the onset of the 21st century provide novel concerns concerning materials, technology, and safety considerations. The primary causes influencing rail deterioration are wear and fatigue, which render railroads unsuitable for service due to poor rail profiles, cracking, spalling, and rail fractures. The degradation of rail is sensitive to both microstructural and macrostructural factors, and there exists a complex interplay of wear mechanisms, wear rates, and the initiation and growth rates of fatigue cracks, all of which influence rail longevity (Eden et al., 2005; Kapoor et al., 2002). Defects like squats and wheelburns manifest even in the most contemporary and meticulously maintained railway networks, and as a general principle, each network generates one such fault annually for every two kilometers. One European railway network experiences around 4,000 rail fractures annually. While such fractures are infrequently perilous when properly treated, they incur substantial repair costs and may disrupt the network (Bhadeshia, 2002). The substitution of such flaws with a short rail segment is costly and often undesirable, as it produces two new discontinuities in the track due to the presence of two aluminothermic welds, hence negating the benefits achieved with lengthy hot-rolled rail.

Considering that the average expense for a repair or brief replacement rail can amount to several thousand euros and that the prevalence of wheel-rail interface defects is expected to rise with the notable increase in traffic levels on most railways, the significance of surface welding is readily apparent. The increasing necessity for restitution, driven by substantial financial obligations, has necessitated study in this domain.

This research aims to demonstrate the potential of surface welding on pearlitic high-carbon steel and the characteristics of the resultant joint, grounded on current theoretical frameworks and reference data. The analysis of the obtained results and conclusions demonstrates the improved qualities of repair welded layers in comparison to the base steel. In restored rail, maximum stresses are generated in the newly deposited layer, which becomes the site for potential crack initiation, hence delaying its occurrence and ensuring secure and dependable operation. This outcome presents more opportunities for more economical and dependable train maintenance in the future. This study demonstrates that repaired rails have superior mechanical properties compared to original rails, attributable to enhanced microstructure and resistance to crack formation.

2. RAIL DEGRADATION

There are many kinds of loadings which can adversely affect the life of rails; amongst these, wear and plastic deformation induced by contact stresses can combine to cause unacceptable changes in the rail head profile. Rails are subjected to complex stress state. There are many stresses that operate in a rail and can influence rail defects and rail failure. As bending and shear stresses arised principally from the gross vehicle load, the rail is also subjected to contact stresses, thermal stresses and residual stresses. Residual stresses in rails are introduced by different mechanisms. Primarily they stem from the manufacturing process, namely from heat treatment and roller straightening (Schleinzer & Fischer, 2000; Schleinzer & Fischer, 2001). A special case of residual stresses is welding residual stresses at rail joints. Since the loading conditions at the tread of a wheel and at the running surface of a rail have a number of features in common the appearance of cracks will also be similar. Cracks may be induced at or below the surface. Surface cracks are initiated due to high traction forces at high speed rails and they will propagate under the influence of a lubricant in an inclined angle in the direction of the motion of the applied load for rails operated in one direction. Transverse branching may then lead to the complete fracture of the rail. Sub- surface cracks are reported to initiate beneath the gauge corner 10–15 mm below the running surface and 6–10 mm from the gauge face (Clayton,1994). They seem to propagate towards the rail surface and to behave like original surface cracks after penetration.

Note, that cracks close to or at the surface are a rather new problem connected with high speed operating. In former times rails experienced enough wear to permanently remove the surface layer containing the new emerging cracks. In order to fulfil the increasing demands for higher axial and dynamic loads modern rail steels tend to exhibit much higher resistance to wearing with the disadvantage that the surface layer removed is not any more large enough to prevent small cracks from extending into the rail (with respect to the development of rail steels (Muders & Rotthauser,2000)).

A typical development of a rail crack is illustrated schematically in Fig. 1 (Ishida &Abe, 1996). Originating from a small surface or sub-surface crack, a dark spot is developing at the surface accompanied by crack growth in an inclined angle below the surface. At a certain point this crack branches into a horizontal and a transverse crack. The transverse crack will extend down into the rail and finally cause its fracture.

Today's rail failures can be divided into three broad groups as follows: those originating from rail manufacturing defects; those originating from defects or damage caused by inappropriate handling, installation and use and those caused by the exhaustion of the rail steel's inherent resistance to fatigue damage (Cannon et al.,2003).

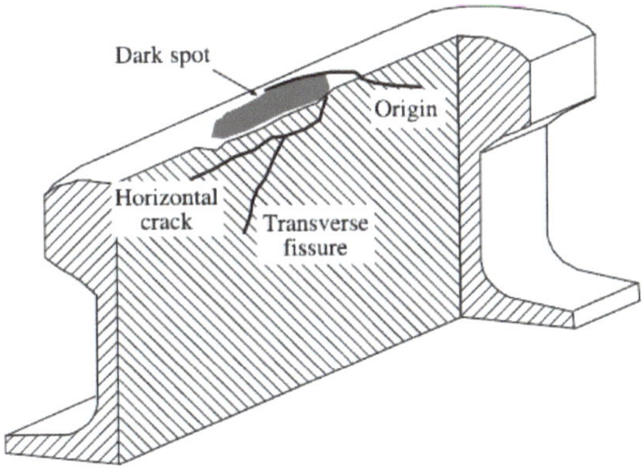

Fig. 1. Typical development of a rail crack (schematically) (Ishida &Abe, 1996).

Rolling contact fatigue (RCF) is likely to be a major future concern as business demands for higher speed, higher axle loads, higher traffic density and higher tractive forces increase. Head checks, gauge-corner cracks and squats are all names for surface-initiated RCF defects. They are caused by a combination of high normal and tangential stresses between the wheel and rail, which cause severe shearing of the surface layer of the rail and either fatigue or exhaustion of ductility of the material. The microscopic crack produced propagates through the heavily deformed (and orthotropic) surface layers of steel at a shallow angle to the rail running surface (about 10°) until it reaches a depth where the steel retains its original isotropic properties. At this stage the crack is a few millimetres deep into the rail head. At this point the crack may simply lead to spalling of material from the rail surface. However, for reasons still not clearly understood, isolated cracks can turn down into the rail, and, if not detected, cause the rail to break. These events appear to be rare, but are highly dangerous since RCF cracks tend to form almost continuously at a given site. Fracture at one crack increases stress in the nearby rail, increasing the risk of further breaks and disintegration of the rail (Cannon et al.,2003).

RCF initiation is not normally associated with any specific metallurgical, mechanical or thermal fault; it is simply a result of the steel's inability to sustain the imposed operating conditions. The problem is known to occur in most of the rail-steel types in common use today.

While wear has been reduced, rolling contact fatigue defects have become more prominent on busy routes where the rails are highly stressed. Although its wear reserve may not be used up, rail may have to be replaced because such defects quickly become critical for safety (Pointner & Frank, 1999). The relationship between RCF and mechanical wear is not well undersood, as for example zero (or minimal) mechanical wear leads to significant microcrack propagation and thus RCF failure. On the other hand, excessive mechanical wear eliminates RCF but leads to unrealistically short rail life (Kapoor et al., 2001).

The pace of rail degradation is influenced by geography; rail head erosion is most pronounced in areas with curved tracks. Figure 2 illustrates the damage to the inner edge of the rail head, resulting from the centrifugal force that seeks to propel the vehicle outward from the track. Such damage can be rectified using surface welding, as illustrated in Fig 2b.

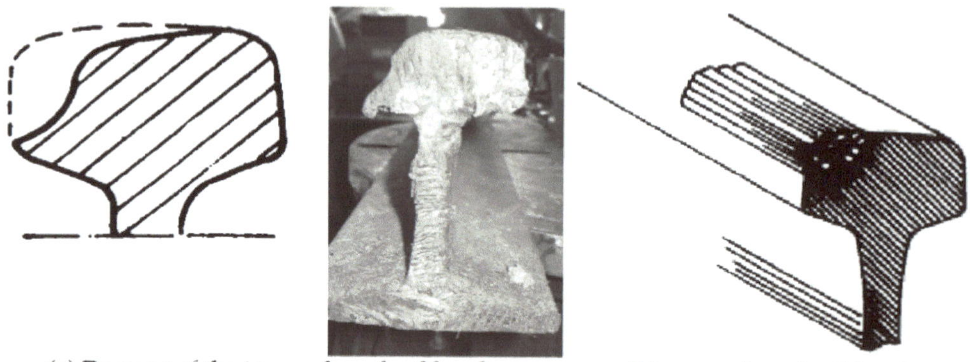

(a) Damage of the inner edge of rail head (b) reparation of rail head

Fig. 2. Rail head degradation (Popovic et al., 2006).

2.1 Fracture control concepts

A few different fracture control concepts are applied in railway systems, and one of them is damage tolerance concept (Zerbst et al., 2005). Within the frame of this concept, the possibility of fatigue crack growth is basically accepted. The aim is to prevent the crack to grow to its critical size during the lifetime of the component, i.e. to estimate number of cycles to critical crack size. In fatigue, crack extension is expressed as a function of stress intensity range ΔK and the crack extension rate, da/dN, whereby da denotes an infinitesimal crack extension due to an infinitesimal number of loading cycles dN. The basic idea is that the largest crack that could escape detection is presupposed as existent. After that, the initial crack can extend due to various mechanisms such as fatigue, stress corrosion cracking, high temperature creep, or combinations of these mechanisms. Such a failure process is visible, and catastrophic rail failure can be prevented by regular examination of the top surface of the railhead. Maintenance methods (lubrication and grinding) help combat the wear and rolling contact fatigue phenomena referred to in local parameters. By applying these methods appropriately, maintenance costs can be reduced (Vitez et al.,2005). Rail grinding prolongs rail service life by preventing the emergence of defects or by delaying their development, preventive grinding to improve the quality of the running surface of newly- laid rails and corrective grinding to remove rail defects that have already developed by reprofiling the rail to optimize wheel/rail contact.

3. RAIL STEELS

Choice of material for rail steels is of fundamental importance. This is because the rail's behaviour in service depends critically on the properties of the metal. Much effort and a considerable amount of research has already been undertaken in the search for the ideal rail steel (Pointner & Frank,1999). In recent years rail steel production has improved as manufacturers have developed steels with increased hardness and better wear resistance.

There are many criteria which determine the suitability of a steel for rail track applications. The primary requirement is structural integrity, which can be compromised by a variety of fatigue mechanisms, by a lack of resistance to brittle failure, by localised plasticity and by excessive wear. All of these depend on interactions between engineering parameters, material properties and the environme nt. The track material must obviously be capable of being manufactured into rails with a high standard of straightness and flatness in order to avoid surface and internal defects which may cause failure. Track installation requires that the steel should be weldable and that procedures be developed to enable its maintenance and repair. Commercial success depends also on material and life time costs.

Since steel has one of the highest values of elastic modulus and shows superb strength, ductility and wear resistance, most modern rails have pearlitic microstructures and carbon- manganese chemistries similar to those produced in rails in 1900. Ordinary rail steels contain about 0.7 wt% of carbon and are pearlitic. Pearlite consists of a mixture of soft ferrite and a hard, relatively brittle iron carbide called cementite, Fig. 3a. Pearlite presumably achieves a high resistance to wear because of the hard cementite and its containment by the more plastic ferrite, but pearlitic steels are not therefore tough. In pealite, altering lamellae of iron and iron carbide are aranged, and lamella spacing has a large effect on hardness.

Naturally cooled standard rails have coarse lamella spacing and relatively low values of about 300 Brinell hardness (HB). Control-cooled premium rails have finer lamella spacing and thus higher hardness of 340-390 HB (Lee & Polycarpou, 2005).

Raising carbon content and refining pearlite spacing increases the hardness of pearlitic steel, and this has been shown to lead to improved wear resistance. Hence rail manufacturers have worked to produce pearlitic steels with higher carbon contents (now achieving approximately 1 wt%) and finer structure (using head-hardening processes). Even though hardness generally has a positive effect on rail wear, there is a limit to the hardness that can be reached with pearlitic steels, and this hardness has been reached in modern rails (Lee & Polycarpou, 2005).

There has been considerable effort devoted to finding alternatives to the pearlitic rails, but with alterable results. In an attempt to develop rail steels with higher hardness and alternative microstructures, several types of bainitic steel were developed. While pearlitic steels obtain their strength from the fine grains of pearlite, bainitic steels (Fig. 3b) derive their strength from ultra-fine structures with a lot dislocations which are harmless but confer high strength (Aglan et al.,2004). Bainitic steel is easy to be cast, welded and inspected by ultrasonic methods. The new generation of bainitic steels achieved higher tensile and fatigue strengths and performed well in service.

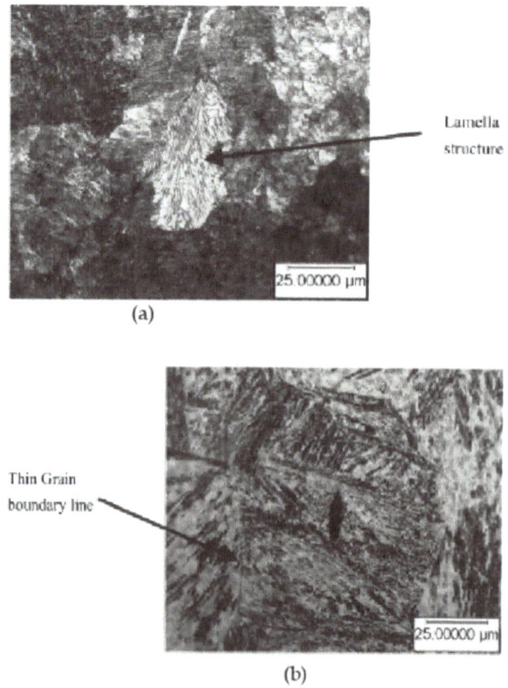

Fig. 3. Optical microstructures of rail steels: (a) pealite; (b) bainite (Aglan et al.,2004).

4. WELDABILITY OF RAILS AND TYPES OF FILLER MATERIALS

Main problem in welding of pearlitic steels is their poor weldability, i.e. susceptibility to welding defects, due to its high carbon equivalent. Since the rail is produced from this type of steel and subjected to complex strain state, leading to its degradation, surface welding is presently the dominant maintenance way to prolong exploitation life. Damaged parts produced from pearlitic high-carbon steel can be surface welded, in spite of their poor weldability, and by properly choice of welding technology, it is possible to get improved structure with dominant properties comparing to the original part (Popovic et al.,2010). To achieve that, it is necessary that obtained morphology corresponds to the new steel generation, i.e. bainitic microstructure.

For surface welding are mostly in use semi-automatic arc welding processes, with flux-cored and self-shielded wires. Basic difference between them is the first requires an external shielding gas, and the second does not. In both cases, core material acts as a deoxidizer, helping to purify the weld metal, generate slag formers and by adding alloying elements to the core, it is possible to increase the strength and provide other desirable weld metal properties (Lee,2001; Sadler,1997). These processes have replaced slowly MMA process and they almost ideal for outdoors in heavy winds. The key strength of these processes lies in the replacement of those aspects of the conventional MMA process that often results in variability in the quality of the repair with automatic and more controlled operations. Although the MMA process is used many industries, it is heavily reliant on the competence of the welder, is time consuming, and is prone to internal defects such as porosity that subsequently grow through fatigue, and if not detected by ultrasonic inspection, result in rail breaks.

The result of flux-cored wire application is higher quality welds, faster welding and maximizing a certain area of welding performance (Popovic et al.,2010). The number of layers in surface welded joint depends of the damage degree, most frequently it's three, sometimes with buffer layer. The buffer layer is applied at the crack sensitive materials, what high carbon steel certainly is (high CE). The function of buffer layer is to slow down the growth of initiated crack with its own plasticity. Constructions, like railways, are exposed to cyclic load and wear in exploatation life, so crack must be initiated. Sometimes in these cases it is necessary to use buffer layer, what besides the good affects, has some drawbacks. Namely, the use of buffer layer significantly slows down surface welding process, due to replacement of wires and settings of other welding parameters. Since, as already noted, for surface welding are mainly in use semi-automatic and automatic processes, it significantly extends the working time. The new classes of flux-cored and self- shielded wires are recently developed, and it is possible to achieve the requested properties of welded joints without buffer layer (Popovic et al.,2011).

5. EXPERIMENTAL PROCEDURE

The material used in present work is pearlitic steel, received in the form of rails, type UIC 860 S49, what is the most common rail type on domestic railroads. It's chemical composition and mechanical properties are given in Table 1.

Table 1. Chemical composition and mechanical properties of base metal.

Chemical composition, %							Tensile strength R_m (N/mm²)	Elongation A_c (%)
C	Si	Mn	P	S	Cu	Al		
0.52	0.39	1.06	0.042	0.038	0.011	0.006	680-830	≥14

The surface welding of the testing plates was perfomed by semi-automatic process. As the filler material, the self-shielded wire (FCAW-S) and flux-cored wires (FCAW) were used, whose chemical compositions and mechanical properties are given in Table 2. The plates were surface welded in three layers; sample 1 with FCAW-S without buffer layer; sample 2 with FCAW with buffer layer (according to Table 2).

Table 2. Chemical composition of filler materials.

Sample No.	Wire designation		Wire diam. mm	Chemical composition							Hardness, HRC
				C	Si	Mn	Cr	Mo	Ni	Al	
Sample 1	OK Tubrodur 15.43 (self-shielded wire)		1.6	0.15	<0.5	1.1	1.0	0.5	2.3	1.6	30-40
Sample 2	1.layer (buffer layer)	Filtub 12B (flux-cored wire)	1.2	0.05	0.35	1.4	-	-	-	-	-
	2. and 3. layer	Filtub dur 12 (flux-cored wire)	1.6	0.12	0.6	1.5	5.5	1.0	-	-	37-42

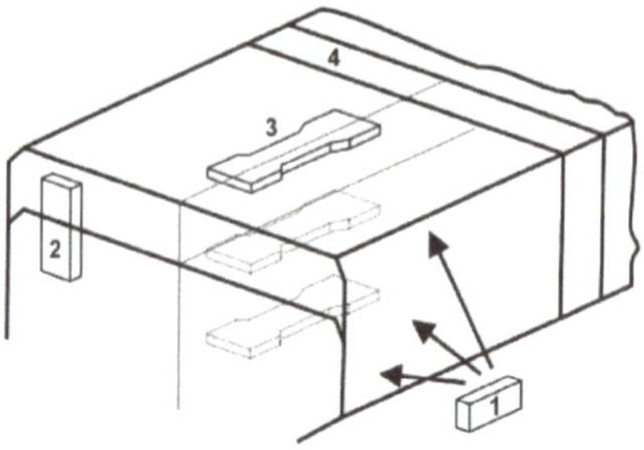

1. specimen for toughness and crack growth resistance estimation
2. specimen for microstructural analysis
3. tensile specimens
4. specimen for hardness measurements

Fig. 4. Specimens from surface welded rail head (Popovic et al.,2010).

Heat input during welding was 10 kJ/cm and preheating temperature was 2300C, since the CE equivalent was CE=0.64 (Popovic et al.,2010). Controlled interpass temperature was 2500C. Sample 1 is surfaced with one type of filler material (self-shielded wire), while for surfacing of sample 2 were used two types of wires, but both flux-cored: one for buffer layer and the second one for last two layers. As shieleded gas for welding of sample 2, CO_2 was used. To evaluate the mechanical properties, specimens for further investigation were cut from surface welded rail head, according to Fig.4.

5.1 Hardness

Hardness measurements were performed using a load of 100Pa. Hardness profiles of surface welded joints are shown in Fig. 5. The lowest hardness is in the base metal (250-300 HV), being the hardness of naturally cooled standard rails(Lee & Polycarpou, 2005; Singh et al., 2001). In HAZ hardness increase is noticable in both samples, due to complex heat treatment and grain refinement (Popovic et al.,2010). In sample 2 comes to a sharp decrease of hardness in first surfaced layer, i.e. in buffer layer. The function of buffer layer is to stop the growth of initiated crack with its own plasticity and reduced hardness. The hardness of II and III welded layers of both samples are the highest and similar, due to influence of alloying elements in filler materials, which shift transformation points to bainitic region4. Maximum hardness level of 350-390 HV is reached in surface welded layers and it provides improvement of mechanical properties and wear resistance.

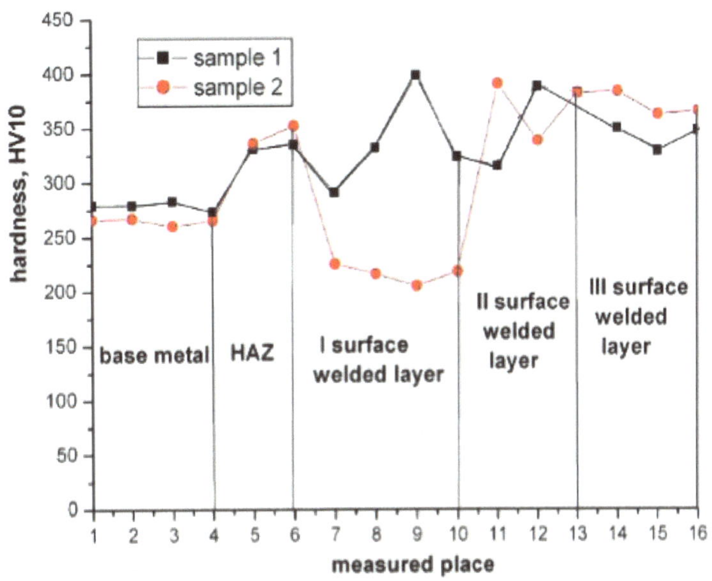

Fig. 5. Hardness profiles along the joint cross-section of samples (Popovic et al.,2011).

5.2 Microstructure

Microstructural analisys of all characteristical zones of welded layer has been done. Heat affected zone (HAZ) also has pearlitic microstructure, but with finer grain, than base metal (Figure 6), so its structure is improved and it is not a critical place in weldment. That is result of thermomechanical treatment of HAZ which is re-heated three times. Structural compatibility between deposite metal and base metal was achieved and martensitic layer wasn't formated.

The greatest differences appear in first layer microstructure, Fig.7. First layer microstructure of sample 1 consists of ferrite, pearlite and bainite, what is result of mixing of low-alloyed filler material with high-carbon base metal. For first layer deposition of sample 2 is used low-carbon wire alloyed with Mn, as a function of buffer layer, so characteristical structure consist of great fraction of ferrite with relatively large primary grains. Beside proeutectoid ferrite, microstructure contains Widmanstatten and acicular ferrite (Popovic et al.,2007).

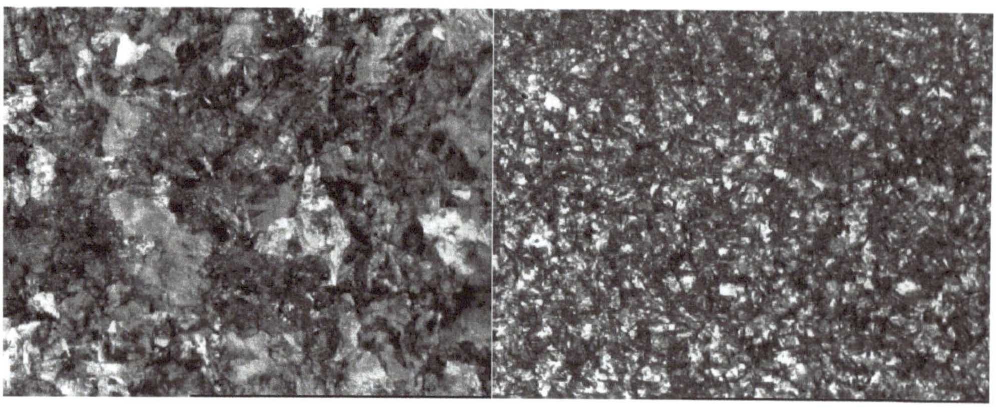

(a) 200x (b) 200x

Fig. 6. Microstructure of a) base metal and b) HAZ of both samples (Popovic et al.,2007).

The second layer microstructure is the most important in surface welded joint, because it has the greatest influence on mechanical and technological properties and exploatation behavior of repaired parts. For this structure is characteristic larger fraction of bainite, consequence to the less mixing with base metal. In second layer of sample 2 occurs fine grain ferritic structure with low content of bainite. This structure has finer grain compare to first layer, what is result of heat treatment and chemical composition (presence of Mo in filler material).

The third layer of sample 1 has some coarser grain structure, with higher content of bainite, compare to previous layer, what is consequence of re-heating absence. For third layer of sample 2 is characteristical bainitic microstructure with small amount of martensite and locally zones of proeutectoid ferrite.

Though used filler materials are different type, alloying concepts, sort of protection, buffer layer, as final result is obtained desirable bainitic microstructure with superior properties compare to base metal (Popovic et al.,2007). Except metallography examination, this is confirmed by other detail tests (Popovic,2006).

	Sample 1	Sample 2
1. layer	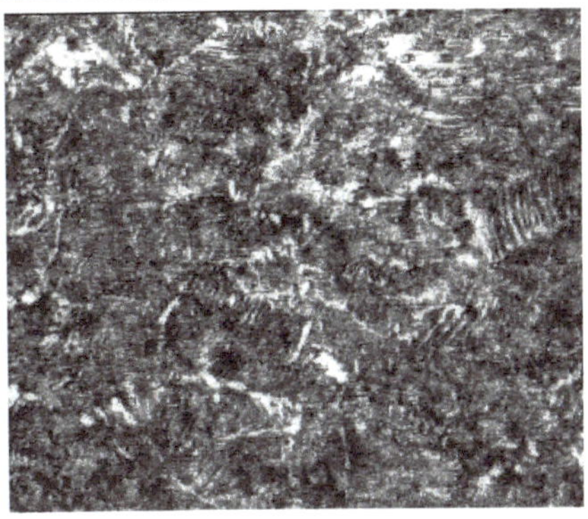 500 x	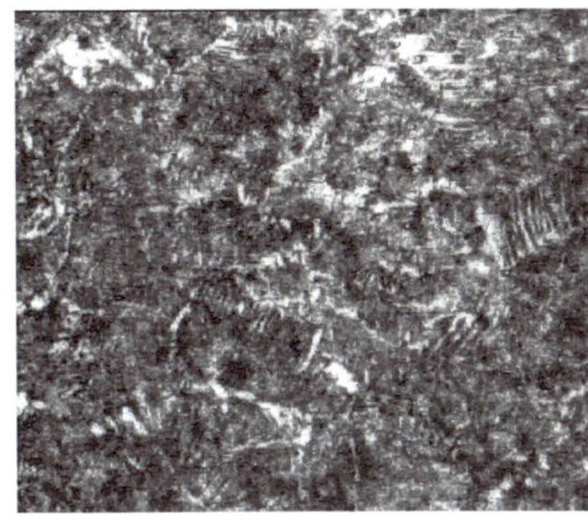 500 x
2. layer	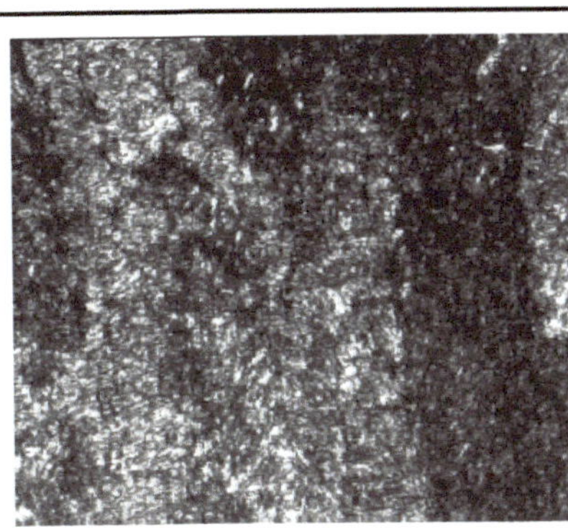 500 x	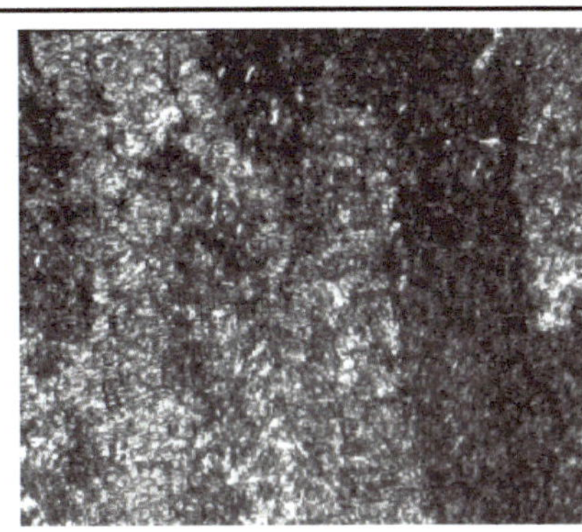 500 x
3. layer	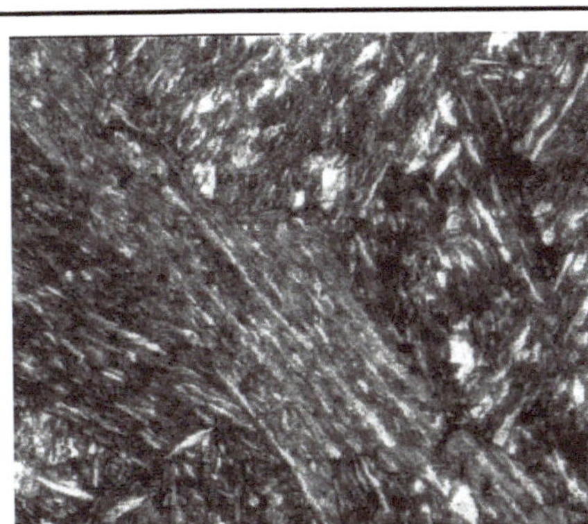 500 x	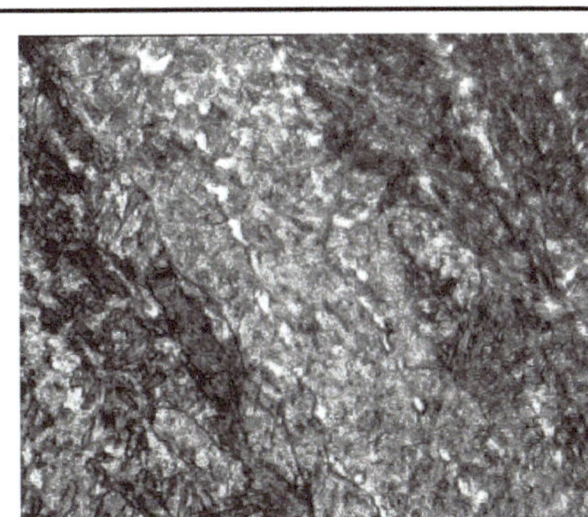 500 x

Fig. 7. Microstructure of all surface welded layers (Popovic et al.,2007).

5.3. Tensile tests

The tensile tests were conducted on a 2 mm thick specimens. The room temperature mechanical properties (ultimate tensile strength, UTS) of the surface welded joint are shown in Figure 8. The basic requirement in welded structures design is to assure the required strength. In most welded structures this is achieved with superior strength of WM compared to BM (overmatching effect), and in tested case this is achieved (Burzic & Adamovic,2008; Manjgo et al.,2010). The highest UTS is in weld metal of sample 2 (1210 MPa), due to solid state strengthening by alloying elements.

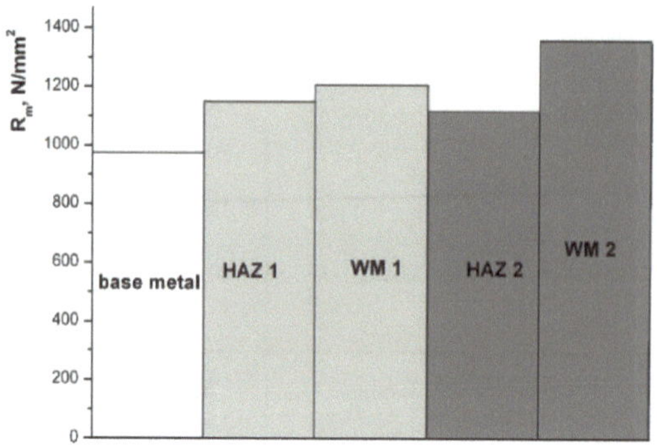

Fig. 8. Ultimate tensile strength of the surface welded joints (Popovic et al.,2011).

5.4 Impact testing

Impact testing is performed according to EN 10045-1, i.e ASTM E23-95, with Charpy V notched specimens, on the instrumented machine SCHENCK TREBEL 150 J. Impact testing results are given in Table 3 for base metal and HAZ at all testing temperatures. Total impact energy, as well as crack initiation and crack propagation energies, for weld metal of both samples at all testing temperatures (200C, -200C and -400C) are presented in Table 4 and in Figure 9.

The total energy of base metal is very low (5 J), due to very hard and very brittle cementite lamellae in pearlite microstructure (Popovic et al.,2011), while the toughness of HAZ is higher (11-12 J) and is similar for both samples at all testing temperatures.

Table 3. Instrumented impact testing results of Charpy V specimens for base metal and HAZ at all testing temperatures.

	Total impact energy, E_u, J		
	20ºC	-20ºC	-40ºC
base metal	5	3	3
sample 1-HAZ	12	11	10
sample 2-HAZ	11	10	9

Table 4. Instrumented impact testing results of Charpy V surface weld metal specimens at all testing temperatures.

	sample 1-WM			sample 2- WM		
	20°C	-20°C	-40°C	20°C	-20°C	-40°C
Total impact energy, E_u, J	29	23	17	34	14	11
Crack initiation energy, E_{in}, J	20	16	15	12	10	10
Crack propagation energy, E_{pr}, J	9	7	2	22	4	1

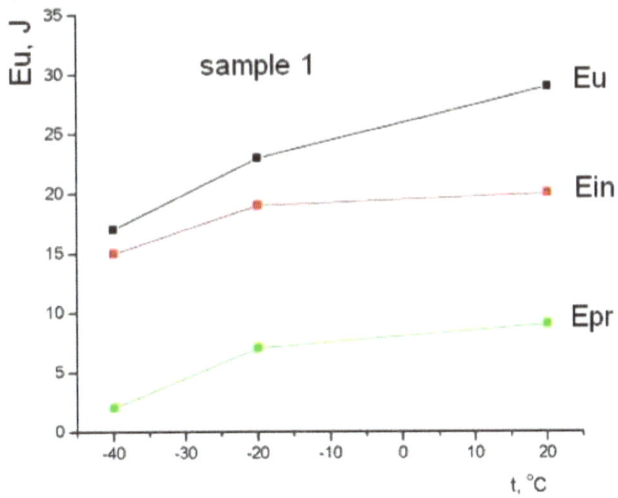

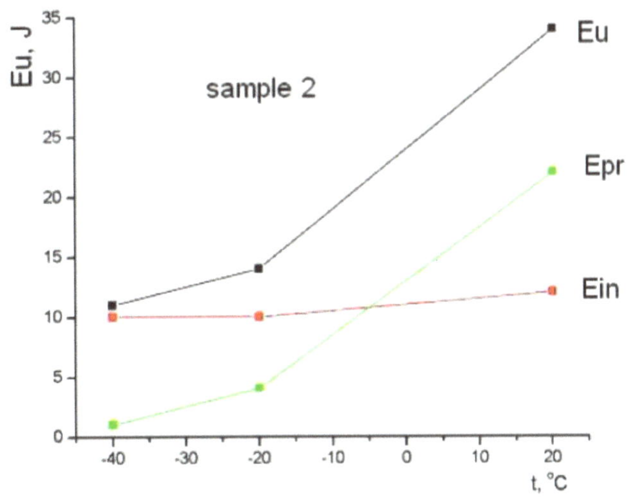

Fig. 9. Dependence total impact energy, crack initiation and crack propagation energy vs.temperature for (a) weld metal of sample 1 and (b)weld metal of sample 2 (Popovic et al.,2011).

Values of total impact energy of samples 1 and 2 at room temperature are significantly higher (29 J and 34 J) than in base metal (5 J), as a consequence of appropriate choice of alloying elements in the filler material. The presence of Ni, Mn and Mo promotes the formation of needled bainitic microstructure and grain refinements, and increases the strength and toughness also(Popovic, 2006). By analyzing the impact energy values of sample 1, a change of toughness in continuity is observed, with no marked drop of toughness, and for all tested temperatures, crack initiation energy is higher than crack propagation energy. This is the reason for the absence of significant decrease of toughness. The highest value of total impact energy was found for the sample 2 at room temperaure (34 J), which is the only case when the initiation energy is lower than propagation energy (12 J and 22 J, respectivelly). This shown practically the buffer layer function. Namely, the initiated crack during propagation comes to plastic buffer layer, which slows down crack further growth. For this reason, the crack propagation energy is the largest part of total impact energy. However, at -200C, significant drop of total impact energy is noticable (14 J) due to losing of buffer layer plastic properties at lower temperatures. The low-carbon wire (0.05%C i 1.4%Mn) has excellent toughness, but and marked rapid drop on S-curve (dependence toughness vs. temperature). Transition temperature of this material above - 200C is confirmed by the obtained impact toughness results. The use of buffer layer is reasonable if the exploatation temperature is above -50C; on the contrary, at lower temperatures, buffer layer is losing its function and the toughess is decreased (Popovic et al.,2011).

Diagrams force-time, obtained by instrumented Charpy pendulum, are given in Figure 10. As can be seen, for the sample 1 the character of diagrams force-time changed little by lower temperature. Namely, this material at room temperature has diagram with marked rapid drop, as consequence of unstable crack growth. After the maximum load, a very fast crack growth is started, and it is confirmed by the low value of crack propagation energy(Grabulov et al.,2008). On the contrary, on the sample 2 diagram at room temperature, the presence of buffer layer is clearly shown. The initiated crack, during its growth, comes to buffer layer which temporary stops the further crack growth and changes crack growth rate. The obtained experimental diagram doesn't belong to any type, according to standard EN 10045-1. This leads to toughness increase, primarily crack propagation energy, and it is also here the only case when the crack initiation energy is lower than crack propagation energy.

t, °C	sample 1	sample 2 (BL)
20°C	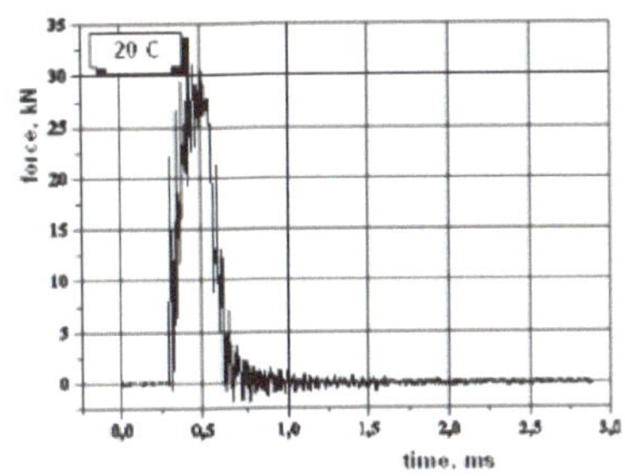	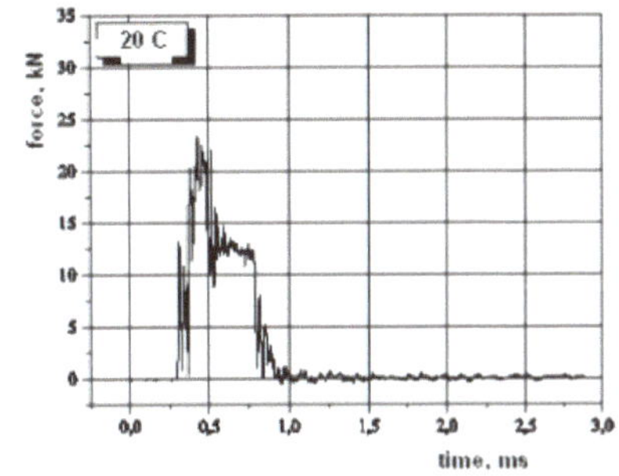
-20°C	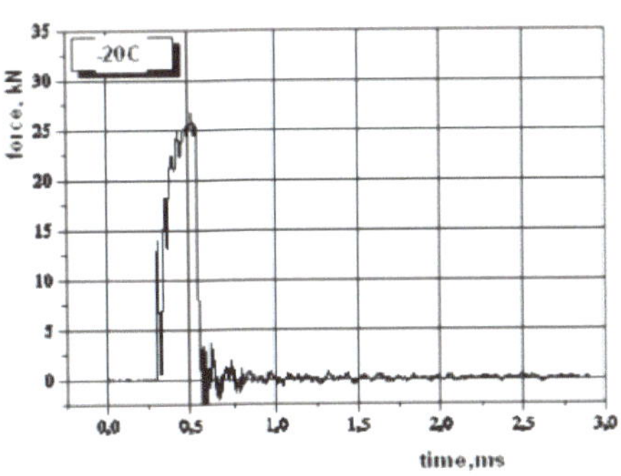	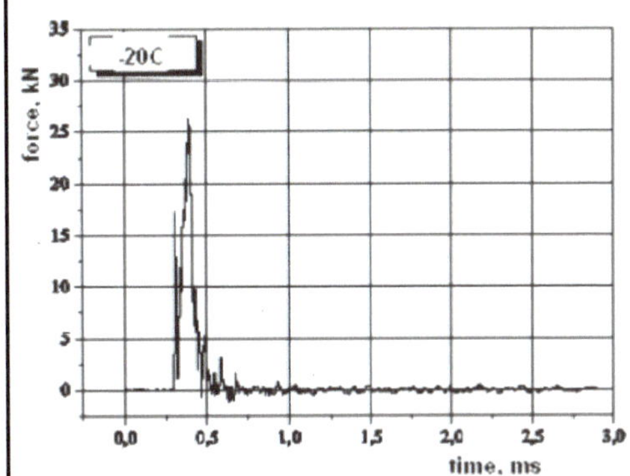
-40°C	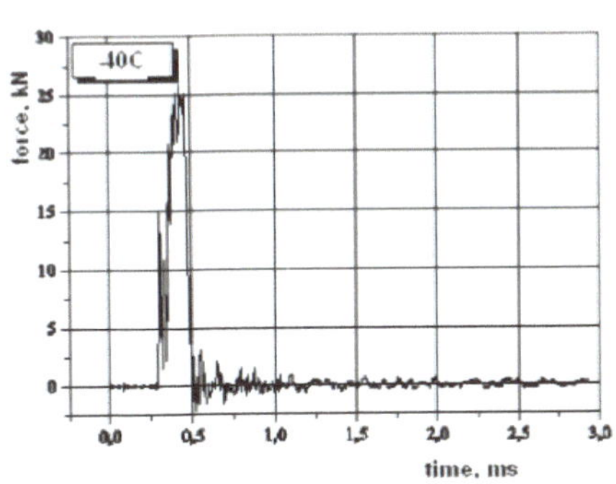	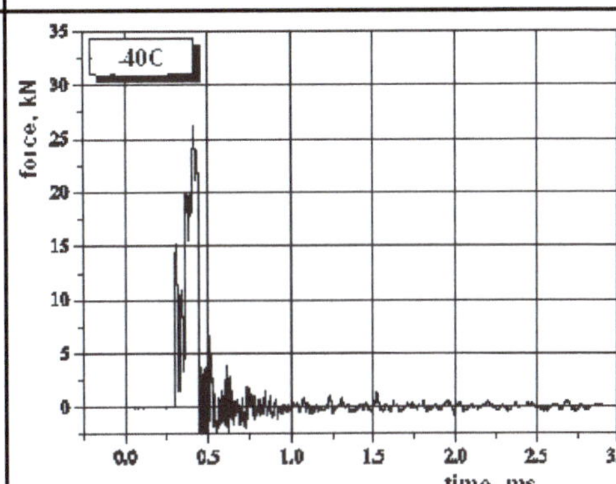

Fig. 10. Diagrams force-time, obtained by instrumented Charpy pendulum for sample 1 and sample 2 (Popovic et al.,2011).

5.5 Crack growth rate

A basic contribution of fracture mechanics in fatigue analysis is the division of fracture process to crack initiation period and the growth period to critical size for fast fracture (Burzic & Adamovic,2008). Fatigue crack growth tests had been performed on the CRACKTRONIC dynamic testing device in FRACTOMAT system, with standard Charpy size specimens, at room temperature, and the ratio R=0.1. A standard 2 mm V notch was located in base metal and in third layer of WM, for the estimation of parameters for BM, WM and HAZ, since initiated crack will propagate through those zones. Crack was initiated from surface (WM) and propagated into HAZ, enabling calculation of crack growth rate *da/dN* and fatigue treshold ΔK_{th}. The results of crack growth resistance parameters, i.e., obtained relationship *da/dN* vs. ΔK for base metal, sample 1 and for sample 2 are given in Figure 11 and 12. Parameters *C* and *m* in Paris law, fatigue threshold ΔK_{th} and crack growth rate values are given in Table 5 for all samples as obtained from relationships given in Figures 11 and 12, for corresponding ΔK values.

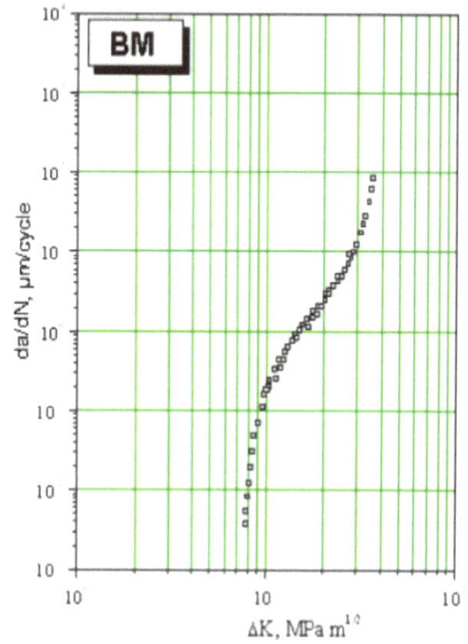

Fig. 11. Diagram da/dN vs. ΔK for base metal.

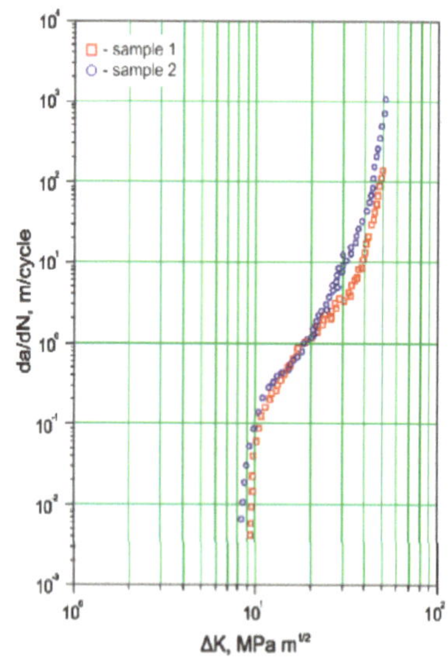

Fig. 12. Diagram da/dN vs. ΔK for sample 1 and sample 2.

The behaviour of welded joint and its constituents should affect the change of curve slope in validity part of Paris law. Materials of lower fatigue-crack growth rate have lower slope in the diagram d*a*/d*N* *vs.* ΔK. For comparison of the properties of surface welded joint constituents the crack growth rates are calculated for different values of stress-intensity factor range ΔK.

Table 5. Parameters C, m, ΔK_{th} and crack growth rate values for all zones of surface welded joints.

	Zone of surface welded joint	Fatigue threshold ΔK_{th}, MPa $m^{1/2}$	Parameter C	Parameter m	Crack growth rate da/dN, m/cycle			
					ΔK=10 $MPam^{1/2}$	ΔK=15 $MPam^{1/2}$	ΔK=20 $MPam^{1/2}$	ΔK=30 $MPam^{1/2}$
Base metal	BM	8.0	$3.31.10^{-11}$	3.28	$6.31.10^{-08}$			
sample 1	WM 1	9,5	$4.45.10^{-13}$	3.74		$1.11.10^{-08}$	-	-
	WM 2		$3.78.10^{-13}$	3.61		-	$1.88.10^{-08}$	-
	HAZ		$4.07.10^{-13}$	3.79		-	-	$1{,}61.10^{-07}$
sample 2	WM 1	8,9	$4.63.10^{-13}$	3.87		$1.65.10^{-08}$	-	-
	WM 2		$3.85.10^{-13}$	3.88		-	$2.07.10^{-07}$	-
	HAZ		$3.76.10^{-13}$	3.93		-	-	$1.18.10^{-06}$

Bearing in mind that weld metal consists of two layers (third layer is used for V notch), as referent values of ΔK were taken: ΔK =10 MPa m1/2 for BM, ΔK =15 MPa m1/2 for WM1, ΔK =20 MPa m1/2 for WM2, and ΔK =30 MPa m1/2 for HAZ. It's important that all selected values are within a middle part of the diagram, where Paris law is applied. The crack growth rate in base metal is 3-4 times higher than in both weld metal layers, i.e. the growth of initiated crack will be slower in weld metal layers. This means that for the same value of stress intensity factor rang ΔK, base metal specimen needs less number of cycles of variable amplitude than weld metal specimen, for the same crack increment.

In all three zones of surface weleded joint (WM2, WM1 and HAZ), sample 2 with buffer layer has higher crack growth rate than sample 1, i.e. the growth of initiated crack will be slower in sample 1. This means that for the same value of stress intensity factor rang ΔK, specimen of sample 2 needs less number of cycles of variable amplitude than specimen of sample 1, for the same crack increment. The maximum fatigue crack growth rate is achieved in HAZ for both samples, when stress intensity factor range approaches to plane strain fracture toughness.

If a structural component is continuously exposed to variable loads, fatigue crack may initiate and propagate from severe stress raisers if the stress intensity factor range at fatigue threshold, ΔK_{th}, is exceeded(Burzic & Adamovic,2008). Fatigue treshold value ΔK_{th} in base metal (ΔK_{th} =8 MPa m1/2) is lower than fatigue treshold value ΔK_{th} in weld metal of both metal. Fatigue treshold value ΔK_{th} for sample 2 (ΔK_{th} =8.9 MPa m1/2) is lower than that for sample 1 (ΔK_{th} =9.5 MPa m1/2). This means that crack in sample 2 will be initiated earlier, i.e. after less number of cycles, than in sample 1.

Values of fatigue threshold and crack growth rates corespond to initiation and propagation energies in impact testing, and in this case, good corelation is achieved (Popovic, 2006). Sample 1 has higher crack initiation energy (20 J) and higher ΔK_{th} ((ΔK_{th} =9.5 MPa m1/2 for sample 1 i ΔK_{th} =8.9 MPa m1/2 for sample 2). With comparation of crack propagation energy and crack growth rate, it is hard to establish the precise analogy, as toughness was estimated for the surface weld metal, whereas crack growth rate for each surface welded layer. Generally, buffer layer didn't show slow the initiated crack growth, with aspect of crack growth rate, while this effect is obvious in the case of toughness, i.e. crack propagation energy (Popovic et al, 2011).

6. CONCISE SUMMARY

Based on the acquired experimental results and their interpretation, the following conclusions are drawn:

The experimental analysis of surface welded connections utilizing various welding techniques has demonstrated, as anticipated, notable disparities in their mechanical characteristics performance. However, in both instances, it was demonstrated that despite the inferior weldability of high carbon steel, successful welding is achievable. Structural compatibility between the deposited metal and the base metal was attained, and a martensitic layer was not formed. The obtained HAZ exhibits superior structure compared to the basic metal.

The filler material is relevant parameter which affects on deposite layer quality. Work with self-shielded wires is more simple, specially for outdoor applications. Both used wires are on high technological level and can be recomended for reparation of high- carbon steel damaged parts. Final microstructure is the result of different influences: type of filler material, heat input, degree of mixture with previous layer and post heat treatment with subsequent surface layer. It is necessary to know all these factors and also to know the way of affect. Though applied wires are with different alloying concepts, result in both cases is that initial pearlitic moprhology is replaced by final desirable bainitic microstructure. It was shown that, by selecting corresponding parameters, it is possible to obtain the morphology of the best properties.

The maximal hardness level of 350-390 HV is reached in surface welded layers of both samples, with equal hardness of base metal (250-300 HV). The main difference appears in the first deposition layer, where as expected, in sample 2 the hardness is significantly lower (buffer layer). The obtained hardness values ensure simultaneously the improvement of mechanical and wear properties, and in the case of a rail, represents maximal hardness preventing the wheel wear (Popovic et al., 2010). Similar results are obtained by tensile testing. Sample 2 has slightly higher ultimate tensile strength (1360 MPa) than sample 1 (1210 MPa) due to solid solution strengthening by alloying elements.

The most improved results are obtained for impact properties. The toughness of base metal is 6-7 times lower than the toughness of weld metal, and more than twice lower than toughness of HAZ. For welding with buffer layer, at -200C, the drop of total impact energy is significant, due to lowering of buffer layer plastic properties at lower temperatures. The transition temperature of this material is above -200C, and it was confirmed by obtained impact toughness results. The use of buffer layer is beneficial for exploatation temperature above -50C. On the contrary, at lower temperatures, buffer layer loses its function and toughess decreases. On the contrary, for sample 1 the change of toughness is continous and without marked drop of toughness. At all tested temperatures, the crack initiation energy is higher than crack propagation energy. This may be the reason for the absence of significant decrease of toughness and that should be kept in mind during design and exploitation.

The results show that base metal is characterized by a lower fatigue threshold than weld metal, i.e. a 3 to 4 times higher crack growth rate. This means that crack will initiate more likely in base metal, and that it requires fewer cycles to reach the critical size. Contrary to a typical welded joint, a surface welded layer is the safest place for crack initiation.

Values of fatigue threshold and crack growth rates corespond to initiation and propagation energies in impact testing. In the case of fatigue treshold and crack initiation energy, good correlation was achieved. Sample 1 has higher crack initiation energy (20 J) and higher ΔK_{th} (9.5 MPa m1/2) than sample 2 (12 J and ΔK_{th} =8.9 MPa m1/2). On the contrary, buffer layer didn't show decrease of initiated crack growth rate, as this effect is obvious in the case of toughness, i.e. crack propagation energy. Since the constructions from high-carbon steel are used at low temperature, and bearing in mind the extended working time, in modern surface welding technologies, the use of buffer layer is not recommended.

The testing findings of the base metal and surface welded layer exhibit the characteristic behavior of two steel microstructures: pearlitic and bainitic, as corroborated by microstructural analysis. It has been demonstrated that, through the judicious selection of filler material and welding process, surface welding of damaged components serves not only as a method of repair but also as a means of enhancing initial qualities..

REFERENCES

1. A.Kapoor et al,(2002). Managing the critical wheel/rail interface, *Railway Gazette International.*, No.1, 25-28, ISSN 0373-5346
2. A.Kapoor et al.,(2001). Tribology of rail transport, In: *Modern Tribology Handbook*, vol.34, 2001.
3. Aglan, H.A., Liu, Z.Y., Hassan, M.F., Fateh, M.,(2004). Mechanical and fracture behavior of bainitic rail steel, *Journal of Materials Processing Technology*,Vol.151,No.1-3, 268-274, ISSN 0924-0136
4. Bhadeshia,H.K.D.H.,(2002). Novel steels for Rails, In:Encyclopedia *of Materials: Science and Technology*, H. K. D. H Bhadeshia, pp. 1-7, Elsevier Science,
5. Burzić, M., Adamović, Ž., (2008). Experimental analysis of crack initiation and growth in welded joint of steel for elevated temperature, *Materials and technology*,Vol.42, No.6, 263-271, ISSN 1580-2949
6. Cannon , D.F., Edel, K.O., Grassie, S.L.,Sawley,K.,(2003). Rail defects: an overview, *Fatigue and Fracture of Engineering Materials and Structures*, Vol.26, No.10, 865-887, ISSN 8756-758X
7. Clayton P,(1994). Tribological aspects of wheel/rail contact: A review of recent experimental research, *Proceedings of the 4th International Conference of contact mechanics and wear of rail/wheel systems*, Vancouver;1994
8. Eden, H.C., Garnham, J.E., Davis, C.L., Influential microstructural changes on rolling contact fatigue crack initiation in pearlitic rail steels, *Materials Science and Technology*,Vol.21, No.6, 623-629, ISSN 0267-0836
9. Grabulov, V.,Blačić I.,Radović A, Sedmak, S., (2008). Toughness and ductility of high strength steels welded joints, *Structural integrity and life*, Vol.8, No.3, 181-190, ISSN 1451-3749
10. Ishida M, Abe N., (1996). Experimental study on rolling contact fatigue from the aspect of residual stress, *Wear*, Vol.191, No.1-2, 65-71, ISSN 0043-1648
11. Lee, K., Polycarpou,A.,(2005). Wear of conventional pearlitic and improved bainitic rail steels, *Wear*, Vol.259,No.1-6, 391-399, ISSN 0043-1648
12. Lee,K. (2001). Increase productivity with optimized FCAW wire, *Welding design & Fabrication*, Vol.74, No.9, 30, ISSN 0043-2253
13. Manjgo, M., Behmen,M., Islamović,F., Burzić, Z., (2010). Behaviour of cracks in microalloyed steel welded joint, *Structural integrity and life*, Vol.10, No.3, 235-238, ISSN 1451- 3749
14. Muders, L., Rotthauser, N., Anspruche an moderne Schienenwerkstoffe, *Proceedings of Internationaled Symposium Schienenfehler*, Brandenburg, Germany, 2000

15. Pointner, P., Frank,N.,(1999). Analysis of rolling contact fatigue helps develop tougher rail steels, *Railway Gazette International*, No.11, 1999, 721-725, ISSN 0373-5346

16. Popovic, O., (2006).Ph.D.Thesis, University of Belgrade, Faculty of Mechanical Engineering Popovic, O., Prokic-Cvetkovic, R., Sedmak, A., Buyukyildrim, G., Bukvic, A., (2011). The influence of buffer layer on the properties of surface welded joint of high-carbon steel, *Materials and technology* Vol.45, No.5, 33-38, ISSN 1580-2949

17. Popovic, O., Prokic-Cvetkovic, R., Sedmak, A., Grabulov, V., Burzic, Z., Rakin, M., (2010). Characterisation of high-carbon steel surface welded layer, *Journal of Mechanical Engineering* Vol.56, No.5, 295-300, ISSN 0039-2480

18. Popovic, O., Prokic-Cvetkovic, R., Sedmak, A., Sijacki-Zeravcic, V., Bakic, G., Djukic, M.,(2007). The influence of filler material on microstructure of high-carbon steel surface welded layer, *Proceedings of 11th International research/ expert conference TMT 2007*, ISBN 978-9958-617-34-8, Hammamet, Tunisia, September 2007, 1491-1494

19. Popović, O., Prokić-Cvetković,R., Grabulov, V., Odanović, Z.,(2006). Selection of the flux cored wires for repair welding of the rails, *Welding and welded structures*, Vol. 51, No. 4, 131-139, ISSN 0354-7965

20. Sadler, H., (1997). Including welding Engineer, *Welding design & Fabrication*, Vol.70, No.6, 74, ISSN 0043-2253

21. Schleinzer, G., Fischer, FD., (2000). Residual stresses in new rails, *Material Science Engineering A*, Vol. 288, No.2, 280-283, ISSN 0921-5093

22. Schleinzer, G., Fischer, FD., (2001). Residual stress formation during the roller straightening of railway rails, *International Journal of Mechanical Science*, Vol.43, No.10, 2281-2295, ISSN 0020-7403

23. Singh, U.P., Roy, B.,Jha,S., Bhattacharyya, S.K.,(2001). Microstructure and mechanical properties of as rolled high strength bainitic rail steels, *Materials Science and Technology*, Vol.17, No.1, 33-38, ISSN 0267-0836

24. Vitez, I., Krumes,D.,Vitez,B.,(2005) UIC-recommendations for the use of rail steel grades, *Metalurgija*, Vol.44, No.2, 137-140, ISSN 0543-5846

25. Zerbst, U., Madler, K., Hintze, H., (2005). Fracture mechanics in railway applications-an overview, *Engineering Fracture Mechanics*, Vol.72, No.2, 163-194, ISSN 0013-7944

Chapter 11

INVESTIGATION OF THIXOTROPIC PLASTIC DEFORMATION IN MAGNESIUM MATRIX COMPOSITES

1. INTRODUCTION: AN OVERVIEW

Magnesium alloys possess numerous advantages in mechanical and physical qualities, including low density, high specific strength, excellent heat conductivity, and effective dampening capabilities. They are extensively developed for automotive, aerospace, electronic, optical instruments, and similar applications. The damping capacity of materials is essential for controlling structural vibrations, reducing noise pollution, and enhancing the fatigue characteristics of components under cyclic stress. Magnesium and its alloys exhibit superior thermodynamic stability, aging stability, and damping capacity; yet, their uses are constrained due to inadequate mechanical characteristics. Creep is a significant property of the mechanical behavior of metals at elevated temperatures, defined as the gradual plastic deformation occurring under constant temperature and sustained load over an extended duration. Creep is a crucial metric for assessing the quality of materials at elevated temperatures in industrial sectors such as the automotive and aerospace industries. The strength and creep resistance of magnesium alloys (AZ91D and AM60) rapidly diminish as the temperature exceeds 150C°. The utilization of magnesium alloys is constrained by limitations such as inadequate strength, hardness, and creep resistance. It is essential to produce magnesium matrix composites (MMCs) that exhibit high strength, toughness, and excellent creep resistance, along with their associated forming process. Particle-reinforced magnesium matrix composites are distinguished by their low cost and straightforward processing, making them a focal point of research in the field of metal matrix composites [Hai et al., 2004]. Nevertheless, magnesium exhibits a low melting point, significant chemical reactivity, and a propensity for flammability, hence complicating the preparation of magnesium matrix composites to some degree. Consequently, it is essential to pursue an improved manufacturing technique for magnesium matrix composites [Zhou et al., 1997]. Powder metallurgy (PM) and casting are prevalent techniques for producing these composite materials. The PM technique requires sophisticated equipment with elevated costs and is incapable of fabricating large-sized and intricate MMC components. It presents risks including powder combustion and detonation. Conversely, the casting method can yield large-sized composites (up to 500 kg) in industrial mass production due to its straightforward procedure and ease of operation, requiring minimal investment in equipment and incurring cheap costs. Numerous researchers are currently examining MMCs produced by the casting process [Kang et al., 1999].

The plastic formability of metal matrix composites (MMCs) is inadequate, necessitating the implementation of an enhanced forming technique. The advancement of semi-solid forming technology has introduced thixoforming as a novel technique for magnesium matrix composites. The technology for creating semi-solid materials offers benefits including reduced deformation resistance and enhanced material mobility [Flemings 1991, Yan et al., 2005]. It consisted of three processes: semi-solid billet production [Yan et al., 2005], partial remelting [Yan et al., 2006], and thixoforming [Yan et al., 2008]. This item examines the fundamental theory of the semi-solid stirring melting production method and the thixoforming process for advanced metal matrix composites (MMCs). The research encompasses the examination of the semi-solid stirring melting manufacturing technique [Yan & Fu et al., 2007; Yan & Lin et al., 2008] and the reheating process [Yan & Zhang et al., 2008; Zhang et al., 2011] for particle-reinforced metal matrix composites (MMCs). A material constitutive relation will be proposed [Yan & Wang et al., 2011]. A finite element model integrated with multiple physical fields will be constructed. The simulation will be eliminated based on the established analysis program. The forming behaviors and deformation laws in thixoforming of particle-reinforced metal matrix composites will be examined by a combination of theoretical analysis and experimental methods [Yan & Huang, 2011]. The results will significantly contribute to establishing the theoretical and practical foundation for the thixoforming process of particle-reinforced metal matrix composites in industrial applications.

2. STUDY ON FABRICATION METHODS AND VARIOUS PROPERTIES FOR MAGNESIUM MATRIX COMPOSITES

2.1 Fabrication methods

The AZ61 alloy was used as the matrix material. The chemical composition of AZ61 was 5.8%7.2%Al, 0.15% Mn, 0.40%1.5%Zn, 0.10% Si, 0.05%Cu, 0.05%Ni, 0.005%Fe, and the rest is Mg. Its solidus temperature was 525^{oC}, and the liquidus temperature was 625^{oC}. The reinforcement was green -SiC particle whose average diameter was 10*m* . A self- manufactured electric resistance furnace was used for melting Mg alloy (shown in Fig.1). The liquid metal was stirred with the mechanical stirrer driven by the timing electrical machine in the melting process. In order to improve the accuracy of controlling temperature, the thermocouple was inserted directly into the liquid metal, and combined with the artificial aptitude modulator BT608 that adopted the industrial micro-processor, whose precision was only 10^{oC}. The MMCs specimens were sampled by the pipette connecting to a vacuum pump in this experiment.

The fabrication processes of SiCp/AZ61 composites were described as follow. AZ61 alloy matrix was heated to melt, gas was gotten rid of and slag was removed. Then SiC reinforcement was added into the molten. There were the three addition processes. In the fully-liquid stirring casting process (about 680^{oC}), SiC particles were introduced into the fully-liquid molten, and then sampled after stirred to reach predetermined time. (2) In the stirring-melt casting process (590^{oC}), SiC particles were introduced at the semi-solid state, then sampled after reached to a fully-liquid temperature of 680^{oC}. (3) In semi-solid stirring casting process, SiC particles were introduced at the semi-solid state, then sampled after stirred to reach predetermined time. In above experiments, their volume fractions of SiC particles were 3%, 6%, and 9% respectively, whose preheated temperature was 500^{oC} with holding time 2h. The stirring rate was 500r/min with holding time 10min. Then the specimen was made to the metallurgical phase sample and corrupted with 0.5% ammonium HF liquor, and its microstructural changes were observed under the optical microscope. Finally, the Vickers hardness was measured in a micro-sclerometer HXS-1000AK.

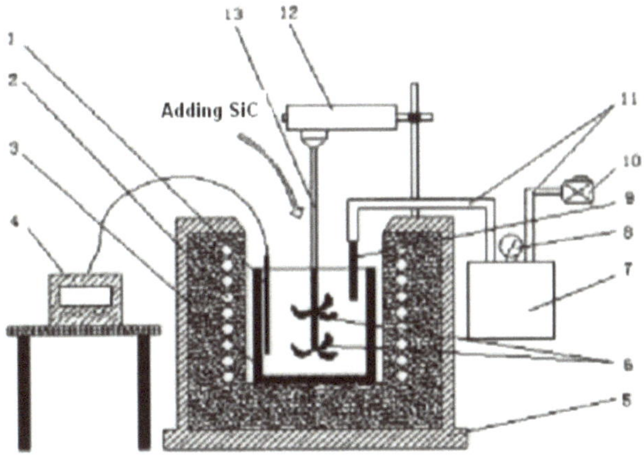

Fig. 1. Schematic diagram of SiCp/AZ61 composites fabricated in stirring casting process 1. thermocouple 2. resistance thread 3. crucible 4. BT608 (artificial aptitude modulator) 5. resistance furnace 6. strring lamina 7. vacuum jar 8. pressure meter 9. pipette 10. vacuum pump 11. guiding windpipe 12. timing electrical machine 13. stirring bar

The microstructures of SiCp/AZ61 composites in three casting processes were shown in Fig.2. The variations of influence of three casting processes on the microstructures of SiCp/AZ61 composites were shown in Fig.2- Fig.4. The distribution of SiC particles was a little uniform in the fully-liquid casting process where a lot of gas cavities and slacks were presented, and SiC particles were easy to sink and float. There were a few gas cavities in the semi-solid casting process where the distribution of SiC particles was inhomogeneous. SiCp/AZ61 composites fabricated by the stirring-melt casting method possessed not only few gas cavities but fairly uniform distribution.

The existence of gas cavities and slacks was attributed to the following factors: (1) Gas was involved in the molten during the mechanical stirring process. (2) Their non-uniform volume shrinkages presented in the composites solidification process due to the differences of their thermal expansion coefficient and heat conduction between matrix and reinforcement. (3) Hydrogen produced in the chemical reactions between Mg and H_2O was dissolved in the molten and formed gas cavities during solidification. (4) The formation of gas cavities was resulted from the particles clustering.

The main problem in the stirring casting process was the inhomogeneous distribution of reinforcement phase. The major reasons were followed as. (1) Due to having the different densities between matrix and reinforcement, SiC particles were settled down. (2) The higher surface tension and poor wettability between SiC and matrix presented, a few SiC particles were floated on the surface of the molten.

SiC particles were introduced at the semi-solid state during the semi-solid stirring casting process where the high viscosity semi-solid alloy can help withstand SiC particles from sinking and floating, but the uniform distribution can not be solved (shown in Fig.3). During the stirring-melt casting process the reinforcement were added at the semi-solid state, and the composites were poured immediately after reached 690^{oC} (liquidus). The quite uniform SiCp/AZ61 composites can be obtained in this method (shown in Fig.4).

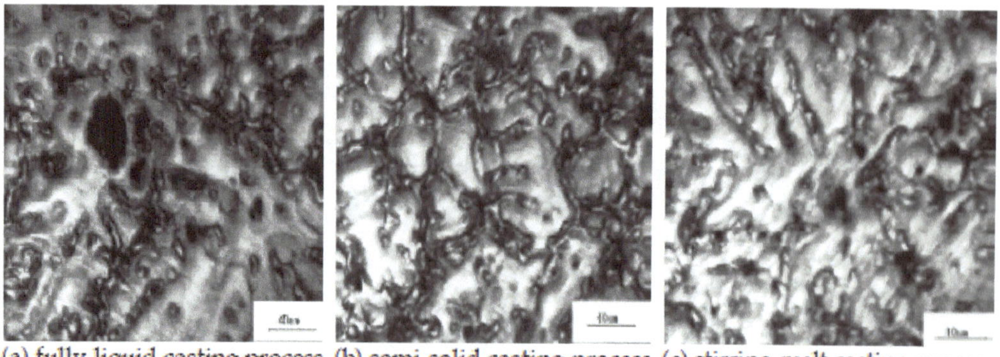

(a) fully-liquid casting process (b) semi-solid casting process (c) stirring-melt casting process

Fig. 2. Microstructures of SiCp/AZ61 composites in three casting processes

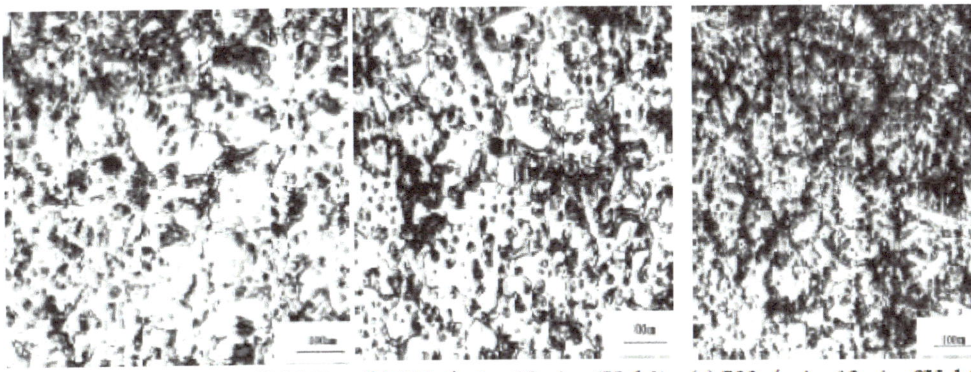

(a) 500r/min, 10min, 3Vol.% (b) 500r/min, 10min, 6Vol.% (c) 500r/min, 10min, 9Vol.%

Fig. 3. Microstructures of SiCp/AZ61 composites with various volume fractions of SiC particles in semi-solid stirring casting process

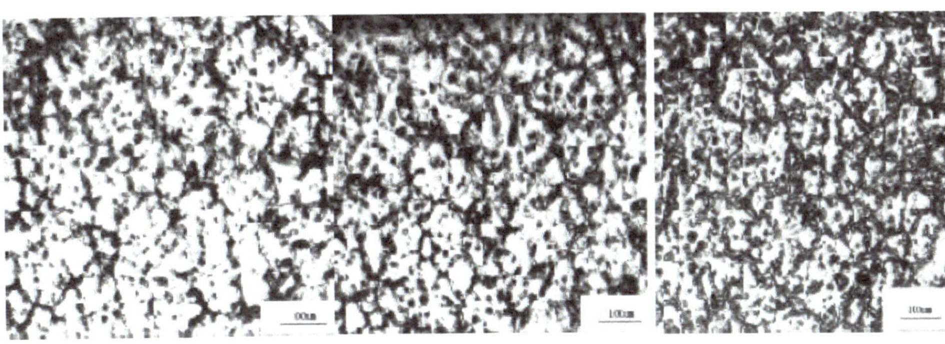

(a) 500r/min, 10min, 3Vol.% (b) 500r/min, 10min, 6Vol.% (c) 500r/min, 10min, 9Vol.%

Fig. 4. Microstructures of SiCp/AZ61 composites with various volume fractions of SiC particles in stirring-melt casting process

2.2 Optimization on stirring melt casting process

In this study, the composites were fabricated by a stirring melt casting method. The effects of volume fraction of SiC particles, stirring temperature and stirring time on the mechanical properties and microstructure of SiCp/AZ61 composites were investigated. The main technological parameters of preparing SiCp/AZ61 composites were optimized, which was helpful for obtaining its good properties.

The effects of volume fraction of SiC particles, stirring temperature and stirring time on the mechanical properties of SiCp/AZ61 composites were investigated by an orthogonal experimental method, in which average particle size and stirring speed were maintained the same. The orthogonal test table with three factors and three levers is shown in Table 1. According to design of the primary experiment, volume fractions of SiC particles were 3%, 6% and 9%, stirring temperatures were 580^{oC}, 587^{oC} and 595^{oC}, and stirring times were 3 min, 5 min and 7 min. Three factors

level

were volume fraction of SiC particles, stirring temperature and stirring time. Two targets were tensile strength and elongation.

Table 1. Factors and levels of test

level	volume fraction of SiC A (*%)	stirring temperature B (ºC)	stirring time C (min)
1	3	580	3
2	6	587	5
3	9	595	7

The effects of volume fraction of SiC particles, stirring temperature and stirring time on the tensile strength and elongation at room temperature of SiCp/AZ61 composites are shown in Table 2. (1) Tensile Strength Analysis.

Table 2. The table of three factors and three levels in the orthogonal experiment

No.	volume fraction of SiC A	stirring temperature B	stirring time C	tensile strength /MPa	elongation /%
1	1	1	1	172	3.8
2	1	2	2	179	4.2
3	1	3	3	163	5.2
4	2	1	2	184	3.5
5	2	2	3	176	4.1
6	2	3	1	189	3.9
7	3	1	3	164	1.3
8	3	2	1	153	1.9
9	3	3	2	170	2.1
I	171.3	173.3	171.3		
II	183.0	169.3	177.7		
III	162.3	174.0	167.7		
R (tensile strength)	20.7	4.7	10.0		
I	4.4	2.9	3.2		
II	3.8	3.4	2.1		
III	1.7	3.7	3.5		
R (elongation)	2.7	0.8	1.4		

The level two (the volume fraction of SiC 6%) was the best among the levels of factor A. The level three (595ºC) was the best among the levels of factor B. The level 2 (5 min) was the best among the levels of factor C. Thus the optimum combination was $A_2B_3C_2$. (2) Elongation analysis. The level two (the volume fraction of SiC 3%) was the best among the levels of factor A. The level 3 (595ºC) was the best among the levels of factor B. The level three (7 min) was the best among the levels of factor C. Thus the optimum combination was $A_1B_3C_3$. (3) Range Comprehensive Analysis. The greater the Range (R), the greater the effect of the lever change of the factor on the test target. This factor was more important. From Table 2, the sequence of tensile strength was R_A>R_C>R_B. So the sequence of primary and secondary in factors of A, B, C was the volume fraction of SiC particles, stirring time and stirring temperature. The sequence of elongation was R_A>R_C>R_B. So the sequence of primary and secondary in factor of A, B, C was also the volume fraction of SiC particles, stirring time and stirring temperature. Besides, from the ranges(R) in table 1, factor A has a more notable impact for tensile strength and elongation, and other factors do not have great impact. After comprehensive analysis, a better combination of factors was $A_2B_3C_2$, namely, the optimum processing plan of SiCp/AZ61 composites in the experimental condition was volume fraction of SiC particles 6%, stirring temperature 595^{oC} and stirring time 5 min.

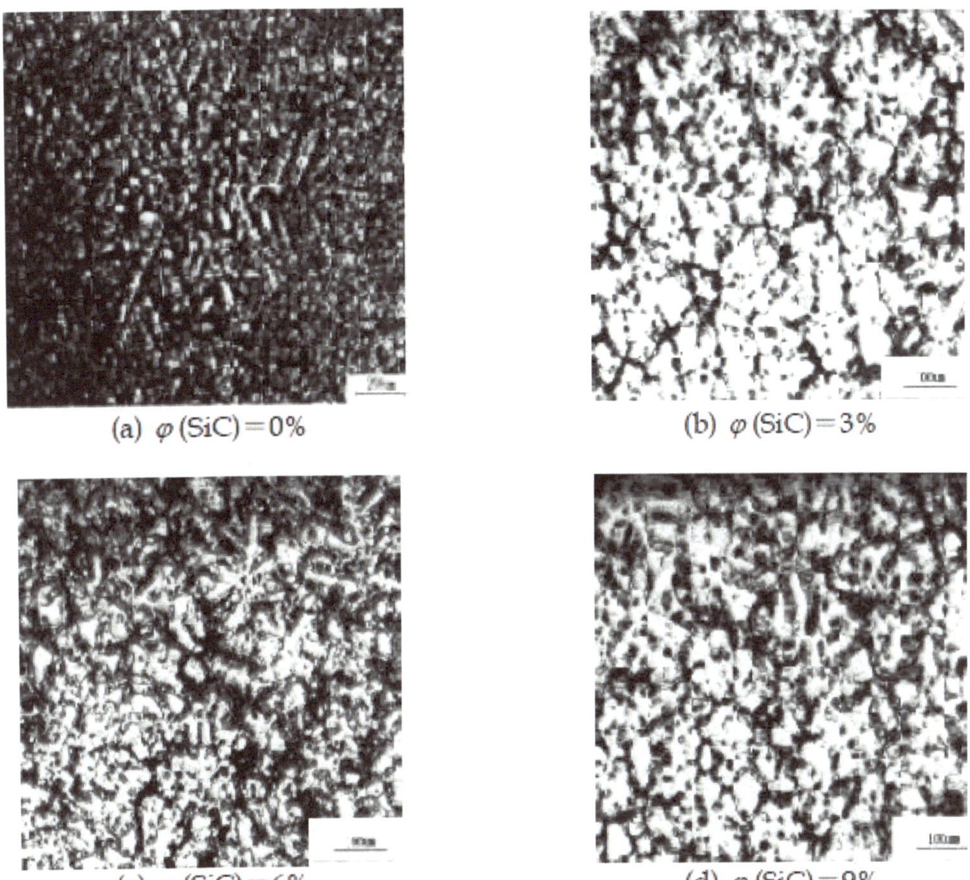

Fig. 5. Microstructures of SiCp/AZ61 composites with various volume fractions of SiC particles

The test results showed that the tensile strength of SiCp/AZ61 composites, that increased as the increasing of volume fraction of SiC particles increasing, and were higher than that of AZ61 (about 165MPa). The tensile strength was up to the maximum 189MPa when the volume fraction of SiC particles was 6%. However, the tensile strength decreased as the volume fraction of SiC particles increased continuously. Comparison with the volume fractions of SiC particles, the change trend of elongation decreased gradually with addition of SiC particles. The reason was a mass of rigid second phase existence in the matrix of SiCp/AZ61 composites, which could improve its rigidity and tensile strength. With the increasing volume fraction of SiC particles, problems of particle packing, agglomerating and clustering were presented in the matrix (Fig. 5), which caused tensile strength to decrease. Decreasing elongation was due to the non-uniform distribution of SiC particles and weak cracks in boundaries between the reinforcement and matrix.

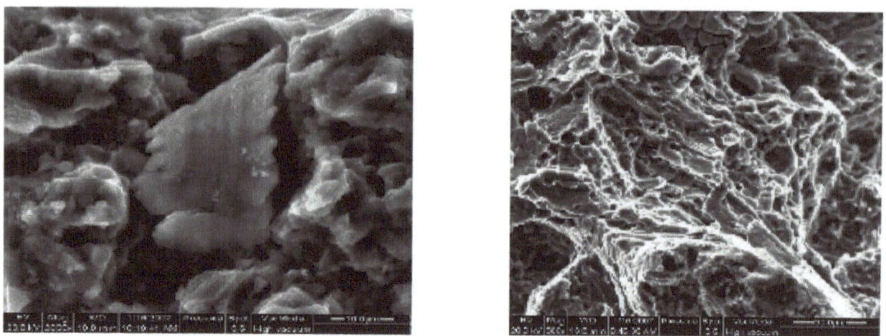

(a) AZ61 matrix (b) SiCp/AZ61 composites

Fig. 6. SEM of tensile fracture surfaces

The fracture morphology of AZ61 matrix at the ambient temperature is shown in Fig.6a. Ductility dimples existed, and cleavage cracks were present in a part. Fig.6b showed the fracture morphology of SiCp/AZ61 composites at the ambient temperature by a better processing plan ($A_2B_3C_2$). Compared with the fracture morphology of the matrix, the fracture morphology of SiCp/AZ61 composites at the ambient temperature were brittle where the fractured SiCp particles were found (Fig. 7a).

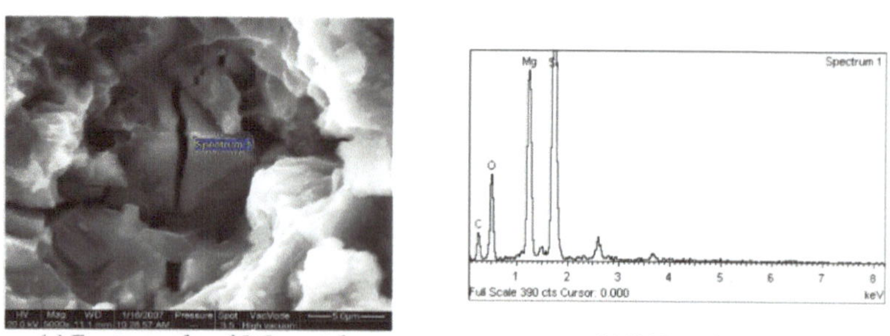

(a) Fractographs of SiC particle (b) EDS analysis

Fig. 7. Fractographs of SiC particle and EDS analysis

3.SEMI-SOLIDISOTHERMALHEATTREATMENTTECHNOLOGYFOR THE PARTIAL REMELTING OF COMPOSITES

In this study, semi-solid isothermal heat treatment technology was used for the partial remelting of composites. The round semi-solid microstructure had been obtained by controlling the reheating processing parameters such as heating temperature and isothermal holding time. The law of microstructural evolution in the remelting process of SiCp/AZ61 composites was investigated, which was expected to offer some theoretical references for the design of thixoforming technology.

The reheating temperatures were taken as 590^{oC}, 595^{oC}, 600^{oC} and 610^{oC} respectively with isothermal temperature heat treatment times of 15min, 30min and 60min. When the scheduled time and temperature were reached, the specimen was taken out and water quenched. Then the specimens were made and etched with 4% nitic acid liquor, and its microstructure change was observed under the optical microscope. The Image-pro Plus software was used to measure the diameter of equal-area of microstructure. The average radius of grain microstructure was then calculated.

According to the Sheil equation (1) and equation (2), the liquid phase volume fraction of the partial remelting structure was calculated (shown in Fig.8).

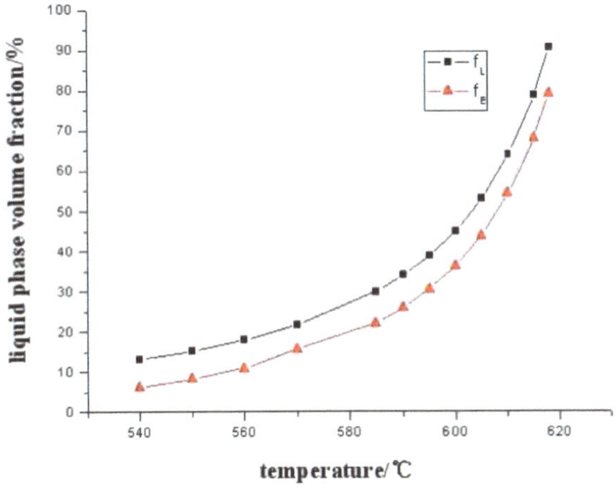

Fig. 8. Relationship between liquidphase volume fraction and temperature

$$f_L = \left(\frac{T_m - T}{T_m - T_L} \right)^{-1/1-K_0} \tag{1}$$

$$f_E = f_S(1 - f_P) - f_P \tag{2}$$

Where *fL* , *fE* and *fP* represent the liquid phase volume fraction of matrix alloy, effective liquid phase volume fractions of composites and enforcing particles. *Tm* and *TL* are melting point of the pure metal and the liquidus temperature of the alloy. K_0 is represented for the coefficient of distribution.

Fig. 9. shows the microstructural evolution of SiCp/AZ61 composites during partial remelting. When the heating temperature reached 590^{oC} with isothermal holding time of 15min, the grain boundaries had almostly been merged and could not be seen clearly. At the same time, SiC particles were inside the grains away from the grain boundaries (Fig. 9a). A separating tendency in the grains of coalescence emerged with the prolongation of isothermal holding time (Fig. 9b). While the holding time reached 60 min, a few grain boundaries became clear. A few globular grains appeared with SiC particles presented in the grain boundaries, but the liquid volume fraction was lower (Fig.9c). When the reheating temperature increased to about 595^{oC} with holding time of 15 min, the grain microstructure evolved quickly, and a globular microstructure appeared, then the eutectic structure began to melt (Fig.9d). The grain boundaries appeared completely with holding time 30 min, and fine globular grains emerged. The effective liquid fraction of SiCp/AZ61 composites was about 31%, and the mean diameter of grains was approximately 60μm (Fig.9e). When the isothermal holding time was further increased to 60min, the grain microstructure was entirely spheroidized, which became more clear and round, and SiC particulate returned to the grain boundaries from interior of grains (Fig.9a,b). At the same time, the mean diameter of grains was about 85μm (Fig.9f). As the reheating temperature increased to 600^{oC} with holding time of 15 min, the microstructural evolution of the sample during remelting was rapid. Some of grains began to spheroidize (Fig.9g). When the holding time reached 30 min, all grains had been spheroidized, whose sizes became relatively fine (Fig.9h). With the prolongation of holding time to 60 min, the grain microstructure tended to spheroidize and increase in size, and the effective liquid fraction was about 37% (Fig.9i). When the reheating temperature was above 610^{oC}, the semi-solid microstructure began to dissolve and disappear. The specimens were susceptible to serious deformation, the liquid flow emerged from the sample, which would prevent semi-solid microstructure from partial remelting (Fig.9j). Therefore the optimal technological parameters of SiCp/AZ61 composites were the reheating temperature of 595^{oC}~600^{oC} and isothermal holding time of 30min~60min.This temperature interval was suitable for semi-solid thixoforming of SiCp/AZ61 magnesium matrix composites.

The microstructures of SiCp/AZ61 composites during partial remelting (Fig.9e, f) were compared with that of AZ61 alloy (Fig.10). It was observed that the microstructures of SiCp/AZ61 composites coalescenced basically before isothermal holding time at the predetermined temperature for 15min, and a separating tendency in the grains didn't appear obviously. After isothermal holding at the predetermined temperature for 25min, the grain microstructure began separating and spheroidizing. However the rate of separation and spheroidization for AZ61 alloy increased. When the reheating temperature reached 595^{oC} with holding time of 0min,the grain microstructure was separated completely, and a few globular grains had appeared. With the prolongation of holding time from 20min to 40min, the mean diameter of the globules was 85μm and 110μm respectively. In addition, compared with AZ61 alloy, the microstructures of SiCp/AZ61 composites were finer during partial remelting due to addition of SiC particulates. Coalescence was restricted since the globules were isolated one with respect to the other by the presence of SiC particulates. At the same time the effective diffusion coefficient of the liquid phase was also reduced because of the presence of reinforced particulates, and during the subsequent isothermal holding process coalescence of α phase was hindered, and Ostwald ripening was also restricted.

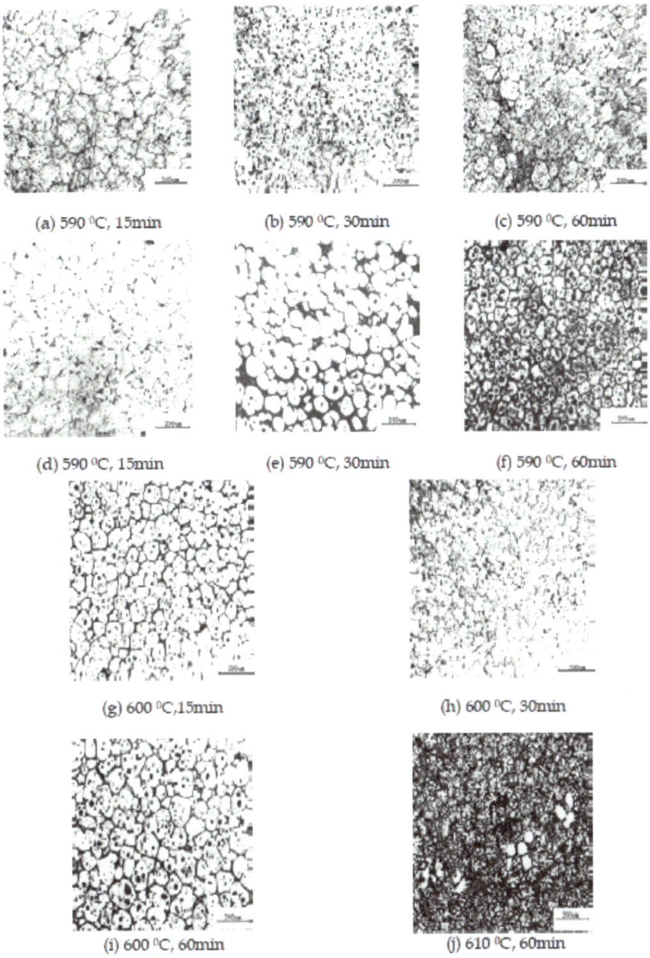

Fig. 9. The microstructural evolution of SiCp/AZ61 composites during partial remelting

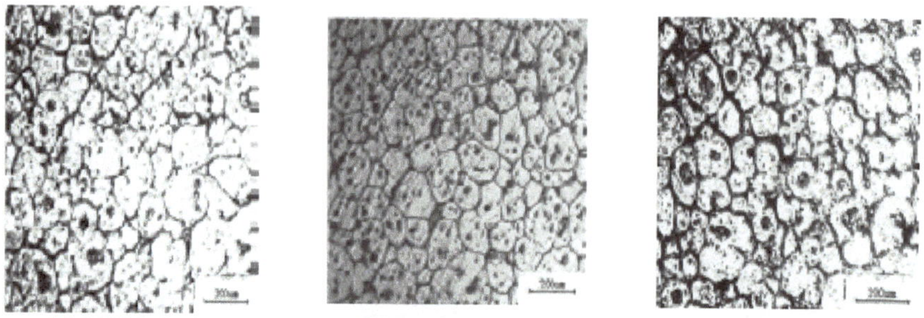

(a) 595 ^{0}C, 0min (b) 595 ^{0}C, 20 min (c) 595 ^{0}C, 40 min

Fig. 10. Microstructures of semi-solid AZ61 magnesium alloy billet during partial remelting

4. THIXOTROPIC COMPRESSION DEFORMATION BEHAVIOR OF COMPOSITES

The characteristics of semi-solid composites deformed mechanism can be understood well only when the relationships between stress and strain are described. So the semi-solid compression tests for SiCp/AZ61 composites were conducted, whose mechanical properties and destruction model were investigated.

The experiments were conducted in a Thermecmastor-Z dynamic material testing machine, whose set-up was shown in Fig.11. The specimen was heated by electromagnetic wave, whose temperature was monitored by thermocouples. The graphite slices were placed between the specimen and the compression heads for reducing the influence of friction on experiment. In order to study and master the characteristic mechanics of semi-solid magnesium matrix composites at high solid volume fractions, the deformation temperatures were taken as 530^{oC}, 545^{oC}, 560^{oC} and 570^{oC} respectively. According to the heating procedure shown in Fig.12, the initial heating rate was 10^{oC}/s; when the specimen temperature reached 500^{oC}, the temperature rate was down to 1^{oC}/s. Then the semi-solid compression experiments were done under the strain rates of 0.1s-1, 0.5 s-1, 5.0 s-1and 10 s-1 respectively, in which the total strain was 0.6.

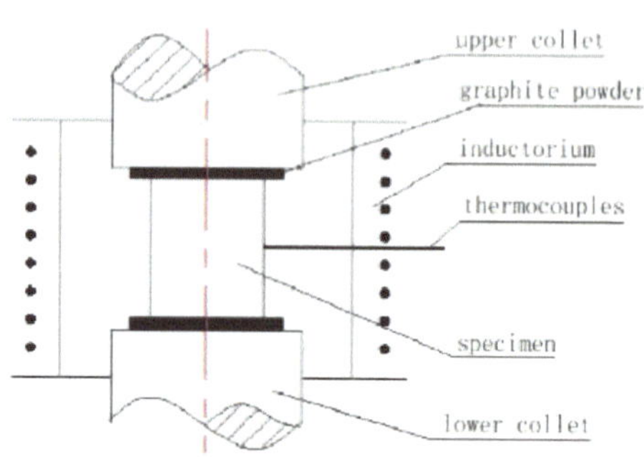

Fig. 11. Schematic diagram for the compressive tests

The stress-strain curves of semi-solid SiCp/AZ61 composites with various volume fractions of SiC particles are shown in Fig.13. The tendency of curves implies that the deformation temperature has a significant effect on the flow stress. It is observed that for a constant strain rate and constant volume fraction of SiC in the composites, the flow stresses and peak stresses decrease with the increasing of deformation temperature, which presents that the thixotropic plastic deformation of the composites is highly sensitive to temperature. The tendency is thought to be the result of variation of volume fractions of solid α phase. When the specimens has high solid volume fractions, the solid grains contact with each other and form a net, sliding and rotation of grains become hard. The plastic deformation of solid particles is the main mechanisms. With the increasing of temperature, the solid volume fractions decrease, and the solid grains are surrounded by liquid phase, which makes the solid grains to slid and rotate easily. Thus the sliding and rotation of grains plays a more significant role in the thixotropic plastic deformation.

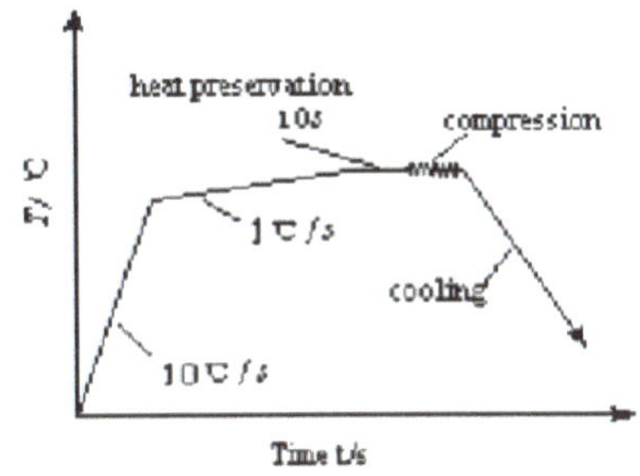

Fig. 12. Heating processing

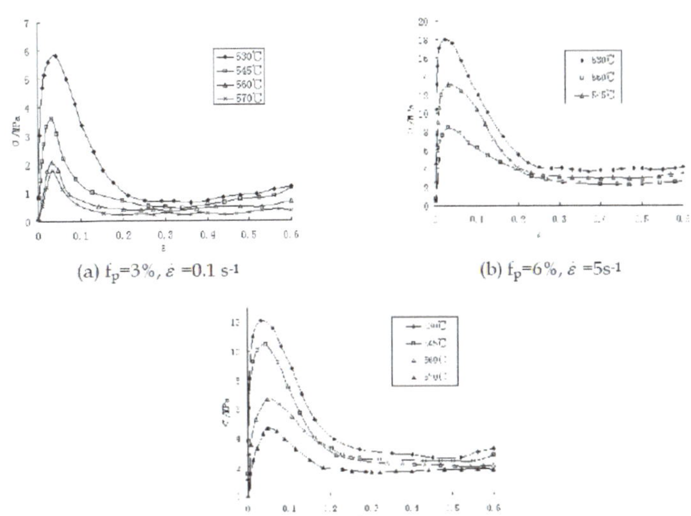

(a) f_p=3%, $\dot{\varepsilon}$ =0.1 s^{-1}

(b) f_p=6%, $\dot{\varepsilon}$ =5s^{-1}

(c) f_p=9%, $\dot{\varepsilon}$ =10s^{-1}

Fig. 13. Curves of stress-strain relation at various temperatures for SiCp/AZ61 composites

Relationships between peak stress and temperature of composites with strain rate of 10s-1 are shown in Fig.14. It can also be seen that the variation of volume fractions of SiC has a significant effect on the peak stress of the composites. When the deformation temperature raises slightly higher than the solid phase line in semi-solid zone, the volume fractions of liquid phase is low in the matrix, and the peak stress decreases rapidly almost as linear form with the increasing of temperature initially, and then slows as the temperature increases further. The results are thought to be of the lower volume fractions of liquid phase. When the specimens has high volume fractions of solid phase , the solid grains contact with each other to form a net, the sliding and rotation of grains become hard. The plastic deformation of solid particles is the main mechanisms. Besides SiC particles increase the resistance to grain boundaries sliding and impose barriers to dislocation motion, which leads to higher resistance for plastic deformation. With the increasing of temperature, the volume fractions of liquid phase increase, and the solid grains are surrounded by liquid phase, which requires smaller force for the solid grains to slid and rotate. Thus the peak flow stress decrease rapidly, and then slowed as the temperature increases further.

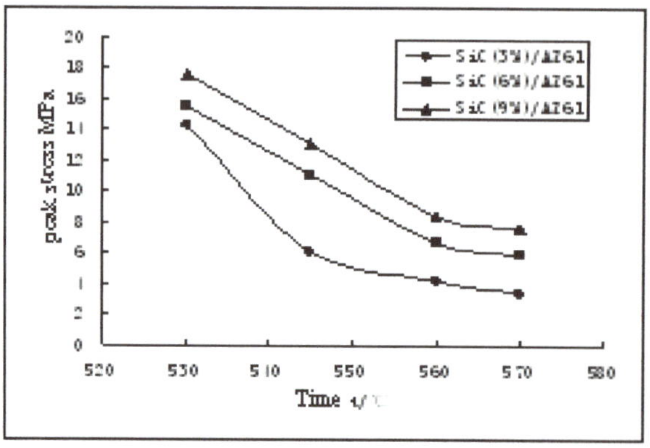

Fig. 14. Relationships between the peak stress and temperature of three SiCp/ AZ61composites in semi-solid thixotropic compression

The relations of stress-strain rate at various strain rates are shown in Fig. 15. It can be seen that the peak stress increases as the strain rate increases at the constant temperature. When the composites are compressed at high strain rate (for example 10 s-1), the liquid phase can not be squeezed out timely, in which the flow stress is very high during the initial deformation stage, and then decreases as the result of high shearing rate. With the further deformation, the liquid is squeezed out and pushed together, in which the solid particles are smashed. During this stage, the grain boundaries sliding and flow become easy and the flow stress decreases rapidly. At lower strain rate the solid grains are surrounded by liquid phase, which requires smaller force for sliding and rotation. So the flow stress is lower.

Fig. 16 presents stress-strain curves of the SiCp/AZ61 composite with different SiC fractions at 545^{oC}and 560^{oC} and constant strain rate of 0.1s-1 and 10s-1. The fractions of SiC have a significant effect on the flow stress. The compression stress increases with the increasing of volume fractions of SiC particles. The reason is that SiC particles are mainly located in the inter-granular and boundary regions in the composites. The SiC particles impose barriers to dislocation motion and resistance to the solid grains sliding during the steady-state compressive deformation. So it increases the resistance for dislocation and grain boundaries sliding with the increasing of volume fractions of SiC particles.

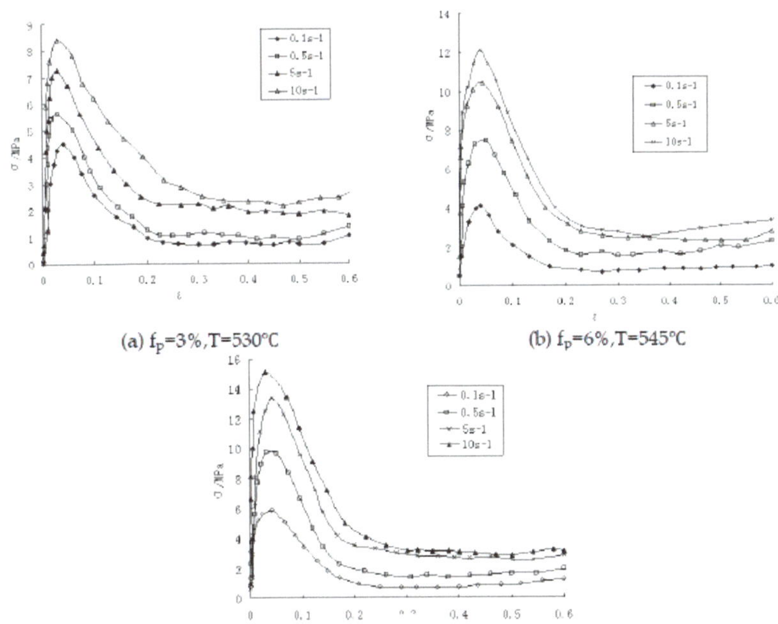

(a) f_p=3%,T=530°C (b) f_p=6%,T=545°C

(c) f_p=9%,T=570°C

Fig. 15. Curves of stress-strain relations at various strain rates for SiCp/AZ61 composites

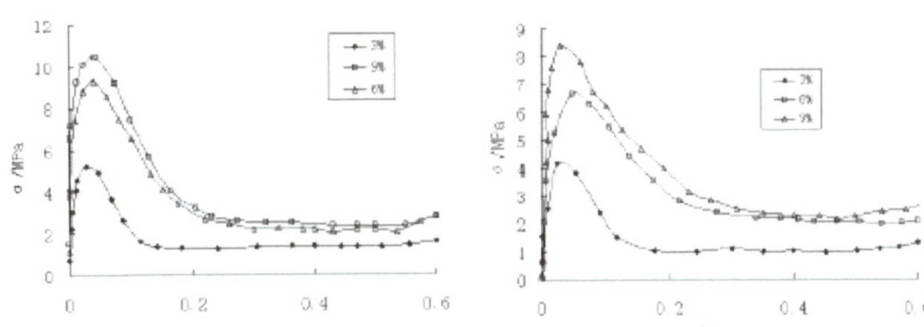

(a) T=545°C, ε =0.1s^{-1} (b) T=560°C, ε =10s^{-1}

Fig. 16. Curves of stress-strain relation at 545^{oC}and 560^{oC} for SiCp/AZ61 composites with different SiC fractions

The appearances of specimens compressed at semi-solid state are shown in Fig. 17. In can be seen that the surface longitudinal cracks of specimen happen at compression ratio of 20% and the liquid phase is squeezed out along the cracks. When the compression ratio reaches up 30%, the volume fractions of liquid phase squeezed out of the surface increases, which generates the mixed liquid-solid outer surfaces. The surface strength is very low and generates easily cracks. When all of liquid is squeezed out, the flow stress starts to ascend. At last the compressed specimen looks like popcorn.

(a) 0% (b) 20% (c) 30% (d) 60%

Fig. 17. Appearances of specimens compressed at semi-solid state

5.CONSTITUTIVEMODELFORTHIXOTROPICPLASTICFORMINGOF COMPOSITES

On the basis of analysis of behavior of thixotropic plastic deformation of composites in compression process, its constitutive model is established. Then the model parameters are determined using the multiple nonlinear regression method.

Based on the experimental analysis of axial compression of composites in semi-solid state, there is a certain non-liner relationship among stress Σ and strain rate z , strain z , temperature T, liquid phase rate *fL* , as well as the volume fraction of reinforcement *f p* [Yan & Wang, 2011]. At the same time Hong Yan present the constitutive relationship of semi- solid magnesium alloy as follow [Yan & Zhou, 2006]:

$$\sigma \propto \exp(1/T)\dot{\varepsilon}^{a_1}\varepsilon^{a_2}(1-\beta f_L)^{a_3} \tag{3}$$

Where Σ - stress, ε - strain, - strain rate, T - temperature, β - geometric parameters(β=1.5), f_L - liquid phase rate.

In the study of deformation behavior of composites under high strain rate [Bao & Lin, 1996] and [Li & Ramesh, 2000] found that the influence of volume fraction of reinforcement on the mechanical behavior of the material was present as following:

$$\sigma(f_p) = \sigma(\varepsilon,\dot{\varepsilon})\cdot g(f_p)\cdot[1+(\alpha\dot{\varepsilon})^m f_p] \tag{4}$$

Where Σ- stress, ε- strain, ε & - strain rate, (ε , ε) - function of strain and strain rate, fp - volume fraction of reinforcement.

So the constitutive model of thixotropic plastic deformation of composites reinforced with particles is proposed.

$$\sigma = \exp(d/T)\cdot\varepsilon^{n}\cdot\dot{\varepsilon}^{m}\cdot[1-\beta f_L]^{a_1}\cdot g(f_p)\cdot[1+(\alpha\dot{\varepsilon})^{m}f_p]^{a_2} \quad (5)$$

Under assumption of $g(f_p)=e^{a+bf_p+cf_p^2}$ the constitutive model is established in the following form.

$$\sigma = \exp(a+bf_p+cf_p^2+d/T)\cdot\varepsilon^{n}\cdot\dot{\varepsilon}^{m}\cdot[1-\beta f_L]^{a_1}\cdot[1+(\alpha\dot{\varepsilon})^{m}f_p]^{a_2} \quad (6)$$

Where a, b, c, d, a_1, a_2 - constant, n—strain hardening index, m—strain rate sensitivity index,β—constant (β=1.5), α - correction coefficient , fp - volume fraction of reinforcement,

f_L - liquid phase rate, $f_L=(\frac{T_M-T_L}{T_M-T})^{\frac{1}{1-K}}$, T_M - the melting point of pure metal, T_L - liquidus temperature of alloy, k - balance coefficient.

The parameters in proposed constitutive model were determined by the multiple nonlinear regression method. The nonlinear equation is transformed into linear one using legarithms for Esq.(3).

$$\ln\sigma = a+bf_p+cf_p^2+d/T+n\ln\varepsilon+m\ln\dot{\varepsilon}+a_1\ln(1-\beta f_L)+a_2\ln[1+(\alpha\dot{\varepsilon})^{m}f_p] \quad (7)$$

Where

$$\begin{aligned} &y=\ln\sigma, X_1=f_p, X_2=f_p^2, X_3=1/T, X_4=\ln\varepsilon, X_5=\ln\dot{\varepsilon}, X_6=\ln(1-\beta f_L),\\ &X_7=\ln[1+(\alpha\dot{\varepsilon})^{m}f_p]\\ &A_0=a, A_1=b, A_2=c, A_3=d, A_4=n, A_5=m, A_6=a_1, A_7=a_2 \end{aligned} \quad (8)$$

Esq.(12) is changed as follow

$$y=A_0+A_1X_1+A_2X_2+A_3X_3+A_4X_4+A_5X_5+A_6X_6+A_7X_7 \quad (9)$$

Table 3 shows the common statistic values. The correlation coefficient R = 0.974, determination coefficient R^2 = 0.949, the adjustment determination coefficient R^2 = 0.931, the Std. Error of the Estimate S =0.0670. As the equation has a number of explained variables, the determination should be based on the adjustment determination coefficient R^2 . As can be seen from the output that $R^2$is close to 1, the fit degree is high. So the representativeness of proposed constitutive model is strong.

Table 3. Model Summary

Model	R	R Square	Adjusted R Square	Error of the Estimate
1	0.974(a)	0.949	0.931	0.0671449560

a Predictors:(Constant), $x_7,x_6,x_4,x_2,x_3,x_5,x_1$ *b Dependent Variable:y*

The analysis is listed in Table 4. The significant test of regression equation is based on the table. Total in Sum of Squares is 835.005, Regression in Sum of Squares and Regression in Mean Square are 413.512 and 59.073 respectively. Residual in Sum of Squares and Residual in Mean Square are 421.492 and 0.451 respectively. The test statistic observations F = 130.902. The concomitant probability p is approximately 0. The linear relationship between variables x and y is significant, which create a linear model.

Table 4. ANOVA (b)

Model		Sum of Squares	df	Mean Square	F	Sig.
1	Regression	413.512	7	59.073	130.902	0.000(a)
	Residual	421.492	934	0.451		
	Total	835.005	941			

a Predictors: (Constant),x7,x6,x4,x2,x3,x5,x1 b Dependent Variable:y

Table 5 shows the regression coefficient analysis. As can be seen from the table and the estimated value of the test results, the corresponding variable regression coefficient A0=- 8.27366, A1=50.158,A2=-296.555,A3 =14253.359,A4 =-0.053,A5=0.242,A6 =2.316,A7 =-0.505.

The concomitant probability p is 0, whose regression is a significant. From comparison of regression coefficients, those indicate that the constitutive model has a significant meaning. The sensitivity coefficient m A5=0.242 of strain rate resulted from regression is good close to the replaced m value.

Table 5. Coefficients (a)

Model		Unstandardized Coefficients	Standardized Coefficients	t	Sig.	
		B	Std. Error	Beta		
1	(Constant)	-8.27366	1.818		-9.728	0.000
	X_1	50.158	7.917	1.303	6.335	0.000
	X_2	-296.555	53.608	-0.934	-5.532	0.000
	X_3	14253.359	1402.791	0.342	10.161	0.000
	X_4	-0.053	0.012	-0.108	-4.601	0.000
	X_5	0.242	0.045	0.474	5.382	0.000

	X_6	2.316	0.553	0.143	4.186	0.000
	X_7	-0.505	0.463	-0.135	-1.091	0.000

a Dependent Variable: y

The analysis of the regression equation is a meaningful, and the following relation is got.

$$y = -8.27366 + 50.158X_1 - 296.555X_2 + 14253.359X_3 - 0.053X_4 + 0.242X_5 + 2.316X_6 - 0.505X_7 \quad (10)$$

From the inverse transform of equations (9) and (10), equation (6) becomes:

$$\sigma = \exp(-8.27366+50.158 f_p - 296.555 f_p^2 + 14253.359 / T) \cdot \varepsilon^{-0.053} \cdot \dot{\varepsilon}^{0.242} \cdot [1 - \beta f_L]^{2.316} \cdot [1 + (2.1 \times 10^4 \dot{\varepsilon})^{0.242} f_p]^{-0.505} \quad (11)$$

Equation (11) is a constitutive relationship of thixotropic plastic forming of SiCp/ AZ61 composites.

Fig.18 is the real stress test - a true strain curves and regression curve of the results of the comparison, Solid line is the experimental curve, dotted line is the calculation of one. The results calculated by multiple non-linear regression method are good agreement with experimental ones. So the proposed constitutive model has the higher forecast precision and practical significance.

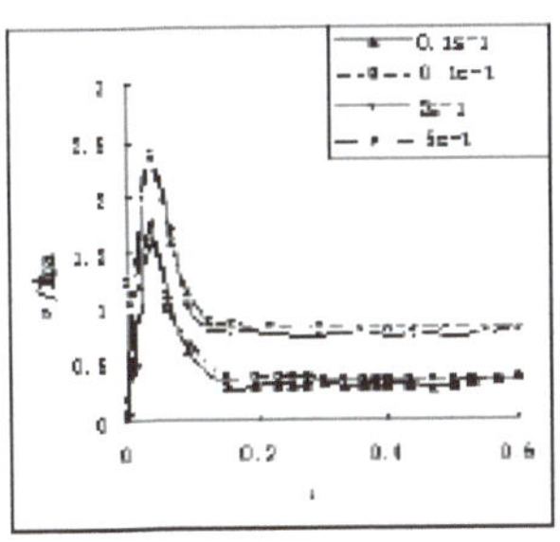

(a) 3vol.% SiCp/AZ61,570°C

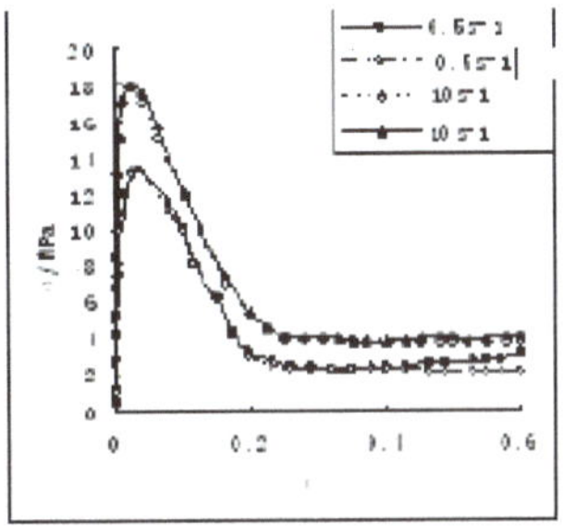

(b) 6vol.% SiCp/AZ61,560°C

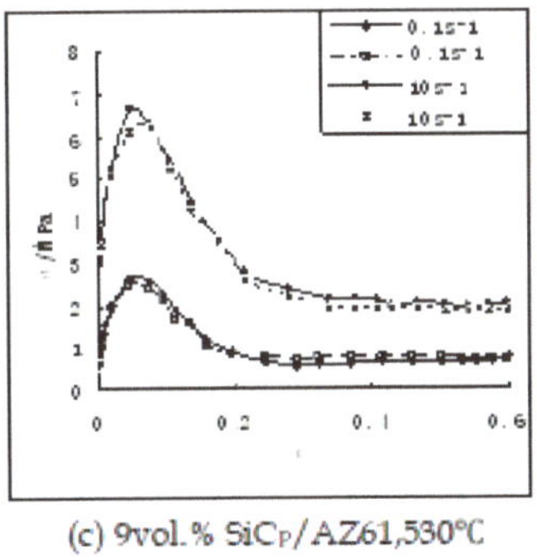

(c) 9vol.% SiCp/AZ61,530°C

Fig. 18. A comparison between true strain—stress curves of the test and regression curves

6.NUMERICALSIMULATIONFORTHIXOTROPICPLASTICFORMING OF COMPOSITES

To investigate thixoforming process with numerical simulation method, which is a nonlinear system, some assumptions are taken as follow: (1) The semi-solid material is assumed as a continuous and incompressible one. (2) The solid grains in semi-solid metal are uniformly distributed in liquid phase, and because of the large deformation in forming, the semi-solid material is considered as an isotropy uniform medium. According to the above assumptions, the material deformation in thixoforming is supposed as a rigid viscoplastic one.

The material adopted in this paper was SiCp/AZ61 composite, and the simulations were performed in thixo-forging and forging. The flow stress model of SiCp/AZ61 composite in thixo-forging is expressed as follow [Yan &Wang, 2011].

$$\sigma = \exp(-8.27366 + 50.158 f_p - 296.55 f_p^2 + 14253.359 / T)$$

$$\varepsilon^{-0.053} \cdot \dot{\varepsilon}^{0.242} \cdot (1 - \beta f_L)^{2.316} \cdot [1 + (2.1 \times 10^4 \dot{\varepsilon})^{0.242} f_p]^{-0.505} \quad (12)$$

where Σ is the stress; ε the strain; z *E* the strain rate; T temperature; β constant(β=1.5); *fp* is Volume fraction of SiC particle; *fL* is liquid volume fraction

For establishing material modal of SiCp/AZ61 composite in forging, true stress-strain curves at various temperature and strain rates were performed by mean of isothermal compression experiments.

In this study, the workpiece is formed by the close-forge method. The experiment set-up was shown in Fig.19. Fig.20 shows the workpiece, whose structure and flow character are complicated. Comparisons between forging and thixo-forging of the workpiece will be done and predicted in advance using numerical simulation. This is an effective method to instruct application of semi-solid forming technology into its practice production.

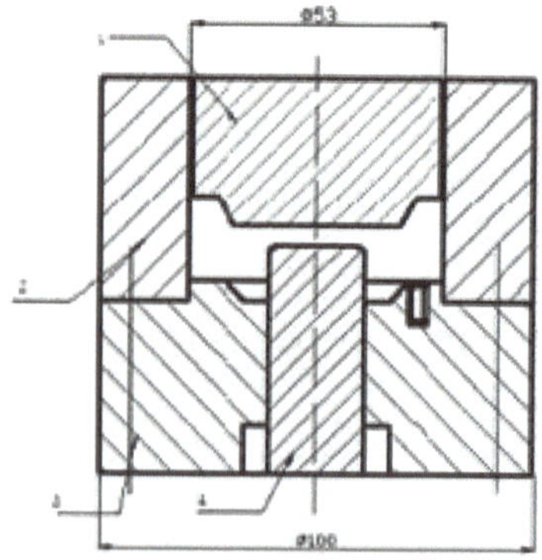

1 up-die 2 sleeve 3 down-die 4 mandril

Fig. 19. Experiment set-up

The same simulated parameters are used to analyze the differences of mechanics properties and flow rule between forging and thixo-forging processes. The materials are normal and semi-solid SiCp/AZ61 composite respectively. Environment temperature is 20^{oC}, warm-up temperature of the die is 320^{oC}. The friction model is constant shearing stress model, whose coefficient is 0.25. Billet size is ø50×18.5mm, which is meshed to 50000 tetrahedron elements. Stroke of up-die is 14mm.

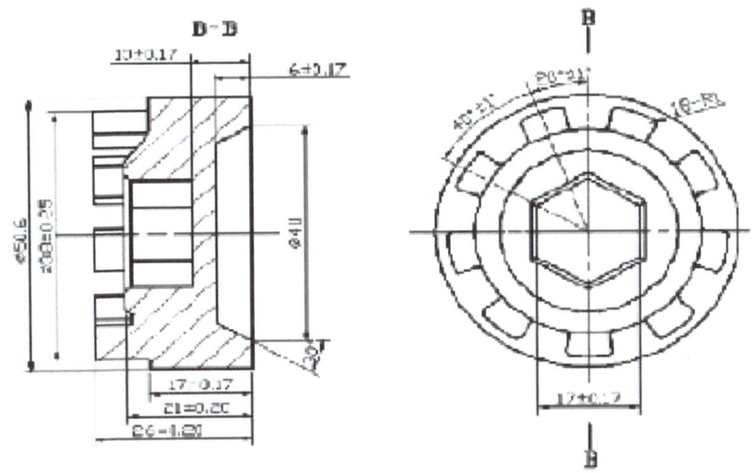

Fig. 20. SiCp/AZ61 composite workpiece

Fig.21 and Fig.22 give the filling stages simulated results in forging and thixo-forging processes respectively. Compared with the two kind of forming processes, it can be concluded that both had the basically identical deformation processes. In the initial stage, the hexagon hole in central section of workpiece was extruded and the rest moved in the rigid motion shown in Fig.21a, Fig.22a. As the stroke increased, metal deformation entered into the second stage, in which metal flowed from central to around in the extrusion pressure, and the cetral protruded and bottom platforms were formed (Seen Fig.21b, Fig.22b). In the last stage, the metal could be filled up claw easily in thixo-forging process, and could not be filled up claw in forging process (Seen Fig.21c, Fig.22c). Therefore, forging was more difficult in filling cavity than thixo-forging.

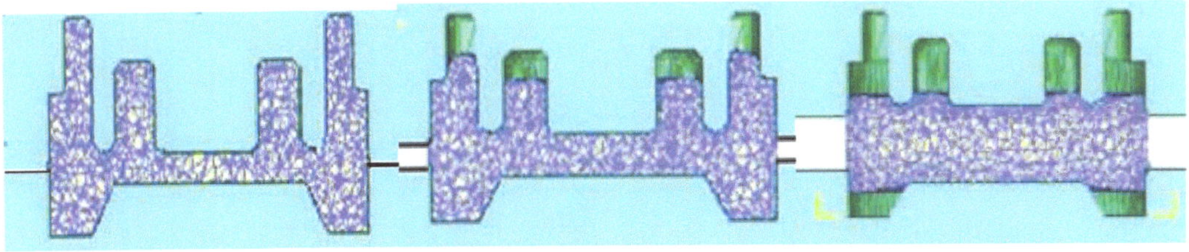

Fig. 21. Filling stages simulated results in thixo-forging process

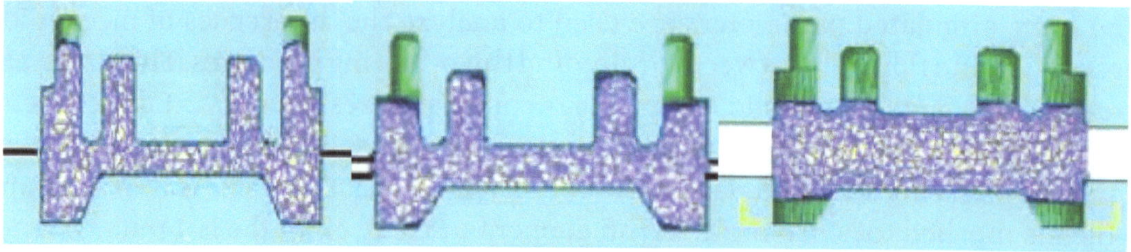

Fig. 22. Filling stages simulated results in forging process

Fig. 23. shows the effective stress distributions at different temperatures in thixo-forging process. The effective stress distribution was more uniform and its value was smaller with the increasing of forming temperature, which was contributed from the excellent fluidity of semi-solid composite.

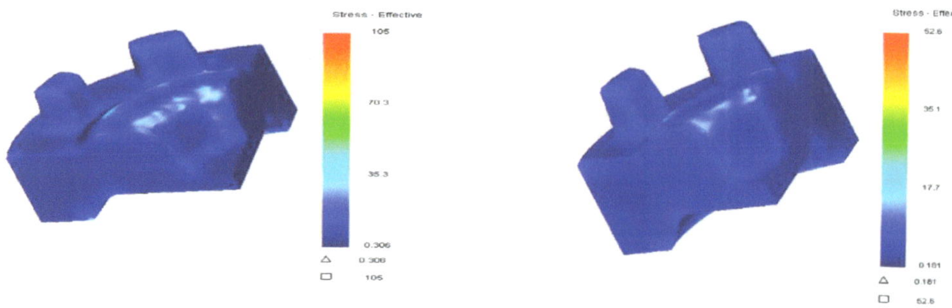

(a) 530°C (b) 560°C

Fig. 23. Effective stress distributions at different temperatures in thixo-forging process

Fig. 24. shows the effective stress distributions at different volume fraction of SiC particle in thixo-forging process. The effective stress was increased with the increasing of volume fraction of SiC particle.

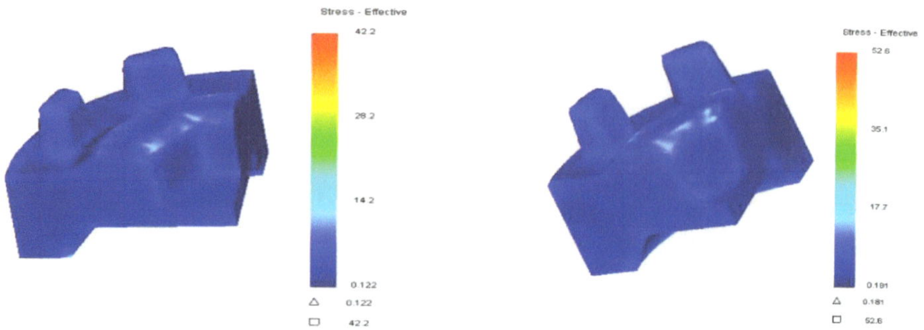

(a) Vol.3% SiC_P/AZ61, 560°C (b) Vol.6% SiC_P/AZ61, 560°C

Fig. 24. Effective stress distributions at different volume fraction of SiC particle in thixo- forging process

Fig. 25 shows temperature distributions at different volume fraction of SiC particle in thixo- forging process. When the volume fraction of SiC particle was 3%, the fluctuation period of temperature was 558~561^{oC}, whose changed value was small. When the volume fraction of SiC particle was 6%, the fluctuation period of temperature was 558~572^{oC}, whose changed value was more greater than that of the former. It could be gained that the temperature distribution in the latter was worse than that in the former.

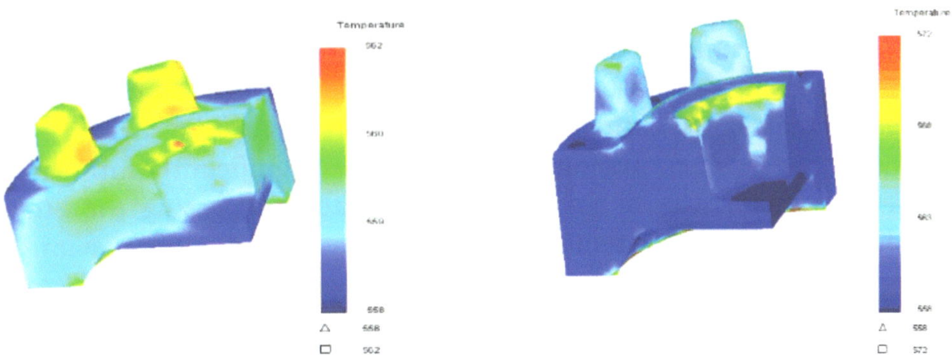

(a) Vol.3% SiC_P/AZ61, 560°C (b) Vol.6% SiC_P/AZ61, 560°C

Fig. 25. Temperature distributions at different volume fraction of SiC particle in thixo- forging process

Fig.26 shows the traditional forging and thixo-forging workpieces of SiCp/AZ61 composite. The thixo-forging has better fill effect and surface finish quality of workpiece than the traditional forging, which could achieve near-end deforming with high quality of workpiece in the former. Those coincide with the simulation results, which indicate that semi-solid SiCp/AZ61 composite has good flow property, and can be used to form complicated workpiece.

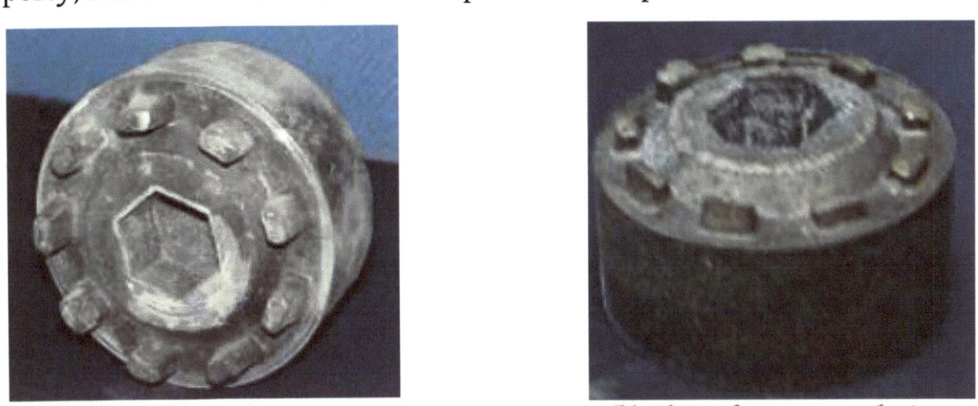

(a) Traditional forging workpiece (b) Thixo-forging workpiece

Fig. 26. Traditional forging and thixo-forging workpieces of composite

7. CONCISE SUMMARY

The microstructural characteristics of magnesium matrix composites were examined across three distinct casting methods. The results demonstrated that SiCp/AZ61 composites produced using the stirring melt casting technique had a relatively uniform distribution of SiC particles and a minimal porosity rate, in contrast to those fabricated through fully liquid stirring casting and semi-solid stirring casting processes. The procedure for fabricating metal matrix composites was exemplary.

The optimal processing parameters for SiCp/AZ61 composites produced by a stirring melt casting method under experimental conditions were a SiC particle volume percentage of 6%, a stirring temperature of 595°C, and a stirring duration of 5 minutes. The volume fraction of SiC particles significantly influenced the mechanical characteristics of SiCp/AZ61 composites, surpassing the effects of stirring duration and stirring temperature, which were the secondary parameters.

Semi-solid isothermal heat treatment technology was employed for the partial remelting of SiCp/AZ61 composites. A refined semi-solid microstructure was achieved, with an equal-area diameter ranging from 60μm to 85μm, and an effective liquid volume percentage of around 31% to 38%. The ideal technological parameters for SiCp/AZ61 composites were a reheating temperature of 595^{oC} to 600^{oC} and an isothermal holding duration of 30 to 60 minutes.

Compression experiments were conducted on semi-solid SiCp/AZ61 magnesium matrix composites. The effects of strain rate, strain, temperature, and volume percentage of SiC particles on flow stress were examined. The findings indicate that the flow stress of semi-solid SiCp/AZ61 composites is influenced by temperature and strain rate. Simultaneously, the flow stress escalates with the augmentation of the volume percentage of SiC particles.

The impact of deformation temperature, strain rate, strain, liquid volume percent, and reinforcement volume fraction on flow stress during the thixotropic plastic deformation process of composites was examined. A novel constitutive model for composites during the thixotropic plastic deformation process was proposed. The constitutive equation for SiCp/AZ61 composites was derived using the multiple nonlinear regression approach, based on data from thixotropic compression tests. The computed results were in good accord with the experimental data. It is utilized to direct the thixotropic plastic deformation process of composites.

Numerical simulation can assist in analyzing the thixoforging process, and the behavior of metal flow has been determined. The effective stress distribution became more uniform and its value decreased with the rise in forming temperature. The effective stress rose with the rising volume percentage of SiC particles. The temperature distribution deteriorated with the increasing volume fraction of SiC particles. An analysis was conducted on the distinctions between traditional forging and thixo-forging methods. Results demonstrated that thixo-forging was superior in cavity filling compared to traditional forging. The intricate workpiece can be completed in a single thixo-forging process. Numerical simulation results are consistent with experimental findings.

REFERENCES

1. Hai, Z. Y., Xing, Y. L. (2004). Review of recent studies in magnesium matrix composites. *Journal of Materials Science*, Vol. 39: 6153-6171
2. Zhou, W., Xu, Z. M. (1997). Casting of SiC reinforced metal matrix composites. *Journal of Materials Processing Technology*, Vol. 63 , No.3: 358-363
3. Kang, C. G., Choi, J. S., Kim, K. H. (1999). The effect of strain rate on macroscopic behavior in the compression forming of semi-solid aluminum alloy. *Materials Processing Technology*, Vol.88, No.1-3: 159-168
4. Zhang, X. Q., Wang, H. W., Liao, L. H. (2004). In situ synthesis method and damping characterization of magnesium matrix composites. *Composites Science and Technology*, Vol. 67:720-727
5. Mordike, B. L. (2002). Development of highly creep resistant magnesium alloys. *Journal of Materials Processing Technology*, Vol. 117, No.3: 391-394
6. Flemings, M. C. (1991). Behavior of metal alloy in the semi-solid state. *Metall Trans A*, Vol.22, No.5: 957-981
7. Yan, H., Xia, J. C. (2005). Theoretical analysis of plastic forming process for semi-solid material. *Materials Science Forum*, Vol.488-489, 389-392
8. Yan, H., Zhang, F. Y. (2005). Structure evolution of AZ61 magnesium alloy in SIMA process. *Transactions of Nonferrous Metals Society of China*, Vol.15, No.3, 560-564
9. Yan, H., Zhang, F. Y. (2006). Microstructural evolution of semi-solid AZ61 magnesium alloy during reheating process. *Solid State Phenomena*, Vol.116-117, 275-278
10. Yan, H., Zhou, B. F. (2006). Thixotrpic deformation behavior of semi-solid AZ61 magnesium alloy during compression process, *Materials Science and Engineering B*, Vol.132, No.1-2, 179-182
11. Yan, H., Zhou, B. F. (2006). Constitutive model of thixotropic plastic forming for semi-solid AZ61 magnesium alloy. *Solid State Phenomena*, Vol.116-117, 577-682

12. Yan, H., Zhou, B. F. (2008). Study on thixo-forging of AZ61 wrought magnesium alloy. *Solid State Phenomena*, Vol.141-143, 577-682
13. Yan, H., Fu, M.F., Zhang, F. Y., Chen, G. X. (2007). Research on properties of SICp/AZ61 magnesium matrix composites in fabrication processes. *Materials Science Forum*, Vols. 561-565: 945-948
14. Yan, H., Lin, L. S. B., Pan, W. (2008). Study of SICp/AZ61 composites. *Solid State Phenomena*, Vols. 141-143 :551-555
15. Yan, H., Zhang, F. Y., Pan, W. (2008). Microstructural evolution of SiCp/AZ61 composites during partial remelting. *Solid State Phenomena*, Vols. 141-143 :545-549
16. Zhang, F. Y., Ye, J. X., Yan, H. (2011). Effects of SiC particle and holding time on semi-solid microstructure of SICp/AZ61 composites. *Advanced Materials Research*, Vols. 152- 153: 628-633
17. Yan, H., Wang, J. J. (2011). Thixotropic compression deformation behavior of SICp/AZ61 magnesium matrix composites. *Transations of Nonferrous Metals Society of China*, Vols. 20: s811-s814
18. Yan, H., Wang, J. J.,Zhang, F. Y. (2011). A constitutive model for thixotropic plastic forming of semi-solid composites. *Advanced Materials Research,* Vols. 154-155:690-693
19. Yan, H., Huang, W. X. (2011). Numerical simulation on thixo-forging of magnesium matrix composite. *Advanced Materials Research,* Vols. 189-193:2535-2538
20. Bao, G., Lin, Z.(1996). High strain rate deformation in particle reinforced metal matrix composites. *Acta Materialia*, Vol.44, No. 3: 1011-1019.
21. Li, Y., Ramesh, K.T., Chin, E.S.C.(2000). The compressive viscoplastic response of an A359/SiCp metal—matrix composite and of the A359 aluminum alloy matrix. *International Journal of Solids and Structures*, Vol. 37, No. 51:7547-7562.
22. Yan, H., Huang, X. ,Hu, Q. (2010). Damping capacity of SICp/AZ61 composites. *Advanced Materials Research*, Vols. 123-125 : 35-38
23. Yan, H., Huang, Z. M. (2011). Study on creep properties of SiCp/AZ61 composites. *Advanced Materials Research*, Vols. 189-193: 4227-4230

12 Chapter

CREATION OF A WINDING MECHANISM FOR AMORPHOUS RIBBON UTILIZED IN TRANSFORMER CORES

1. INTRODUCTION: AN OVERVIEW

In recent years, the scope of applications for soft magnetic materials has expanded considerably due to advancements in amorphous and nanocrystallized systems. Specific ferromagnetic alloys can be acquired in vitreous phases using quick quenching methods; some undergo partial crystallization via specific heat treatments, resulting in structures comprised of grains of 10 to 40 nanometers in length, encased in a vitreous phase. One quick quenching approach is melt-spinning, which produces amorphous metal strips that are subsequently twisted into rolls.

The further use of the wound rolls involves the configuration of electric transformer cores, demonstrating significant enhancements in overall performance, increased efficiency, and reduced environmental effect. Historically, these cores have been fabricated using grain-oriented and non-grain-oriented silicon steel sheets, ferrite sheets, and Ni-Fe and Co-Fe alloy sheets produced through traditional casting methods, necessitating multiple mechanical and thermal processes, some of which incur significant costs (Gelinas, 2000). The production of nano-structured magnetic packages can be achieved by directly employing melt-spun strips subjected to various heat treatments, allowing for the adjustment of the hysteresis cycle. Moreover, its applications can be broadened to intricate geometries by incorporating a milling phase subsequent to the melt-spinning process, resulting in refined elemental powder particles (Nowacki, 2006; Byoung et al., 2007), whose dimensions can be adjusted through the regulation of milling duration (Dobrzanskia et al., 2004). The implications of utilizing soft magnetic alloys influence not only transformer cores but also AC motors (Pagnola et al., 2009; Pagnola, 2009). These novel amorphous and nano-crystallized materials are presently marketed at prices up to three times higher than those of traditional materials (Condes, 2008).

Magnetic cores dissipate energy via two distinct mechanisms: hysteresis (energy loss during the reorientation cycle of magnetic domains) and eddy currents (also known as Foucault currents). These losses may reach between 5% and 15% of the total energy produced, varying according to the production method utilized. Independent research and several writers corroborate that these losses can be diminished by about 80% compared to those observed in devices constructed with conventional steel (De Cristofaro, 1998; Douglas, 1988; Richardson, 1990). Table 1 and Figure 1 illustrate the significantly reduced losses and, more importantly, the quantity of energy conserved. The LSA executed the Melt-Spinning technique via the project titled "Advanced Technology Magnetic Materials Production" (PICT-2007-02018), with the objective of minimizing energy losses to specified values. Amorphous ribbons, akin to FINEMET®, were produced during preliminary testing; these ribbons were 1mm in width and 20μm in thickness, and were rapidly quenched on a copper wheel in an air atmosphere, achieving a cooling rate of 106 K/sec (Muraca et al., 2009).

Table 1. Core losses in regular Fe-Si cores and Amorphous alloys cores refer to Fe-Si (100%).

Power [kVA]	Core Losses [W]		Saving percentage [%]	Manufacturer
	Fe - Si	Amorphous		
10	40	13,5	66	Osaka Transformer
10	40	11	72	Westinghouse
15	50	14	72	Allied and MIT
25	85	28	67	General Electric
25	85	16	81	Prototype Allied

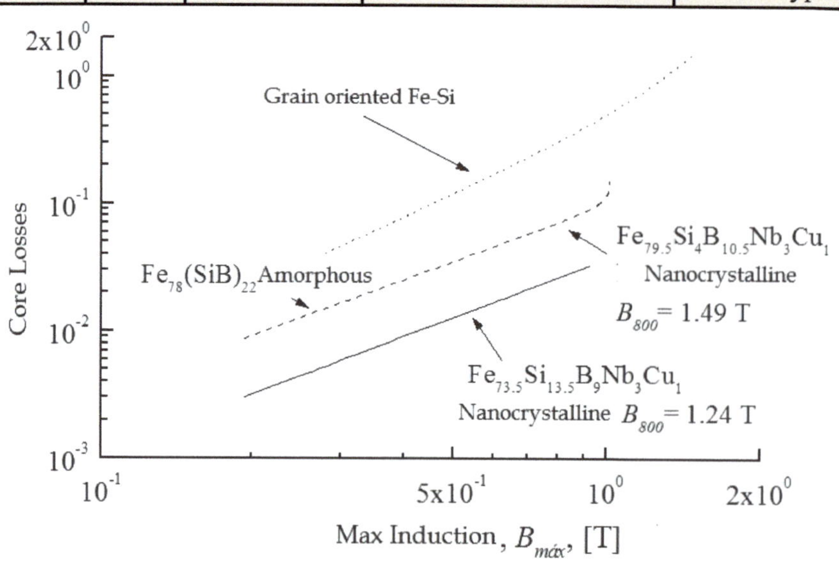

Fig. 1. Core losses with different alloys (Dobrzanskia et al., 2004).

2. MELT-SPINNING

One of the most common rapid quenching techniques to produce amorphous metals is the one called melt-spinning. Using this technique, the molten alloy is jetted on the surface of a high speed spinning copper wheel through a nozzle. The casting wheel acts as a heat sink reaching one million degrees per second cooling rate (Praisner et al., 1995) necessary to achieve the vitreous phase instead of a crystalline structure (see figure 2). In figure 3 it is shown a diagram of the melt-spinning apparatus, where it can be seen the small and weak linkage between the ribbon and the casting wheel.

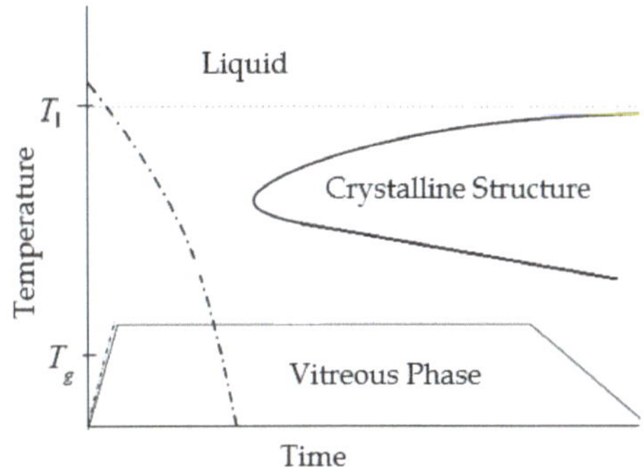

Fig. 2. Cooling procedure to avoid crystalline structure. (Moya, 2009).

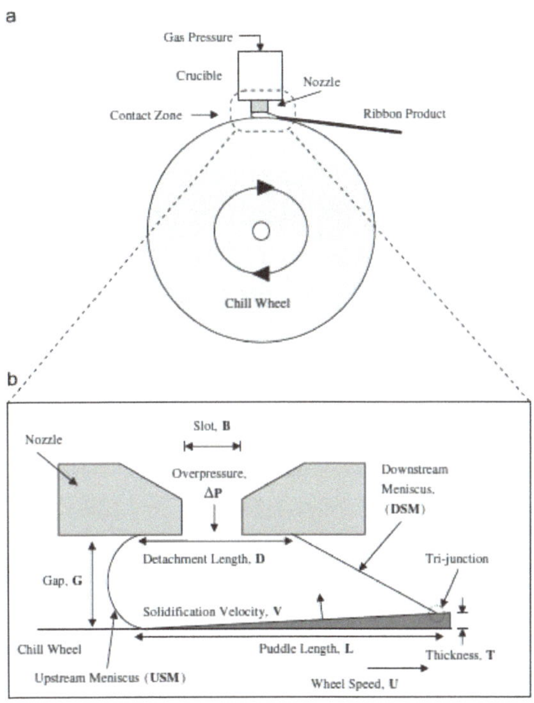

Fig. 3. (a) Schematic of a melt-spinning apparatus, (b) Blow up of the contact zone. (Theisen et al., 2010).

The amorphous alloy is obtained from a crystalline alloy, called mother alloy, which has the same chemical composition as the amorphous one. The way to get to the mother alloy is melting the proper quantity of the different components into an induction heater several times in order to insure a homogeneous alloy. Afterwards, the alloy is introduced into a quartz crucible with an induction coil which heats the alloy over the melting point; then, an argon over-pressure expulses the alloy through the nozzle on the high speed spinning wheel. As a result, a continuous amorphous ribbon is obtained; its thickness (≈ 20 – 100 μm) is a function of the injection pressure, the gap between the nozzle and the wheel and the cooling rate. Depending on the alloy and its corrosion susceptibility, the process should be in a controlled atmosphere, in a vacuum chamber or even in environmental conditions.

3. WINDING SYSTEM

The winding mechanism designed is capable of working at high winding speeds (an order of magnitude higher than those used in paper winding and steel-making, see table 2), it also insures the quality of the product as it´s has been solidified at the wheel without changing its surface roughness generated in the previous stage, since any aspect that has influence on its surface integrity during this stage has a direct impact on the magnetic package´s performance. This winding system is assembled next to the cylindrical sleeve by the casting wheel seen in Figure 4.

Two problems hold the design back at the first stage of the process:

1. Thread of the strip into the winding reel.
2. Tension control of the roll.

Both of these issues mainly appear because of the intrinsic characteristics of the melt- spinning technique. Due to the speed of the process and the fact that the strip has no fixed point at the casting wheel the solutions given are rarely similar to those found in regular winding machines. First of all, an automatic threading system was designed due to the impossibility to count on the proper time to thread the strip into the winding reel by a human (~ 5 to 10 seconds). This time implies an excessive collection of material (250 to 500m) by the casting wheel that can be wrecked by its own weight or successive folding.

With regard of the tension control, it's critical not only because of the typical problems in every wound roll, but also because an over-tension can separate the strip from the casting wheel where the material is still in a liquid state. Therefore, two zones were established in the machine, a free-tension zone and another one where it is controlled up to a set-point determined by the tension profile.

Figure 5 and 6 shows the proposed design, where it can be seen a set of guiding belts (1), that generate an air flow capable of dragging the strip from the casting wheel up to the pinch rollers (2). These rollers are powered by an asynchronous motor and variable frequency drive, where the strip purely rolls over them; the control of their speed and the winding velocity of the reel are the manipulated variables of the tension control system. Between the guiding belts and the pinch rollers, there are a set of idle rollers (8 & 9), which provide the system a stock of material in order to prevent an unwanted detachment of it at the solidification meniscus. Next to the pinch rollers there is a deflector (4) that simply guides the strip towards the winding reel (5). At last, the winding reel is surrounded by a wrapping belt that ensures the thread of the strip into the reel during the startup.

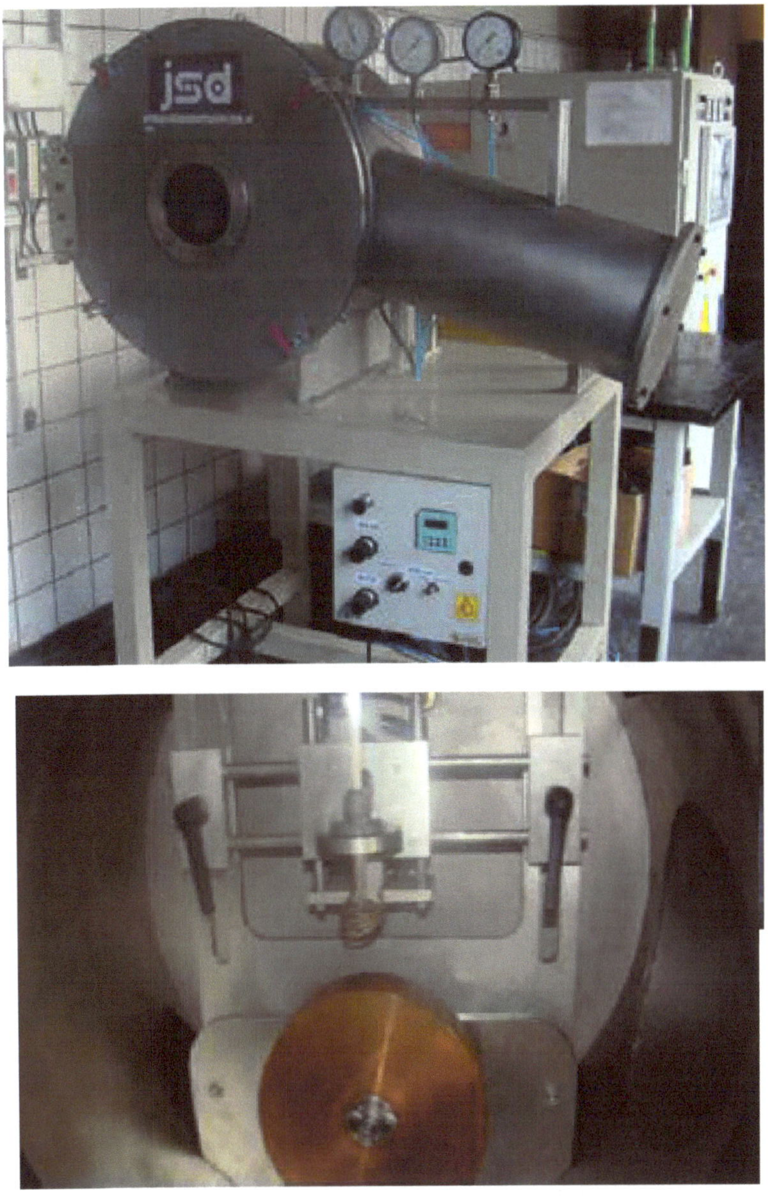

Fig. 4. Melt-spinning Equipment developed at LSA. Crucible and copper wheel.

Table 2. State of the art - Winding parameters found in different industries.

Material	Winding Speed [m/s]	Winding Tension [Mpa]	Source
Paper	≈ 8	< 10	Liu, 2009
Steel	≈ 8	15 - 75	Liu, 2009
Plastic Films	≈ 15	< 4	Lee et al., 2002
Magnetic Tape	< 5	< 4	Liu, 2009
Amorphous strip	25 – 50	15 - 30	Own development

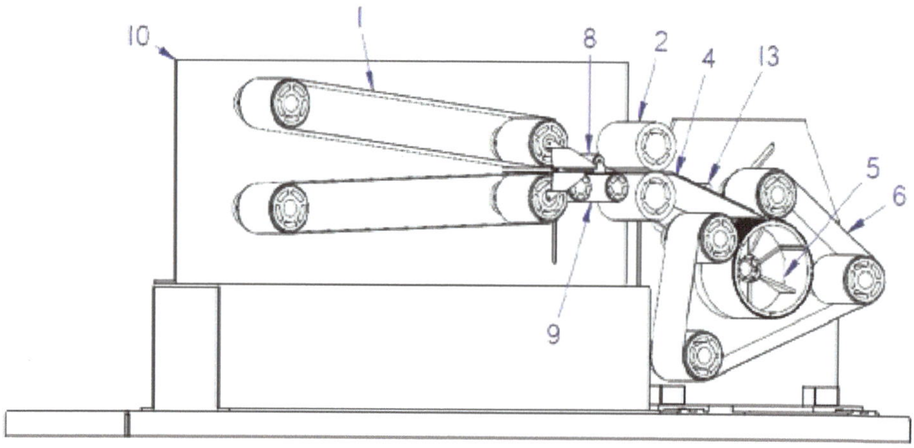

Fig. 5. Winding machine at startup: 1. Guiding belts; 2. Pinch rollers; 3. Initial deflector; 4. Main deflector; 5. Winding reel; 6. Wrapping belts; 7.Crosslide; 8. Stocking system – dancer roller; 9. Stocking system - Fixed roller; 10. Chassis; 11. Main motor; 12. Slide guides; 13.Tension sensor.

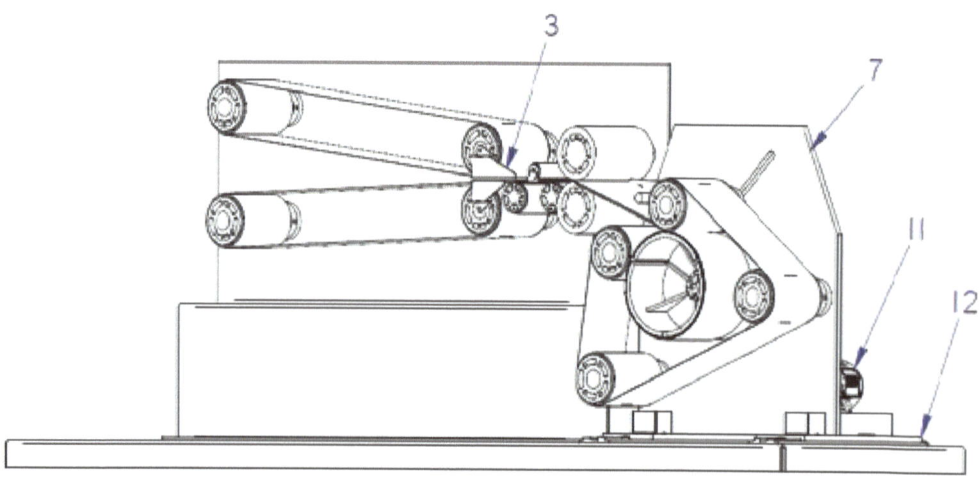

Fig. 6. Winding machine at startup.

3.1 Winding tension profile

Several winding stress models have been developed in order to find the proper winding tension profile (Li et al., 2009; Liu, 2009; Lee et al., 2002). Following the model proposed by Liu, every new wound lap is considered a collection of concentric laps of web material. In every one of them it is formulated the differential equations of internal equilibrium to find out stress, strain, displacement and pressures developed during the winding. Finally, the profiles shown in figure 7 were obtained.

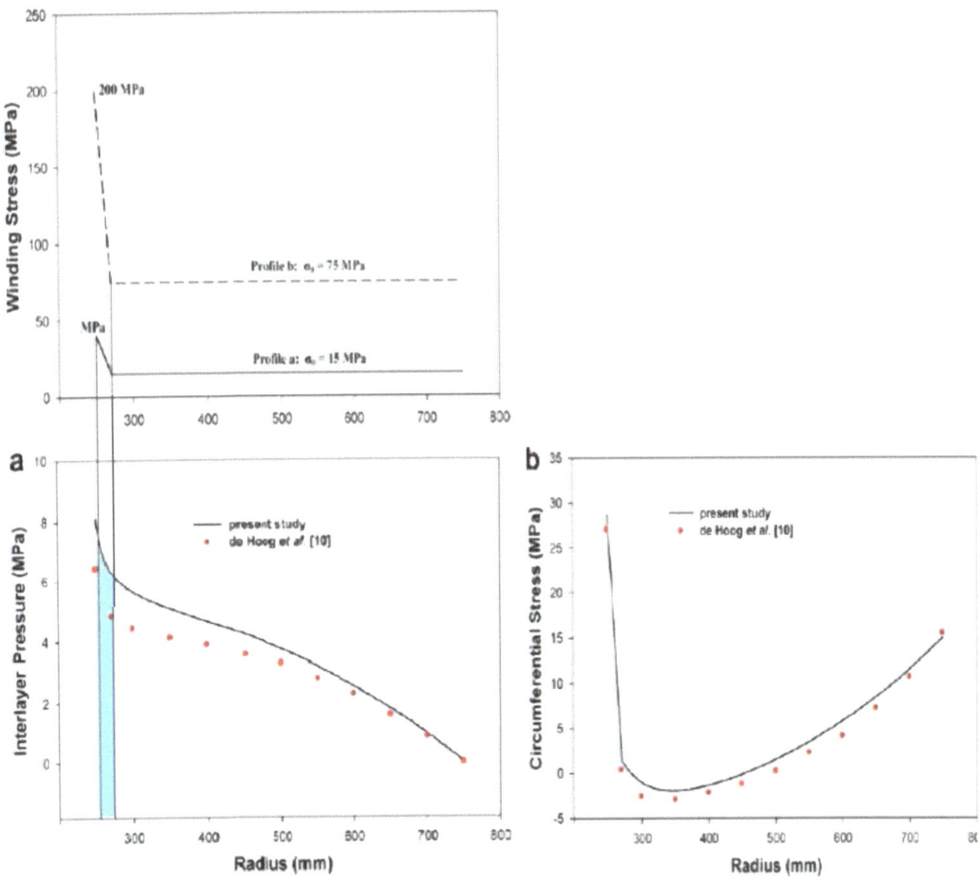

Fig. 7. Interlayer pressure (a) and circumferential stress (b) for the winding stress profile in the above figure (Liu, 2009).

As it can be seen, the winding stress profile starts at a higher value and then it starts decreasing through a ramp towards the regime value. During this ramp the roll is setting up its linkage with the winding reel, which will determine the end of the startup stage and the beginning of the working regime stage. To insure that the threading is complete the friction force generated by the internal pressure times the surface of every wound lap must be higher than the inertial force plus the winding tension. From now on, the roll is fixed to the winding reel and no slip between them will be found.

$$\mu_s \cdot \int_{r_o}^{r'} (p_i \cdot S) \cdot dr = F_{friction} > F_{inertia} + T_{winding} \quad (1)$$

Once established the tension profile, it can be calculated the power necessary for the winding reel motor and for the pinch rollers motor. On one hand, the winding reel motor (main motor) is going to take most of the torque necessary for the winding, on the other hand the pinch rollers motor will be working almost as a brake because upstream it is a free- tension zone and downstream the tension is provided by the main motor. This is why the power of the main motor is considered to take this torque times a service factor to make up for the startup situation. Taking into account these considerations and the dimensions of the strip and the reel, it is needed a 5HP AC motor for the winding reel.

3.2 Startup

When the casting wheel starts throwing the first cuts of amorphous strip, the guiding belts drag it towards the pinch rollers (2) where a first thread is done. At this point, it´s the first contact with the winding machine and it is found pure rolling friction between the strip and the pinch rollers, where these jog the strip forward with its own tangential speed. As a consequence, it is needed a high precision in the mounting of this rollers in order to preserve the clearance between them; if it is bigger than the designed one the strip would slip between them. But if it is smaller, the material would be damage by an operation similar to a laminate. Due to the constant contact between the strip and these rollers, it is recommended to pay close attention to the hardness during the material selection of the rollers in order not to damage the surface quality of the strip; several options appear like copper, brass or even PTFE (TEFLON®) inserts. Next to the pinch rollers (2) the strip is guided towards the winding reel (5) by the main deflector (4). The winding reel is spinning to a higher tangential speed than the speed of the strip in order to ensure the threaded. Additionally, the reel is surrounded by a wrapping belt (6) which guarantees the strip to follow the profile of the reel until there is enough friction between the successive wounds of strip, so to create a bond within the reel and the strip, but this won´t happen until several wounds of ribbon had already been rolled over the reel

3.3 Working regime

Once insured the threaded, the wrapping belts are completely removed, as can be seen in Figure 8, to look after the surface integrity of the strip. This action is performed by a pneumatic actuated scissor mechanism. Meanwhile, the spinning speed of the reel is reduced because during the threaded it was significantly higher;

also, from this moment on, the dancer roller (8) can freely move in the vertical axis, and the tension sensor (13) is disposed as it is shown in Figure 8. From now on, the winder is at a working regime and we must proceed to the tension control of the roll.

The basic principle of the tension control is the small difference between the pinch roller´s speed and winding speed (represented by the tangential speed of the reel) which is slightly higher (He et al., 2010). The structure of the device for controlling the tension is shown in Figure 10. The tension measurement is used to tune up the spinning speed of the reel, by an asynchronous motor, variable frequency drive and encoder. The control set point establishes a tighter roll at the beginning and looser at the end, known as taper tension control (Good et al., 2008). With this system it is intended to obtain an optimum tension of the roll without inflicting any damage to the material.

A storage system is incorporated in order to prevent flaws on the tension control system, such as response time, slipping of the strip on the pinch rollers, lack of precision on electric and electronic components. Every difference between the pinch rollers and the casting wheel speed, overcomes into an over-tension of the strip or an excessive storage of material, which may cause a possible detachment of the solidification meniscus.

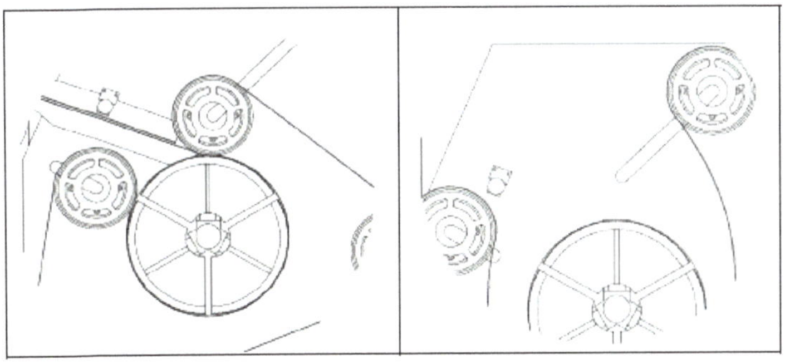

Fig. 8. Wrapping belts and tension sensor at startup condition (3.1) in working regime (3.2).

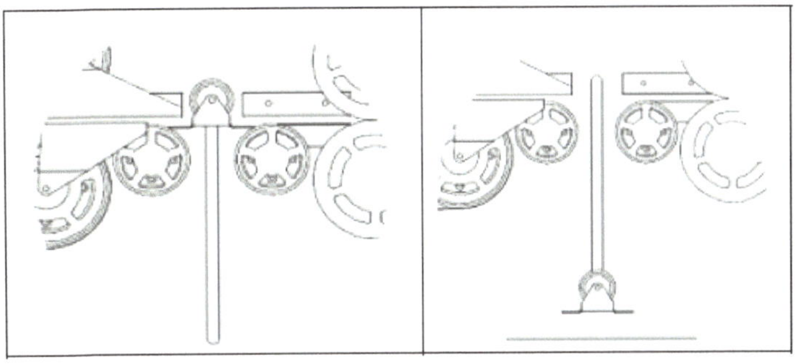

Fig. 9. Stocking system at startup and during the working regime.

The mechanism is composed by 3 idle rollers (Figure 9), two of them (9) are fixed, while the centered one (8) can move along the vertical axis forcing the strip to take a larger profile instead of the straight line from the startup. Sensing the position of this roller, it is tuned up the pinch rollers speed, only when the stock is excessive or insufficient. So, this variation will be considered as a transitory regime for the tension control system, in order to assume the pinch rollers′ speed as a constant. With this system a little stock of ribbon is created in order to absorb every produced over-tension.

The winding reel (5) is provided with a mandrel for a quick demounting of the finished roll. Moreover, the reel along with the wrapping belt is mounted on a cross slide (7) which can be moved over the sliding guides (12) in a cross direction, as can be seen in Figure 5 and 6. With this system several angular defects during the tuned-up before the startup of the equipment are corrected.

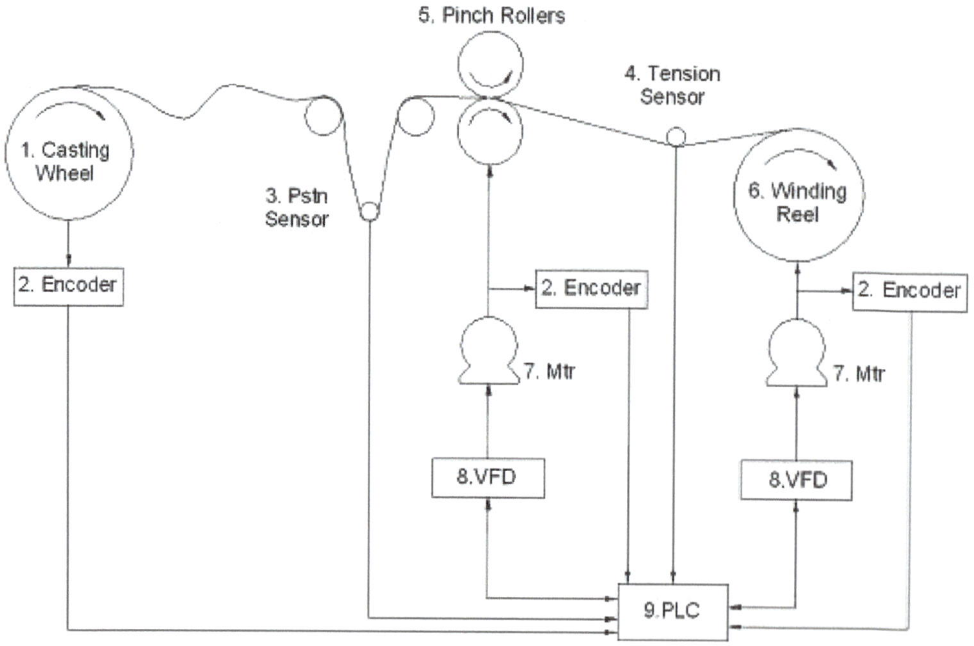

Fig. 10. Control System Structure: 1. Casting Wheel; 2. Encoder; 3. Position Sensor; 4. Tension Sensor; 5. Pinch Rollers; 6. Winding Reel; 7. Asynchronous Motor; 8. Variable Frequency Drive; 9. PLC.

4. RESULTS AND DISCUSSION

The initial investment and operating cost for the 25-30 lifetime-years of the transformer will be called Total Cost of Ownership. Within the Operating Cost of the device is included the cost of the dissipated electric energy at the windings (Cu) and at the core (Fe). Consequently, had the core losses been diminished (by using amorphous metal cores), the Total Cost of Ownership of a device produced by this technology would be reduced in comparison to those produced by traditional technology; and would profit a considerable economic gain to the owner of this machine. From this point of view, the implementation of an accurate winding system as the one proposed, presents an optimum solution to the formerly described process, not only to behoove the handling of the final product, but also to simplify the post-melt spinning heat treatments, such as isotermic annealing which is used to obtain nano-crystalized ribbons (Muraca . et al. 2009).

An accurate control of the material and design of the nozzle´s orifices (Saito, 2010; Kurokawa, et al. 1999), working pressures of the chamber and ejection temperature are crucial to prevent unwanted flaws (Saito, 2010; Marashi et al 2009) which include the absence of the formation of the strip on the casting wheel as can be seen in Figure 11. For this reason it is highly recommended to ensure the working conditions described along this article.

(a)

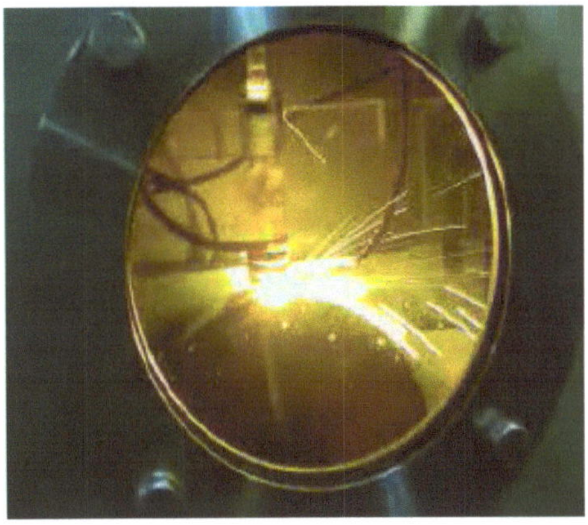

(b)

Fig. 11. (a) Ejection of molten material on the wheel in a non-operational regime, (b) Ejection of molten material on the wheel in a operational regime; both photos on its own equipment in LSA.

5. CONCISE SUMMARY

This work proposes the construction of a winding mechanism for amorphous strips utilized in magnetic transformer cores, with general dimensions detailed in the figure in Figure 12, and its assembly with the custom-designed equipment is entirely feasible.

The designs of components and parts are informed by our own experience in constructing this equipment and our research on the production of micro and nano-materials (Ozols et al., 1999; Pagnola, 2009; Muraca et al., 2009), alongside technical insights from other authors regarding the fabrication of various products for industrial magnetic packages, as illustrated in Figure N. 13. Croat (1992) and Kurokawa et al. (1999) were regarded.

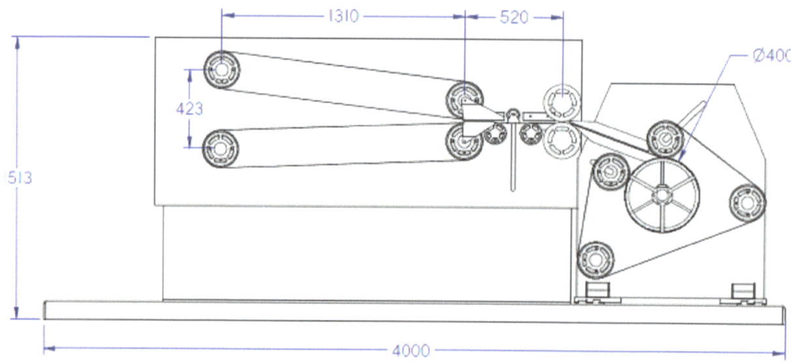

Fig. 12. General drawing of mechanical parts.

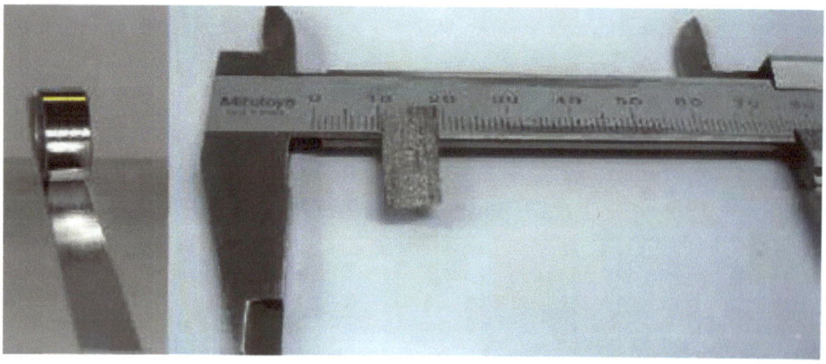

Fig. 13. Fe_{78} $Si_{13}B_9$ amorphous strips used in magnetic cores, and industrial magnetic package.

REFERENCES

1. Bedell J.R., Method of and apparatus for casting metal strip employing a localized conditioning shoe, United States Patent: 4649984, March 17, 1987
2. Budzyn B., Carlson C., Two piece casting wheel, United States Patent: 4537239, August 27, 1985.

3. Byoung-Gi M., Yong Sohn K., Won-Wook P., Taek-Dong L., Effect of milling on the soft magnetic behavior of nanocrystalline alloy cores, Materials Science and Engineering A, pp. 449 451, 2007.

4. ConDes (Consultora Conexiones para el Desarrollo), Mercado de Materiales Magnético, en el marco proyecto FAN (www.con-des.com.ar), 2008.-

5. Croat J., Energy product rare earth-iron magnet alloys, United States Patent: 5172751, December 22, 1992.

6. De Cristofaro N., Amorphous Metals in Electric-Power Distribution Applications, Materials Research Society, MRS Bulletin, Vol. 23, No. 5, 1998, pp. 50-56.

7. Dobrzanskia L.A. , Nowosielskia R., Koniecznya J., Wysłockib J. , Przybyłb A., Properties and structure of the toroidal magnetically soft cores made from the amorphous strips, powder, and composite materials, Journal of Materials Processing Technology, 2004, pp.157-158.

8. Douglas J., Transformers with lower losses, Power Engineering Review, IEEE, Vol. 8, No. 3, 1988, pp. 12-13.

9. Frissora A.., Kojak A., Chilled casting wheel, United States Patent: 4502528, March 5, 1985. Gelinas C., Soft magnetic Powders for AC Magnetic Applications. Euro PM2000,Soft Magnetic Materials Workshop, Berlin, 2000, pp. 1-8.

10. Good J., Roisum D., Winding: machines, mechanics and measurements, DEStech Publications Inc., ISBN: 978-1-932078-69-5, 2008.

11. He F., Zhao H. & Wang Q., Constant Linear Speed Control of Motor Based on Fuzzy PI in Strip Winding System, DBTA 2010, 978-1-4244-6977-2, pp 1-4.

12. Kurokawa K., Suhara S., Matsukawa T., Ishizuka H., Sato T., Method of manufacturing a wide metal thin strip, United States Patent: 5908068,June 1, 1999.

13. Kurokawa, Method of manufacturing a wide metal thin strip, United States Patent: 5908068, June 1, 1999.

14. Lee Y., Wickert J., Stress field width axisymetric wound rolls, Journal of applied mechanics, ASME, Vol 69, March 2002.

15. Li S., Cao J., A hybrid approach for quantifying the winding process and material effects on sheet coil deformation, Journal of Engineering Materials and technology, ASME,Vol. 126, July 2004.

16. Liu M., A nonlinear model of center-wound rolls incorporating refined boundary conditions, ELSEVIER, Computers and structures 87 (2009) 552-563.

17. Marashi S P H, Abedi A., Kaviani S., Aboutalebi S H., Rainforth M. and Davies H A, Effect of melt-spinning roll speed on the nanostructure and magnetic properties of stoichiometric and near stoichiometric Nd–Fe–B alloy ribbons, J. Phys. D: Appl. Phys. 42 11, 2009.
18. Muraca D., Silveyra J., Pagnola M., Cremaschi M., Nanocrystals magnetic contribution to FINEMET-type soft magnetic materials with Ge addition, Journal of Magnetism and Magnetic Materials, 321, 2009, pp. 3640–3645.
19. Moya J., "Vidrios Metálicos y Aleaciones Nanocristalinas: Nuevos Materiales de Estructura Avanzada", Cuadernos de Facultad n. 4, UBA 2009
20. Nowacki J., Polyphase sintering and properties of metal matrix composites, Journal of Materials Processing Technology 175, 2006, pp 316 323.
21. Ozols A., Sirkin H. and Vicente E.E., Segregation in Stellite powders produced by the plasma rotating electrode process Materials Science and Engineering: A Volume 262, Issues 1-2, 1999, pp. 64-69.
22. Pagnola M., Tesis Doctoral UBA: Desarrollo de Composites Ferromagnéticos, 2009.
23. Pagnola M., Saccone F., Ozols A., Sirkin H,, Improvement to approximation of second order function of hysteresis in magnetic materials, COMPEL: The International Journal for Computation and Mathematics in Electrical and Electronic Engineering, Volume 28 Issue 6, 2009, pp. 1579-1589.
24. Praisner T., Chen J., Tseng A., An experimental study of process behavior in planar flow melt spinning, Metallurgical and materials transactions B, Volume 26B, December 1995.
25. Richardson B., Amorphous metal cored transformers-justifying their use, IEE Colloquium on Magnetic Ribbons and Wires in Power, Electronic and Automotive Applications, 1990, pp.1-4.
26. Saito T., Electrical resistivity and magnetic properties of Nd–Fe–B alloys produced by Melt spinning technique, Journal of Alloys and Compounds Volume 505, Issue 1, 2010, pp. 23-28.
27. Theisen E., Davis M., Weinstein S., SteenP., Transient behavior of the planar-flow melt Spinning, ELSEVIER, Chemical engineering science 65 (2010) 3249-3259.
28. Wang C., Yan M., Surface quality, microstructure and magnetic properties of Nd2(Fe,Zr,Co)14B/Fe alloys prepared by different melt-spinning equipments, Materials Science and Engineering B 164, 2009, pp. 71-75.

13

Chapter

ANALYSIS OF FREE VIBRATION IN CENTRIFUGALLY STIFFENED NON-UNIFORM TIMOSHENKO BEAMS

1. INTRODUCTION: AN OVERVIEW

Rotating beam-like structures are extensively utilized across various engineering disciplines and are particularly significant for modeling components such as wind turbine blades, helicopter rotors, robotic manipulators, turbo-machinery, and airplane propellers. The governing differential equations of motion for free vibration of a non-uniform rotating Timoshenko beam, with general elastic restrictions at the ends, are resolved using the differential quadrature method (Bellman & Roth, 1986; Felix et al., 2008, 2009). The equations of motion are formulated to incorporate the influences of shear deformation, rotating inertia, hub radius, elastic end restraints, and non-uniform cross-sectional area change of the beam. Banerjee has created a framework utilizing Hamilton's principle to account for the centrifugal stiffening that occurs in rapidly rotating structures due to the presence of centrifugal force from rotational motion (Banerjee, 2001). The proposed approach accommodates a wide array of conditions for resolution. Specific instances with classical constraints can be derived for the limiting values of the rigidities. Step changes in cross-section are also taken into account (Naguleswaran, 2004).

The natural vibration frequencies and mode forms of rotating beams have garnered significant interest and attention. A significant number of academics have investigated the dynamic behavior of rotating uniform or tapered Euler-Bernoulli beams. (Yang et al., 2004; Özdemir & Kaya, 2006; Lin & Hsiao, 2001). Banerjee calculated the dynamic stiffness matrix of a rotating Bernoulli-Euler beam utilizing the Frobenius approach in power series, incorporating an axial force at the outboard end of the beam alongside the conventional centrifugal force (Banerjee, 2000).

Few studies have addressed the issue of rotating beams while considering rotary inertia, shear deformation, and their combined effects, along with hub radius and elastically constrained ends (Bambill et al., 2010). In scenarios where rotating inertia and shear deformation effects are negligible, a study utilizing Euler–Bernoulli beam theory is applicable. Timoshenko theory facilitates the characterization of vibrations in short beams, sandwich composite beams, or higher modes of narrow beams (Rossi et al., 1991; Seon et al., 1999). Banerjee et al. (2006) examined the free bending vibrations of rotating tapered Timoshenko beams utilizing the dynamic stiffness approach. (Ozgumus & Kaya, 2010) employed the Differential Transform Method for the investigation of free vibrations in a rotating, tapered Timoshenko beam.

The finite element method was employed by Hodges and Rutkowski in 1981. Vinod et al. (2007) conducted a study on spectral finite element formulation for a spinning beam subjected to brief impact. Gunda and Ganguli (2008) devised a novel beam finite element, deriving its basis functions from the precise solution of the governing static homogeneous differential equation of a stiff string, which emerged from an approximation in the rotating beam equation. Singh et al. (2007) employed Genetic Programming to develop an approximation model of rotating beams. (Gunda et al., 2007) presented a low degree of freedom model for the dynamic analysis of rotating tapered beams, utilizing a computationally efficient superelement constructed from a combination of polynomial and Fourier series form functions. Kumar and Ganguli (2009) investigated rotating beams whose eigenpair, frequency, and mode shape correspond to those of uniform non-rotating beams for a specific mode. A noteworthy article (Ganesh & Ganguli, 2011) introduced a physics-based basis function for the vibration analysis of high-speed rotating beams utilizing the finite element approach. The basis function generated shape functions that are contingent upon the element's position within the beam, as well as its material, geometric attributes, and rotational velocity.

The current study aims to offer answers for practical engineering scenarios and may also serve as a baseline for comparing other numerical models. The suggested differential quadrature method provides an effective and precise approach for solving linear and nonlinear partial differential equations. It was utilized by Bellman in the 1970s. He employed this method to determine the natural frequencies of transverse vibrations in a spinning cantilever beam. Bellman and Casti, 1971. Various writers have employed the differential quadrature method, acknowledging its efficacy in addressing such challenges (Bert & Malik, 1996; Shu & Chen, 1999; Choi et al., 2000; Liu & Wu, 2001; Shu, 2000).

Numerical findings are derived for the natural frequencies of transverse vibration and the mode forms of rotating beams, accounting for elastic restrictions and non-uniform variations in cross-sectional area. Several of those examples have been resolved utilizing the finite element method, and the resultant data are in remarkable concordance..

2. THEORY

Figure 1 shows the rotating tapered beam considered in the present paper. The beam could have step jumps in cross section and rotates at speed η . The X -axis coincides with the centroidal axis of the beam, the Y -axis is parallel with the axis of rotation and the Z -axis lies in the plane of rotation. L is the length of the beam, L_k is the length of the segment k and L_d is the length of the last segment of the beam. The displacement in the Y direction is denoted as w and the section rotation is denoted as ψ . Only displacements in the $X - Y$ plane are taken into account and the Coriolis effects are not considered.

The centrifugal force of a beam element at a distance $Rk + xk$ expressed as from the axis of rotation can be

$$dFk = \eta^2 (Rk + xk)dm \tag{1}$$

where $dm = \rho Ak(xk)dxk$ is its mass, with ρ the mass density of material, and $Ak(xk)$, is the cross-sectional area at xk . Figure 2. The centrifugal force $Nk(xk)$ generated by η is

$$dNk(xk) = \eta^{2+} (Rk + xk) Ak(xk)dxk \tag{2}$$

The total axial force at the cross section located at $Rk + xk$ is

$$\bar{N}_k(\bar{x}_k) = \bar{\eta}^2 \rho \int_{\bar{x}_k}^{L_k} (\bar{R}_k + \bar{x}_k) A_k(\bar{x}_k) d\bar{x}_k + \bar{F}_{k+1} = \bar{\eta}^2 \rho \left(\bar{R}_k \int_{\bar{x}_k}^{L_k} A_k(\bar{x}_k) d\bar{x}_k + \int_{\bar{x}_k}^{L_k} A_k(\bar{x}_k) \bar{x}_k \, d\bar{x}_k \right) + \bar{F}_{k+1} \tag{3}$$

$Fk+1$ is the outboard force at the end of the segment k, due to the adjacent segments k+1 to d.

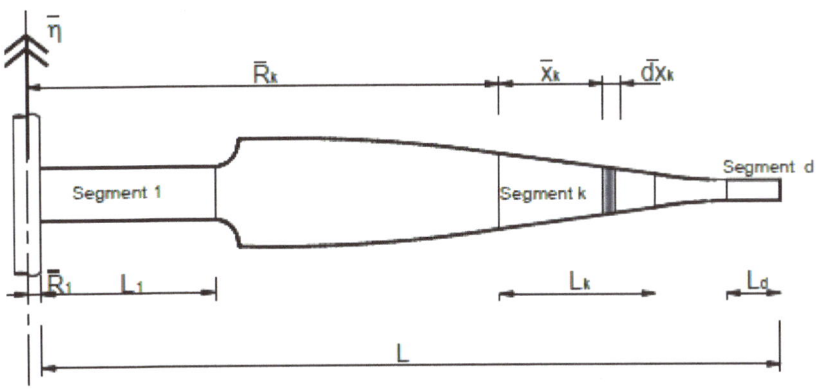

Fig. 1. Rotating beam model

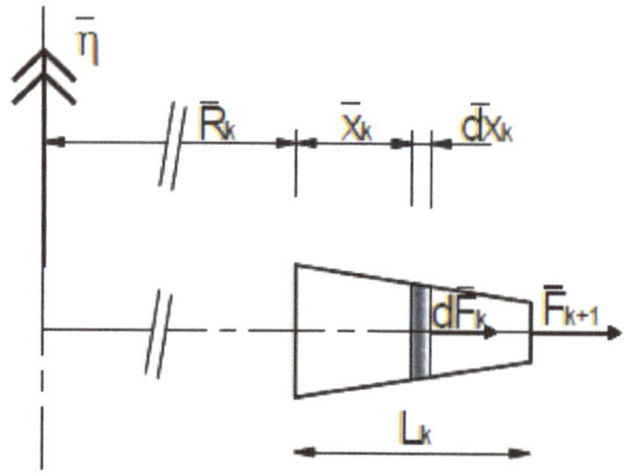

Fig. 2. Rotating beam segment *k* of length L_k

Finally, the tensile force can be written as

$$\bar{N}_k(\bar{x}_k)=\bar{\eta}^2\rho\left(\bar{R}_k V_k(L_k)+\Phi_k(L_k)-\bar{R}_k V_k(\bar{x}_k)-\Phi_k(\bar{x}_k)\right)+\bar{F}_{k+1} \quad (4)$$

with

$$V_k(\bar{x}_k)=\int_0^{\bar{x}_k} A_k(\bar{x}_k)\,d\bar{x}_k\ ;\ \Phi_k(\bar{x}_k)=\int_0^{\bar{x}_k} A_k(\bar{x}_k)\,\bar{x}_k\,d\bar{x}_k \quad (5a, b)$$

The expressions for shear force and bending moment at an instant *t* in the rotating beam are

$$\bar{Q}_k^*(\bar{x}_k,t)=\bar{N}_k(\bar{x}_k)\frac{\partial\bar{w}_k(\bar{x}_k,t)}{\partial\bar{x}_k}+\kappa GA_k(\bar{x}_k)\left(\frac{\partial\bar{w}_k(\bar{x}_k,t)}{\partial\bar{x}_k}-\bar{\psi}_k(\bar{x}_k,t)\right) \quad (6)$$

$$\bar{M}_k^*(\bar{x}_k,t)=EI_k(\bar{x}_k)\frac{\partial\bar{\psi}_k(\bar{x}_k,t)}{\partial\bar{x}_k} \quad (7)$$

where *Ik* (*xk*) is the second moment of area of the beam cross-section; *t* the time; *wk* (*x* ,*t*) the

transverse displacement; $\psi_k(x,t)$ the section rotation; *E* the Young's modulus; *v* the

Poisson's ratio; $G = E / 2(1 + v)$ the shear modulus and *k* is the shear factor.

The governing differential equations of motion of a rotating Timoshenko beams (Banerjee, 2001) are:

$$\frac{\partial\bar{Q}_k^*(\bar{x}_k,t)}{\partial\bar{x}_k}=\rho A_k(\bar{x}_k)\frac{\partial^2\bar{w}_k(\bar{x}_k,t)}{\partial t^2}$$

$$\bar{Q}_k^*(\bar{x}_k,t)-N_k(\bar{x}_k)\frac{\partial\bar{w}_k(\bar{x}_k,t)}{\partial\bar{x}_k}+\frac{\partial\bar{M}_k^*(\bar{x}_k,t)}{\partial\bar{x}_k}+\rho I_k(\bar{x}_k)\bar{\eta}^2\bar{\psi}_k(\bar{x}_k,t)=\rho I_k(\bar{x}_k)\frac{\partial^2\bar{\psi}_k(\bar{x}_k,t)}{\partial t^2} \quad (8a,b)$$

Assuming simple harmonic oscillation

$$\bar{w}_k(\bar{x}_k,t)=\bar{W}_k(\bar{x}_k)e^{i\omega t}\quad;\quad \bar{\psi}_k(\bar{x}_k,t)=\bar{\Psi}_k(\bar{x}_k)e^{i\omega t} \tag{9a,b}$$

where *w* is the circular frequency in radian per second. The bending moment and the shear force are expressed as

$$\bar{Q}_k^*(\bar{x}_k,t)=\bar{Q}_k(\bar{x}_k)e^{i\omega t};\quad \bar{M}_k^*(\bar{x}_k,t)=\bar{M}_k(\bar{x}_k)e^{i\omega t} \tag{10a,b}$$

where

$$\bar{Q}_k(\bar{x}_k)=\left(\bar{N}_k(\bar{x}_k)+\kappa GA_k(\bar{x}_k)\right)\frac{d\bar{W}_k(\bar{x}_k)}{d\bar{x}_k}-\kappa GA_k(\bar{x}_k)\bar{\Psi}_k(\bar{x}_k);\ \bar{M}_k(\bar{x}_k)=EI_k(\bar{x}_k)\frac{d\bar{\Psi}_k(\bar{x}_k)}{d\bar{x}_k} \tag{11a,b}$$

Substituting equations (9-10) into equations (8), the equations of motion for the free vibration of the segment *k* of the rotating beam result in:

$$\begin{gathered}-\frac{d\bar{Q}_k(\bar{x}_k)}{d\bar{x}_k}=\rho A_k(\bar{x}_k)\omega^2\bar{W}_k(\bar{x}_k)\\ -\bar{Q}_k(\bar{x}_k)+\bar{N}_k(\bar{x}_k)\frac{d\bar{W}_k(\bar{x}_k)}{d\bar{x}_k}-\frac{d\bar{M}_k(\bar{x}_k)}{d\bar{x}_k}-\rho I_k(\bar{x}_k)\bar{\eta}^2\bar{\Psi}_k(\bar{x}_k)=\rho I_k(\bar{x}_k)\omega^2\bar{\Psi}_k(\bar{x}_k)\end{gathered} \tag{12a,b}$$

Replacing equations (11) into equations (12), the differential equations of motion become:

$$\begin{gathered}-\frac{d\bar{N}_k(\bar{x}_k)}{d\bar{x}_k}\frac{d\bar{W}_k(\bar{x}_k)}{d\bar{x}_k}-\bar{N}_k(\bar{x}_k)\frac{d^2\bar{W}_k(\bar{x}_k)}{d\bar{x}_k^2}-\kappa GA_k(\bar{x}_k)\left(\frac{d^2\bar{W}_k(\bar{x}_k)}{d\bar{x}_k^2}-\frac{d\bar{\Psi}_k(\bar{x}_k)}{d\bar{x}}\right)-\\ \kappa G\frac{dA_k(\bar{x}_k)}{d\bar{x}_k}\left(\frac{d\bar{W}_k(\bar{x}_k)}{d\bar{x}_k}-\bar{\Psi}_k(\bar{x}_k)\right)=\rho A_k(\bar{x}_k)\omega^2\bar{W}_k(\bar{x}_k)\\ -\kappa GA_k(\bar{x}_k)\left(\frac{d\bar{W}_k(\bar{x}_k)}{d\bar{x}_k}-\bar{\Psi}_k(\bar{x}_k)\right)-EI_k(\bar{x}_k)\frac{d^2\bar{\Psi}_k(\bar{x}_k)}{d\bar{x}_k^2}-\\ E\frac{dI_k(\bar{x}_k)}{d\bar{x}_k}\frac{d\bar{\Psi}_k(\bar{x}_k)}{d\bar{x}_k}-\rho I_k(\bar{x}_k)\bar{\eta}^2\bar{\Psi}_k(\bar{x}_k)=\rho I_k(\bar{x}_k)\omega^2\bar{\Psi}_k(\bar{x}_k)\end{gathered} \tag{13a,b}$$

The term $pIk(xk)\eta^2\psi k(xk)$ included in equation (13.b) was introduced by Banerjee, 2001.

This term generates more realistic results especially for high rotational speeds, η^2. The conditions for displacements and forces between adjacent segments, *k* and *k*+1, are:

$$\bar{W}_k(L_k)-\bar{W}_{k+1}(0)=0;\quad \bar{\Psi}_k(L_k)-\bar{\Psi}_{k+1}(0)=0 \tag{14a,b}$$

$$\bar{Q}_k(L_k)-\bar{Q}_{k+1}(0)=0;\quad \bar{M}_k(L_k)-\bar{M}_{k+1}(0)=0 \tag{15a,b}$$

Figure 3 shows the beam elastically restrained at both ends.

The boundary conditions of the beam at its ends are, for the first segment k=1, at $x_1 = 0$:

$$\bar{Q}_1(0) - \bar{K}_{W1}\ \bar{W}_1(0) = 0;\ \bar{M}_1(0) - \bar{K}_{\Psi 1}\ \bar{\Psi}_1(0) = 0 \tag{16a, b}$$

and for the last segment k=d, at $xd = Ld$:

$$\bar{Q}_d(L_d) - \bar{K}_{Wd}\ \bar{W}_d(0) = 0;\ \bar{M}_d(L_d) - \bar{K}_{\Psi d}\ \bar{\Psi}_d(0) = 0 \tag{17a,b}$$

The four spring constants are denoted as: $KW1$, KWd, $K\psi 1$, $K\psi d$.

The expressions and parameters in dimensionless form are defined as follows:

$$\Omega^2 = \frac{\rho A_1(0)}{EI_1(0)} L^4 \omega^2;\ \eta^2 = \frac{\rho A_1(0)}{EI_1(0)} L^4 \bar{\eta}^2;$$

$$x = \frac{\bar{x}_k}{L_k};\ l_k = \frac{L_k}{L};\ r_k^2 = \frac{I_k(0)}{A_k(0)};\ s_k = \frac{L}{r_k};\ R_k = \frac{\bar{R}_k}{L};\ W_k(x) = \frac{\bar{W}_k(\bar{x}_k)}{L_k};\ \Psi_k(x) = \bar{\Psi}_k(\bar{x}_k);$$

$$a_k(x) = \frac{A_k(\bar{x}_k)}{A_k(0)};\ b_k(x) = \frac{I_k(\bar{x}_k)}{I_k(0)};\ v_k(x) = \frac{V_k(\bar{x}_k)}{l_k A_k(0)};\ \phi_k(x) = \frac{\Phi_k(\bar{x}_k)}{l_k^2 A_k(0)};$$

$$N_{k+1} = \frac{\bar{F}_{k+1}}{EA_k(0)};\ N_k(x) = \frac{\bar{N}_k(\bar{x}_k)}{EA_k(0)};\ Q_k(x) = \frac{\bar{Q}_k(\bar{x}_k)}{EA_k(0)};\ M_k(x) = \frac{L_k}{EI_k(0)} \bar{M}_k(\bar{x}_k);$$

and $K_{Wj} = \bar{K}_{Wj} \dfrac{L}{EA_1(0)}$; $K_{\Psi j} = \bar{K}_{\Psi j} \dfrac{L}{EI_1(0)}$; with j =1 or j=d.

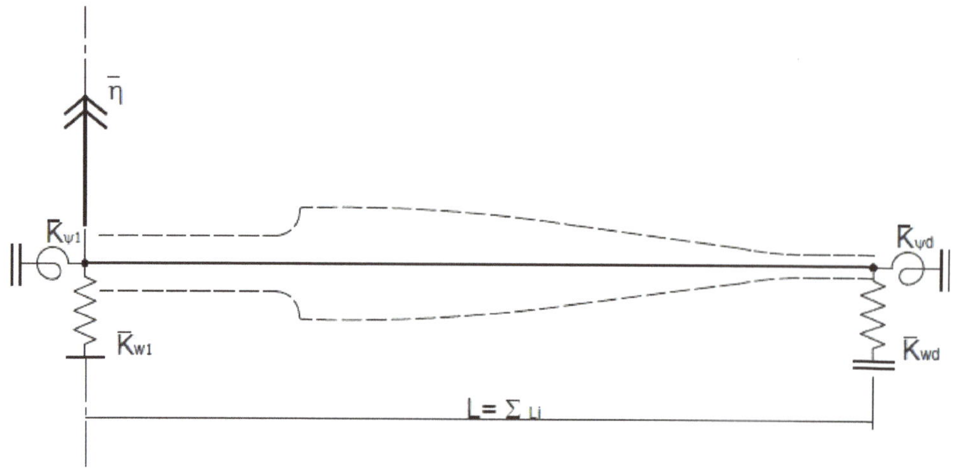

Fig. 3. Elastic restraints of the rotating beam

In each segment k of the beam, x varies between 0 and 1.

The axial force, the shear force and the bending moment in the adimensional form become:

$$N_k(x)=\eta^2\frac{l_k^2}{s_1^2}(R_k v_k(1)+\phi_k(1)-R_k v_k(x)-\phi_k(x))+N_{k+1};\ \text{with } s_1=\frac{L}{r_1};\ r_1^2=\frac{I_1(0)}{A_1(0)} \tag{18}$$

$$Q_k(x)=\left(N_k(x)+\frac{\kappa}{2(1+\nu)}a_k(x)\right)\frac{dW_k(x)}{dx}-\frac{\kappa}{2(1+\nu)}a_k(x)\Psi_k(x) \tag{19}$$

$$M_k(x)=b_k(x)\frac{d\Psi_k(x)}{dx} \tag{20}$$

And the equations of motion in dimensionless form are:

$$\eta^2 a_k(x)(R_k+x)\frac{dW_k(x)}{dx}-\frac{s_1^2}{l_k^2}N_k(x)\frac{d^2W_k(x)}{dx^2}-\frac{\kappa}{2(1+\nu)}\frac{s_1^2}{l_k^2}a_k(x)\left(\frac{d^2W_k(x)}{dx^2}-\frac{d\Psi_k(x)}{dx}\right)-$$
$$-\frac{\kappa}{2(1+\nu)}\frac{s_1^2}{l_k^2}\frac{da_k(x)}{dx}\left(\frac{dW_k(x)}{dx}-\Psi_k(x)\right)=\Omega^2 a_k(x)W_k(x) \tag{21}$$

$$-s_1^2\frac{\kappa}{2(1+\nu)}s_k^2 a_k(x)\left(\frac{dW_k(x)}{dx}-\Psi_k(x)\right)-\frac{s_1^2}{l_k^2}b_k(x)\frac{d^2\Psi_k(x)}{dx^2}-\frac{s_1^2}{l_k^2}\frac{db_k(x)}{dx}\frac{d\Psi_k(x)}{dx}-$$
$$-\eta^2 b_k(x)\Psi_k(x)=\Omega^2 b_k(x)\Psi_k(x) \tag{22}$$

The equations (14), which satisfy continuity of displacement and rotation, can be expressed in dimensionless form as follows:

$$l_k W_k(1)-l_{k+1}W_{k+1}(0)=0;\ \Psi_k(1)-\Psi_{k+1}(0)=0 \tag{23a,b}$$

and the equations (15) of compatibility of the bending moment and the shear force, result in the following adimensional equations:

$$\alpha_k Q_k(1)-\alpha_{k+1}Q_{k+1}(0)=0;\ \frac{\beta_k}{l_k}M_k(1)-\frac{\beta_{k+1}}{l_{k+1}}M_{k+1}(0)=0$$

or

$$\alpha_k\left[\left(N_k(x)+\frac{\kappa}{2(1+\nu)}a_k(x)\right)\frac{dW_k(x)}{dx}-\frac{\kappa}{2(1+\nu)}a_k(x)\Psi_k(x)\right]_{x=1}-$$
$$-\alpha_{k+1}\left[\left(N_{k+1}(x)+\frac{\kappa}{2(1+\nu)}a_{k+1}(x)\right)\frac{dW_{k+1}(x)}{dx}-\frac{\kappa}{2(1+\nu)}a_{k+1}(x)\Psi_{k+1}(x)\right]_{x=0}=0; \tag{24a,b}$$

$$\frac{\beta_k}{l_k}b_k(x)\frac{d\Psi_k(x)}{dx}\bigg|_{x=1}-\frac{\beta_{k+1}}{l_{k+1}}b_{k+1}(x)\frac{d\Psi_{k+1}(x)}{dx}\bigg|_{x=0}=0$$

where $\alpha_k = \frac{A_k(0)}{A_1(0)}$; $\beta_k = \frac{I_k(0)}{I_1(0)}$.

The boundary conditions at the end closest to the axis of rotation, segment 1, x=0, are:

$$Q_1(0) - K_{W1} l_1 W_1(0) = 0 ; \left(N_1(0) + \frac{\kappa a_1(0)}{2(1+\nu)} \right) \frac{dW_1(x)}{dx}\bigg|_{x=0} - \frac{\kappa a_1(0)\Psi_1(0)}{2(1+\nu)} - K_{W1} l_1 W_1(0) = 0$$

$$M_1(0) - K_{\varphi 1} l_1 \Psi_1(0) = 0 ; \; b_1(0)\frac{d\Psi_1(x)}{dx}\bigg|_{x=0} - K_{\varphi 1} l_1 \Psi_1(0) = 0 \tag{25a,b}$$

and at the other end of the rotating beam, segment d , x=1, they are:

$$Q_d(1) - K_{Wd}\frac{l_d}{\alpha_d} W_d(1) = 0; \left(N_d(1) + \frac{\kappa a_d(1)}{2(1+\nu)} \right) \frac{dW_d(x)}{dx}\bigg|_{x=1} - \frac{\kappa a_d(1)}{2(1+\nu)}\Psi_d(1) - \frac{K_{Wd} l_d}{\alpha_d} W_d(1) = 0$$

$$M_d(1) - \frac{K_{\varphi d} l_d}{\beta_d}\Psi_d(1) = 0 ; \; b_d(1)\frac{d\Psi_d(x)}{dx}\bigg|_{x=1} - \frac{K_{\varphi d} l_d}{\beta_d}\Psi_d(1) = 0 \tag{26a,b}$$

where Nd (1) is an outboard force at the end of the beam, farthest from the axis of rotation, that is equal to zero in the present study.

3. DIFFERENTIAL QUADRATURE METHOD, DQM

In order to obtain the DQM analog equations from the governing equations of the rotating beam, the beam segment domain is discretized in a grid of i points, using the Chebyshev – Gauss - Lobato expression, (Shu, 2000). (See Fig. A.1 in Appendix A)

Equations (18, 19, 20) assumed the form:

$$N_k(x_i) = \eta^2 \frac{l_k^2}{s_1^2}\left(R_k v_k(1) + \phi_k(1) - R_k v_k(x_i) - \phi_k(x_i) \right) + N_{k+1} \tag{27}$$

$$Q_k(x_i) = \left(N_k(x_i) + \frac{\kappa}{2(1+\nu)} a_k(x_i) \right) \sum_{j=1}^{n} A_{ij}^{(1)} W_{kj} - \frac{\kappa}{2(1+\nu)} a_k(x_i)\Psi_{ki} \tag{28}$$

$$M_k(x_i) = b_k(x_i) \sum_{j=1}^{n} A_{ij}^{(1)} \Psi_{kj} \tag{29}$$

The equations of motion (21) and (22) become:

$$\begin{aligned} &\left(\eta^2 a_k(x_i)(R_k + x_i) - \frac{\kappa}{2(1+\nu)} \frac{s_1^2}{l_k^2} \frac{da_k(x_i)}{dx} \right) \sum_{j=1}^{n} \left(A_{ij}^{(1)} \right) W_{kj} - \\ &-\left(\frac{s_1^2}{l_k^2} N_k(x_i) + \frac{\kappa}{2(1+\nu)} \frac{s_1^2}{l_k^2} a_k(x_i) \right) \sum_{j=1}^{n} \left(A_{ij}^{(2)} \right) W_{kj} + \frac{\kappa}{2(1+\nu)} \frac{s_1^2}{l_k^2} a_k(x_i) \sum_{j=1}^{n} A_{ij}^{(1)} \Psi_{kj} + \\ &+ \frac{\kappa}{2(1+\nu)} \frac{s_1^2}{l_k^2} \frac{da_k(x_i)}{dx} \Psi_{ki} = \Omega^2 a_k(x_i) W_{ki} \end{aligned} \tag{30}$$

$$-\frac{\kappa}{2(1+\upsilon)} s_1^2 s_k^2 a_k(x_i) \sum_{j=1}^{n} A_{ij}^{(1)} W_{kj} - \frac{s_1^2}{l_k^2} b_k(x_i) \sum_{j=1}^{n} A_{ij}^{(2)} \Psi_{kj} +$$
$$+\left(\frac{\kappa}{2(1+\nu)} s_1^2 s_k^2 a_k(x_i) - \eta^2 b_k(x_i) \right) \Psi_{ki} - \frac{s_1^2}{l_k^2} \frac{db_k(x_i)}{dx} \sum_{j=1}^{n} A_{ij}^{(1)} \Psi_{kj} = \Omega^2 b_k(x_i) \Psi_{ki} \tag{31}$$

where the $A^{(1)}$ and $\overset{A}{(2)}$ ij are the weighting coefficients of linear algebraic equations. (See Appendix A.1 for more details).

Finally, the conditions (23) and (24) are replaced by:

$$l_k W_{kn} - l_{k+1} W_{(k+1)1} = 0 ; \ \Psi_{kn} - \Psi_{(k+1)1} = 0 ; \tag{32a,b}$$

$$\alpha_k \left(\left(N_k(1) + \frac{\kappa}{2(1+\nu)} a_k(1) \right) \sum_{j=1}^{n} A_{nj}^{(1)} W_{kj} - \frac{\kappa}{2(1+\nu)} a_k(1) \Psi_{kn} \right)$$
$$-\alpha_{k+1} \left(\left(N_{k+1}(0) + \frac{\kappa}{2(1+\nu)} a_{k+1}(0) \right) \sum_{j=1}^{n} A_{1j}^{(1)} W_{(k+1)j} - \frac{\kappa}{2(1+\nu)} a_{k+1}(0) \Psi_{k1} \right) = 0; \tag{33a,b}$$
$$\frac{\alpha_k}{l_k} b_k(1) \sum_{j=1}^{n} A_{nj}^{(1)} \Psi_{kj} - \frac{\alpha_{k+1}}{l_{k+1}} b_{k+1}(0) \sum_{j=1}^{n} A_{1j}^{(1)} \Psi_{(k+1)j} = 0$$

and the boundary conditions (25) and (26) replaced by:

$$\left(N_1(0) + \frac{\kappa}{2(1+\nu)} a_1(0) \right) \sum_{j=1}^{n} A_{1j}^{(1)} W_{1j} - \frac{\kappa}{2(1+\nu)} a_1(0) \Psi_{11} - l_1 K_{W1} W_{11} = 0 ;$$
$$K_{\Psi 1} \Psi_{11} - \frac{b_1(0)}{l_1} \sum_{j=1}^{n} A_{1j}^{(1)} \Psi_{1j} = 0 \tag{34a,b}$$

$$\left(N_d(1) + \frac{\kappa}{2(1+\nu)} a_d(1) \right) \sum_{j=1}^{n} A_{nj}^{(1)} W_{dj} - \frac{\kappa}{2(1+\nu)} a_d(1) \Psi_{dn} - l_d K_{Wd} W_{dn} = 0;$$
$$K_{\Psi n} \Psi_{dn} - \frac{b_d(1)}{l_d} \sum_{j=1}^{n} A_{nj}^{(1)} \Psi_{dj} = 0 \tag{35a,b}$$

The DQM linear equation system is used to determine the natural frequencies and mode shapes of the rotating beam.

The number of terms taken in the summations had been studied for many situations and the system has acceptable convergence by n= 21 terms. (See Table 1)

4. FINITE ELEMENT METHOD, MEF

An independent set of results for the natural frequencies, was also obtained by a finite element code. (Bambill et al., 2010). The finite element model employed in the analysis has 3000 beam elements of two nodes in the longitudinal direction (Rossi, 2007). See Table 2. This number of elements was proved to be enough with a convergence analysis.

The beam model also takes into account the shear deformation (Timoshenko beam's theory) and the increase in bending stiffness induced by the centrifugal force.

The term $\rho\ Ik\ (xk)\eta^{2}\ \psi k\ (xk)$ of equation (13.b) was not included in the finite element formulation. Probably for this reason some small differences between both sets of numerical results (DQM and FEM) begin to appear when the rotational speed η increases.

5. NUMERICAL RESULTS

In the following examples some calculations were performed over elliptical cross sections. ($\kappa = 0.886364$). Without loss of generality, one may choose to keep constant width $e_k=e$ and vary the height $hk\ (x)$ in each segment of the beam. The area and the second moment of area of the cross section of the beam will be $A_k(x) = \frac{\pi\ eh_k(x)}{4}$, $I_k(x) = \frac{\pi\ eh_k^3(x)}{64}$, and for this particular situation there are:

$$a_k(x) = \frac{h_k(x)}{h_k(0)};\ b_k(x) = \left(\frac{h_k(x)}{h_k(0)}\right)^3$$

The following formula is proposed to a quadratic variation of the height in each segment of beam:

$$hk\ (x) = c0\ k + c1k\ x + c2\ k\ x^2$$

And the slope is the derivative of this function

$$h_k'(x) = \frac{dh_k(x)}{dx} = c_{1k} + 2c_{2k}\,x$$

where $c0k$, $c1k$ and $c2\ k$ are constants, which are defined by the heights and slopes at both ends of each segment k. The heights and slopes at each end are identified with the subscript A for x=0: hAk ; $h'Ak$ and with the subscript B for x=1: hBk ; $h'Bk$.

If the segment of the beam shows a linear variation of height , $c_2k = 0$ and

$$hAk = c0k\ ;\ hBk = c0k + c1k\ ;\ h'Ak = h'Bk = c1k$$

As it can be seen in Table 1, the frequency coefficients calculated by the Differential Quadrature Method, DQM, using a summation with $n \geq 19$ (i= 1, 2, 3, ..., n) points, show none significant improvement.

Table 1. Convergence analysis of the DQM, for a two-span rotating Timoshenko beam elastically restrained al both ends, with a quadratic variation of height.

n	Ω_1	Ω_2	Ω_3	Ω_4	Ω_5
5	15.6861	29.2939	49.1602	63.9792	112.610
7	15.1981	28.9907	46.9070	64.9219	88.8670
9	14.9057	29.5079	47.4960	64.7054	87.4079
11	14.8340	29.6332	47.6579	64.7247	87.6724
13	14.8281	29.6467	47.6811	64.7310	87.7047
15	14.8291	29.6464	47.6820	64.7319	87.7079
17	14.8295	29.6460	47.6816	64.7320	87.7080
19	14.8296	29.6459	47.6815	64.7320	87.7080
21	14.8296	29.6459	47.6815	64.7320	87.7080

The frequency coefficients in Table 1, correspond to a beam of two segments, rotating at speed $\eta = 10$, whose characteristics are: elliptical cross section; $\nu = 0.3$; $\kappa = 0.886364$; R_1=0; $l_1/L = l_2/L = 1/2$; $s_1 = \sqrt{300}$; $h_{B1}/h_{A1} = 1/2$; $h'_{B1} = 0$; $h_{A2}/h_{B1} = 1/2$; $h_{B2}/h_{A2} = 1/2$; $h'_{A2} = 0$; $K_{W1} = 10$; $K_{\psi 1} = 5$; $K_{Wd} = 0.1$; $K_{\psi d} = 1$.

In Table 2 the values obtained for the natural frequency coefficients using the finite element method are presented for $\eta = \sqrt{\rho A_0 / EI_0}\, L^2 \bar{\eta} = 0$ and $\eta = 10$. The number of elements is increased from 10 to 3000.

The model of the rotating beam of Table 2 has the following characteristics: one segment; rectangular cross section; $\nu = 0.3$; $\kappa = 10(1+\nu)/(12+11\nu)$ =0.849673; R_1=0; $s_1 = \sqrt{300}$; $h_B/h_A = 1/4$; h'_B=0; $K_{W1} \to \infty$; $K_{\psi 1} \to \infty$; $K_{Wd} = 0$; $K_{\psi d} = 0$.

In the first examples it is assumed a perfect clamped condition at the axis of rotation, given by: $K_{W1} \to \infty$ and $K_{\psi 1} \to \infty$. (Tables 3, 4 and 5).

Table 3 presents the effect of the rotational speed parameter η on the natural frequency coefficients of a rotating cantilever beam of one segment, ($KW1 \to \infty$; $Kw\,1 \to \infty$; $KWd = 0$; $Kw\,d = 0$). The results correspond to a linear variation of height and a comparison is made with (Barnejee, 2006) when Banerjee´s parameter is n=1. As it can be observed the agreement is excellent.

Table 2. Convergence analysis of the frequency coefficients $\Omega_i = \sqrt{\rho A_0 / EI_0}\, L^2 \omega_i$ using MEF.

	$\eta = 0$				
Number of elements	Ω_1	Ω_2	Ω_3	Ω_4	Ω_5
10	3.38628165	11.7689336	26.5951854	46.6658427	71.0448001
100	3.37398143	11.7248502	26.4438604	46.1408176	69.5136708
1000	3.37385398	11.7243988	26.4423706	46.1357196	69.4986357

2000	3.37385302	11.7243954	26.4423593	46.1356810	69.4985219
3000	3.37385284	11.7243946	26.4423572	46.1356739	69.4985008
$\eta = 10$					
10	11.6074237	25.8805102	44.0407905	66.3753084	92.6859627
100	11.6098042	25.7094320	43.5638284	65.4674874	90.8491237
1000	11.6098077	25.7074626	43.5585908	65.4579769	90.8301746
2000	11.6098078	25.7074476	43.5585511	65.4579049	90.8300310
3000	11.6098078	25.7074448	43.5585437	65.4578915	90.8300044

Table 3. Frequency coefficients $\Omega_i = \sqrt{\rho A_1(0)/EI_1(0)}\, L^2 \omega_i$ **for a one-span beam, l1 /L = 1 ;** $s_1 = \sqrt{1000}$; $h_B / h_A = 1/2$; $K_{W1} \to \infty$; $K_{\psi 1} \to \infty$; $K_{Wd} = 0$; $K_{\psi d} = 0$.

η		Ω_1	Ω_2	Ω_3	Ω_4	Ω_5
0	DQM	3.82377	18.3171	47.2638	90.4468	147.992
	(Barnejee,2006)	3.82379	18.3173	47.2648	90.4505	148.002
2	DQM	4.43680	18.9365	47.8706	91.0589	148.609
	(Barnejee,2006)	4.43680	18.9366	47.8717	91.0625	148.619
4	DQM	5.87874	20.6850	49.6446	92.8693	150.444
	(Barnejee,2006)	5.87877	20.6851	49.6456	92.8730	150.454
6	DQM	7.65512	23.3091	52.4622	95.8054	153.450
	(Barnejee,2006)	7.65514	23.3093	52.4632	95.8090	153.460
8	DQM	9.55392	26.5435	56.1584	99.7601	157.555
	(Barnejee,2006)	9.55396	26.5437	56.1595	99.7638	157.564
10	DQM	11.5015	30.1825	60.5628	104.608	162.668
	(Barnejee,2006)	11.5015	30.1827	60.5639	104.612	162.677

All the calculations performed for the following Tables and Graphics used $R_1 = 0$; and $\nu = 0.30$; $\kappa = 0.886364$ (elliptical cross section).

The DQM results are determined using $n = 21$ in each segment of the beam, and the MEF results were obtained with 3000 elements.

The beam considered in Table 4 has one segment and is elastically restrained at its outer end. The parameter of rotation speed η is taken equal to 10. The Table presents the frequency coefficients for the first five mode shapes which correspond to different sets of elastically boundary conditions given by the spring constant parameters KWd and $K\psi d$. The other details of the beam are specified in the legend of the table.

The beam model considered in Table 5 has two segments of equal length and similar conditions and parameters as Table 4.

Table 4. First natural frequencies $\Omega_i = \sqrt{\rho A_1(0)/EI_1(0)}\, L^2 \omega_i$ for a one-span rotating Timoshenko beam, with elliptical cross section and quadratic height variation along the axis. $\nu = 0.3$; $s_1 = \sqrt{300}$; $h_B/h_A = 1/2$; $h'_B = 0$; $K_{W1} \to \infty$; $K_{\psi 1} \to \infty$; $\eta = 10$.

Kψ d	KWd	Method	Ω_1	Ω_2	Ω_3	Ω_4	Ω_5
0	0	DQM	11.2148	27.6174	50.0089	77.5866	108.472
		FEM	11.2375	27.6743	50.0711	77.6432	108.523
	0.1	DQM	15.4254	32.5178	52.8516	79.0733	109.357
		FEM	15.4438	32.5548	52.9087	79.1298	109.408
	1	DQM	18.0157	40.3494	65.6538	92.2848	119.836
		FEM	18.0465	40.3841	65.6882	92.3208	119.877
	10	DQM	18.3978	41.6361	69.0216	99.4111	131.859
		FEM	18.4315	41.6757	69.0611	99.4484	131.893
	→∞	DQM	18.4417	41.7750	69.3474	100.033	132.894
		FEM	18.4757	41.8151	69.3878	100.071	132.929
	0	DQM	11.3941	29.3678	53.0174	81.0192	112.024
		FEM	11.4148	29.4104	53.0660	81.0662	112.068
	0.1	DQM	15.6233	32.8965	55.0307	82.2247	112.825
		FEM	15.6400	32.9308	55.0763	82.2710	112.868
1	1	DQM	19.2962	41.3980	65.9339	92.3365	120.822
		FEM	19.3219	41.4295	65.9674	92.3723	120.859
	10	DQM	19.9179	43.3987	70.5662	100.622	132.723
		FEM	19.9463	43.4345	70.6034	100.658	132.756
	→∞	DQM	19.9899	43.6199	71.0558	101.509	134.136
		FEM	20.0187	43.6562	71.0937	101.546	134.170
	0	DQM	11.4913	30.3954	55.1815	84.0260	115.628
		FEM	11.5115	30.4328	55.2229	84.0663	115.665
	0.1	DQM	15.7621	33.1503	56.5688	84.8630	116.210
		FEM	15.7780	33.1835	56.6092	84.9031	116.247
10	1	DQM	20.4765	42.5961	66.2635	92.3899	121.730
		FEM	20.4994	42.6248	66.2963	92.4255	121.765
	10	DQM	21.3539	45.5548	72.8197	102.560	134.141
		FEM	21.3795	45.5875	72.8543	102.594	134.174
	→∞	DQM	21.4553	45.8807	73.5609	103.912	136.261
		FEM	21.4813	45.9139	73.5962	103.947	136.294
	0	DQM	11.5091	30.5860	55.6105	84.6706	116.454
		FEM	11.5291	30.6228	55.6510	84.7101	116.491

	0.1	DQM	15.7905	33.2010	56.8768	85.4233	116.975
		FEM	15.8064	33.2340	56.9165	85.4625	117.012
→∞	1	DQM	20.7557	42.9193	66.3549	92.4035	121.943
		FEM	20.7782	42.9475	66.3875	92.4392	121.978
	10	DQM	21.6961	46.1510	73.5285	103.223	134.643
		FEM	21.7214	46.1832	73.5626	103.257	134.675
	→∞	DQM	21.8045	46.5039	74.3452	104.735	137.026
		FEM	21.8302	46.5368	74.3801	104.769	137.059

Table 5. First natural frequencies $\Omega_i = \sqrt{\rho A_1(0) / EI_1(0)}\, L^2 \omega_i$ for a two-span elastically restrained rotating Timoshenko beam, with elliptical cross section and quadratic height variation along the axis. $l1 / L = 1/2$ $l2 / L = 1/2$, $hB1 / hA1 = 1/2$, $h'B1 = 0$, $hA2 / hB1 = 1/2$, $hB2 / hA2 = 1/2$, $h'A2 = 0$, $KW1 \to\infty$ $K\psi 1 \to\infty$, $\eta = 10$.

Kψ d	KWd	Method	Ω_1	Ω_2	Ω_3	Ω_4	Ω_5
0	0	DQM	11.8651	24.5717	40.8347	59.8775	81.1573
		FEM	11.8796	24.5914	40.8559	59.9110	81.1826
	0.1	DQM	15.2667	30.3903	49.8409	67.9228	90.5546
		FEM	15.2858	30.4064	49.8638	67.9498	90.5743
	1	DQM	15.6938	31.4585	52.2093	72.0062	98.7038
		FEM	15.7140	31.4754	52.2362	72.0330	98.7265
	10	DQM	15.7412	31.5756	52.4458	72.4214	99.4803
		FEM	15.7616	31.5927	52.4732	72.4482	99.5038
	→∞	DQM	15.7466	31.5887	52.4718	72.4669	99.5627
		FEM	15.7669	31.6059	52.4993	72.4937	99.5862
	0	DQM	11.9142	25.1342	42.8878	62.4877	85.6040
		FEM	11.9288	25.1532	42.9079	62.5196	85.6258
	0.1	DQM	16.2121	31.6526	50.5459	67.9979	90.8436
		FEM	16.2314	31.6672	50.5682	68.0245	90.8635
1	1	DQM	16.9952	33.8476	55.0090	75.3283	102.166
		FEM	17.0160	33.8634	55.0372	75.3520	102.190
	10	DQM	17.0842	34.0961	55.4704	76.2542	103.723
		FEM	17.1052	34.1120	55.4993	76.2779	103.748
	→∞	DQM	17.0942	34.1238	55.5205	76.3541	103.882
		FEM	17.1152	34.1398	55.5496	76.3778	103.907
	0	DQM	11.9157	25.1505	42.9498	62.5733	85.7690

		FEM	11.9302	25.1695	42.9699	62.6051	85.7907
	0.1	DQM	16.2528	31.7152	50.5831	68.0018	90.8571
		FEM	16.2721	31.7297	50.6053	68.0283	90.8770
10	1	DQM	17.0528	33.9729	55.1728	75.5622	102.430
		FEM	17.0737	33.9886	55.2011	75.5857	102.453
	10	DQM	17.1437	34.2281	55.6450	76.5200	104.034
		FEM	17.1648	34.2440	55.6741	76.5435	104.059
	→∞	DQM	17.1539	34.2566	55.6962	76.6231	104.197
		FEM	17.1750	34.2725	55.7255	76.6466	104.222
	0	DQM	11.9158	25.1524	42.9569	62.5831	85.7880
		FEM	11.9304	25.1713	42.9770	62.6149	85.8097
	0.1	DQM	16.2575	31.7225	50.5875	68.0023	90.8587
		FEM	16.2768	31.7370	50.6097	68.0288	90.8786
→∞	1	DQM	17.0595	33.9876	55.1921	75.5901	102.461
		FEM	17.0804	34.0033	55.2205	75.6136	102.485
	10	DQM	17.1506	34.2436	55.6656	76.5517	104.071
		FEM	17.1717	34.2595	55.6947	76.5752	104.096
	→∞	DQM	17.1609	34.2722	55.7169	76.6551	104.234
		FEM	17.1820	34.2881	55.7461	76.6786	104.259

Next Tables, 6 to 10, correspond to beams of two segments, elastically restrained at both ends and any particular details are expressed in each legend.

Table 6. First natural frequencies $\Omega_i = \sqrt{\rho A_1(0) / EI_1(0)}\, L^2\, \omega_i$ for a two-span elastically restrained rotating Timoshenko beam, with elliptical cross section and quadratic height variation along the axis. $l1 / L = 1/2$ $l2 / L = 1/2$, $hB1 / hA1 = 1/2$, $h'B1 = 0$, $hA2 / hB1 = 1/2$, $hB2 / hA2 = 1/2$, $h'A2 = 0$, $KW1 \to \infty$, $K\psi 1 = 0.1$, $\eta = 10$.

Kψ d	KWd	Method	Ω_1	Ω_2	Ω_3	Ω_4	Ω_5
0	0	DQM	9.98841	21.2706	37.3110	54.9224	77.7336
		FEM	10.0246	21.3074	37.3506	54.9699	77.7658
	0.1	DQM	12.4181	26.7466	45.4389	63.7717	87.4947
		FEM	12.4641	26.7810	45.4827	63.8054	87.5227
	1	DQM	12.7051	27.6834	47.3154	67.8701	94.9448
		FEM	12.7526	27.7193	47.3634	67.9035	94.9788
	10	DQM	12.7370	27.7869	47.5065	68.2901	95.6398
		FEM	12.7847	27.8230	47.5548	68.3236	95.6747

	→∞	DQM	12.7406	27.7985	47.5276	68.3362	95.7137
		FEM	12.7883	27.8347	47.5760	68.3697	95.7487
	0	DQM	10.0086	21.6505	39.0007	57.4853	82.1930
		FEM	10.0451	21.6876	39.0412	57.5291	82.2227
	0.1	DQM	13.1047	28.0775	46.2626	63.9615	87.6768
		FEM	13.1521	28.1101	46.3049	63.9941	87.7052
1	1	DQM	13.6220	29.9765	49.8847	71.5084	98.2767
		FEM	13.6718	30.0120	49.9325	71.5384	98.3117
	10	DQM	13.6812	30.1918	50.2664	72.4302	99.6646
		FEM	13.7312	30.2278	50.3147	72.4604	99.7012
	→∞	DQM	13.6879	30.2159	50.3082	72.5299	99.8074
		FEM	13.7379	30.2519	50.3566	72.5601	99.8441
	0	DQM	10.0092	21.6615	39.0505	57.5689	82.3537
		FEM	10.0457	21.6987	39.0910	57.6127	82.3835
	0.1	DQM	13.1336	28.1416	46.3059	63.9714	87.6854
		FEM	13.1811	28.1741	46.3481	64.0039	87.7138
10	1	DQM	13.6618	30.0922	50.0321	71.7561	98.5256
		FEM	13.7116	30.1278	50.0799	71.7860	98.5607
	10	DQM	13.7222	30.3131	50.4233	72.7081	99.9556
		FEM	13.7723	30.3491	50.4716	72.7381	99.9923
	→∞	DQM	13.7290	30.3379	50.4661	72.8107	100.102
		FEM	13.7792	30.3739	50.5145	72.8408	100.139
	0	DQM	10.0093	21.6628	39.0562	57.5786	82.3722
		FEM	10.0458	21.6999	39.0968	57.6224	82.4020
	0.1	DQM	13.1369	28.1491	46.3110	63.9726	87.6864
		FEM	13.1844	28.1817	46.3532	64.0051	87.7149
→∞	1	DQM	13.6664	30.1057	50.0495	71.7856	98.5555
		FEM	13.7163	30.1414	50.0973	71.8155	98.5906
	10	DQM	13.7270	30.3273	50.4418	72.7411	99.9903
		FEM	13.7771	30.3634	50.4901	72.7711	100.027
	→∞	DQM	13.7338	30.3521	50.4847	72.8441	100.137
		FEM	13.7840	30.3882	50.5331	72.8741	100.174

Table 7. First natural frequencies $\Omega_i = \sqrt{\rho A_1(0) / EI_1(0)}\, L^2 \omega_i$ for a two-span elastically restrained rotating Timoshenko beam, with elliptical cross section and quadratic height variation along the axis. $l1 / L = 1/2$ $l2 / L = 1/2$, $hB1 / hA1 = 1/2$, $h'B1 = 0$, $hA2 / hB1 = 1/2$, $hB2 / hA2 = 1/2$, $h'A2 = 0$, $KW1 = 10$, $K\psi 1 = 10$, $\eta = 10$.

Kψ d	KWd	Method	Ω_1	Ω_2	Ω_3	Ω_4	Ω_5
0	0	DQM	11.3734	23.4059	39.2570	56.9363	78.4787
		FEM	11.3904	23.4273	39.2820	56.9721	78.5041
	0.1	DQM	14.4551	28.9705	47.6260	65.1400	87.8678
		FEM	14.4773	28.9890	47.6542	65.1664	87.8894
	1	DQM	14.8320	29.9641	49.6507	69.1005	95.0494
		FEM	14.8553	29.9837	49.6827	69.1264	95.0757
	10	DQM	14.8738	30.0733	49.8536	69.5057	95.7100
		FEM	14.8972	30.0931	49.8861	69.5316	95.7371
	→∞	DQM	14.8785	30.0856	49.8761	69.5501	95.7801
		FEM	14.9019	30.1054	49.9086	69.5761	95.8073
	0	DQM	11.4126	23.8883	41.1048	59.4030	82.7965
		FEM	11.4297	23.9092	41.1294	59.4361	82.8192
	0.1	DQM	15.3087	30.2367	48.3683	65.2804	88.0547
		FEM	15.3312	30.2539	48.3957	65.3061	88.0766
1	1	DQM	15.9925	32.2677	52.2109	72.5363	98.2033
		FEM	16.0166	32.2865	52.2436	72.5592	98.2306
	10	DQM	16.0702	32.4974	52.6080	73.4315	99.5067
		FEM	16.0945	32.5165	52.6413	73.4545	99.5353
	→∞	DQM	16.0790	32.5231	52.6513	73.5282	99.6404
		FEM	16.1032	32.5422	52.6847	73.5513	99.6691
	0	DQM	11.4138	23.9023	41.1598	59.4837	82.9526
		FEM	11.4309	23.9232	41.1844	59.5168	82.9753
	0.1	DQM	15.3450	30.2988	48.4074	65.2877	88.0636
		FEM	15.3676	30.3159	48.4348	65.3134	88.0854
10	1	DQM	16.0434	32.3866	52.3585	72.7730	98.4379
		FEM	16.0676	32.4055	52.3914	72.7958	98.4653
	10	DQM	16.1228	32.6224	52.7651	73.6979	99.7796
		FEM	16.1471	32.6415	52.7985	73.7208	99.8083
	→∞	DQM	16.1317	32.6487	52.8094	73.7975	99.9164
		FEM	16.1560	32.6679	52.8428	73.8204	99.9453

	0	DQM	11.4139	23.9039	41.1661	59.4929	82.9706
		FEM	11.4310	23.9248	41.1908	59.5260	82.9933
	0.1	DQM	15.3493	30.3061	48.4120	65.2886	88.0646
		FEM	15.3718	30.3232	48.4394	65.3142	88.0865
→∞	1	DQM	16.0493	32.4005	52.3759	72.8012	98.4661
		FEM	16.0735	32.4194	52.4088	72.8240	98.4934
	10	DQM	16.1289	32.6371	52.7836	73.7295	99.8122
		FEM	16.1532	32.6562	52.8170	73.7524	99.8409
	→∞	DQM	16.1378	32.6635	52.8280	73.8295	99.9493
		FEM	16.1622	32.6827	52.8615	73.8524	99.9782

Table 8. First natural frequencies $\Omega_i = \sqrt{\rho A_1(0) / EI_1(0)}\, L^2 \omega_i$ for a two-span elastically restrained rotating Timoshenko beam, with elliptical cross section and quadratic height variation along the axis. $l1 / L = 1/2\; l2 / L = 1/2$, $hB1 / hA1 = 1/2$, $h'B1 = 0$, $hA2 / hB1 = 1/2$, $hB2 / hA2 = 1/2$, $h'A2 = 0$, $KW1 = 10$, $K\psi 1 = 5$, $\eta = 10$.

Kψ d	KWd	Method	Ω_1	Ω_2	Ω_3	Ω_4	Ω_5
0	0	DQM	11.0954	22.8658	38.6771	56.1987	78.0667
		FEM	11.1149	22.8898	38.7052	56.2367	78.0931
	0.1	DQM	14.0189	28.3688	46.9190	64.5736	87.5306
		FEM	14.0444	28.3902	46.9506	64.6011	87.5532
	1	DQM	14.3726	29.3408	48.8766	68.5581	94.6820
		FEM	14.3992	29.3634	48.9120	68.5850	94.7094
	10	DQM	14.4118	29.4478	49.0737	68.9659	95.3401
		FEM	14.4386	29.4706	49.1095	68.9930	95.3684
	→∞	DQM	14.4162	29.4598	49.0954	69.0107	95.4100
		FEM	14.4430	29.4826	49.1314	69.0377	95.4383
	0	DQM	11.1299	23.3178	40.4680	58.6771	82.4016
		FEM	11.1495	23.3414	40.4960	58.7122	82.4254
	0.1	DQM	14.8296	29.6459	47.6815	64.7320	87.7080
		FEM	14.8556	29.6659	47.7122	64.7587	87.7309
1	1	DQM	15.4691	31.6287	51.4150	72.0511	97.8483
		FEM	15.4967	31.6508	51.4509	72.0751	97.8766
	10	DQM	15.5418	31.8531	51.8028	72.9497	99.1495
		FEM	15.5696	31.8754	51.8392	72.9737	99.1791
	→∞	DQM	15.5499	31.8782	51.8452	73.0468	99.2832

		FEM	15.5778	31.9005	51.8817	73.0708	99.3129
	0	DQM	11.1309	23.3309	40.5211	58.7582	82.5578
		FEM	11.1505	23.3545	40.5491	58.7932	82.5816
	0.1	DQM	14.8640	29.7082	47.7217	64.7403	87.7163
		FEM	14.8900	29.7282	47.7523	64.7669	87.7393
10	1	DQM	15.5170	31.7461	51.5610	72.2905	98.0837
		FEM	15.5447	31.7682	51.5970	72.3143	98.1121
	10	DQM	15.5912	31.9764	51.9582	73.2187	99.4233
		FEM	15.6191	31.9987	51.9947	73.2426	99.4531
	→∞	DQM	15.5996	32.0021	52.0016	73.3186	99.5601
		FEM	15.6275	32.0245	52.0381	73.3426	99.5900
	0	DQM	11.1310	23.3324	40.5272	58.7675	82.5758
		FEM	11.1506	23.3560	40.5552	58.8025	82.5996
	0.1	DQM	14.8680	29.7155	47.7264	64.7412	87.7173
		FEM	14.8940	29.7355	47.7570	64.7679	87.7402
→∞	1	DQM	15.5225	31.7598	51.5783	72.3191	98.1119
		FEM	15.5503	31.7819	51.6143	72.3428	98.1403
	10	DQM	15.5970	31.9908	51.9765	73.2506	99.4560
		FEM	15.6249	32.0132	52.0131	73.2745	99.4857
	→∞	DQM	15.6053	32.0166	52.0200	73.3510	99.5931
		FEM	15.6333	32.0391	52.0566	73.3749	99.6230

Table 9. First natural frequencies $\Omega_i = \sqrt{\rho A_1(0) / EI_1(0)}\, L^2 \omega_i$ for a two-span elastically restrained rotating Timoshenko beam, with elliptical cross section and quadratic height variation along the axis. $l1/L = 1/2$ $l2/L = 1/2$, $hB1/hA1 = 1/2$, $h'B1 = 0$, $hA2/hB1 = 1/2$, $hB2/hA2 = 1/2$, $h'A2 = 0$, $KW1 = 10$, $K\psi 1 = 1$, $\eta = 10$.

Kψ d	KWd	Method	Ω_1	Ω_2	Ω_3	Ω_4	Ω_5
0	0	DQM	10.3650	21.7083	37.5771	54.9536	77.3897
		FEM	10.3937	21.7398	37.6121	54.9961	77.4183
	0.1	DQM	12.9366	27.1514	45.6367	63.6520	86.9652
		FEM	12.9736	27.1805	45.6754	63.6816	86.9900
	1	DQM	13.2417	28.0899	47.4960	67.6756	94.0576
		FEM	13.2800	28.1204	47.5385	67.7048	94.0875
	10	DQM	13.2755	28.1934	47.6848	68.0876	94.7110
		FEM	13.3141	28.2242	47.7276	68.1169	94.7416

	→∞	DQM	13.2794	28.2050	47.7057	68.1329	94.7804
		FEM	13.3179	28.2358	47.7485	68.1621	94.8111
	0	DQM	10.3892	22.1044	39.2705	57.4666	81.7463
		FEM	10.4182	22.1360	39.3061	57.5054	81.7724
	0.1	DQM	13.6565	28.4599	46.4416	63.8427	87.1268
		FEM	13.6946	28.4875	46.4789	63.8714	87.1519
1	1	DQM	14.2060	30.3646	50.0205	71.2606	97.2421
		FEM	14.2462	30.3947	50.0627	71.2866	97.2727
	10	DQM	14.2687	30.5803	50.3961	72.1634	98.5382
		FEM	14.3091	30.6108	50.4388	72.1896	98.5700
	→∞	DQM	14.2757	30.6044	50.4372	72.2610	98.6715
		FEM	14.3161	30.6350	50.4800	72.2872	98.7035
	0	DQM	10.3900	22.1160	39.3204	57.5486	81.9025
		FEM	10.4190	22.1475	39.3560	57.5874	81.9286
	0.1	DQM	13.6869	28.5231	46.4839	63.8527	87.1344
		FEM	13.7250	28.5506	46.5212	63.8813	87.1595
10	1	DQM	14.2479	30.4798	50.1652	71.5043	97.4786
		FEM	14.2881	30.5099	50.2075	71.5301	97.5092
	10	DQM	14.3119	30.7011	50.5501	72.4364	98.8132
		FEM	14.3524	30.7317	50.5928	72.4624	98.8452
	→∞	DQM	14.3191	30.7259	50.5922	72.5368	98.9498
		FEM	14.3596	30.7565	50.6350	72.5629	98.9819
	0	DQM	10.3901	22.1173	39.3262	57.5580	81.9205
		FEM	10.4190	22.1488	39.3618	57.5968	81.9466
	0.1	DQM	13.6904	28.5305	46.4889	63.8539	87.1353
		FEM	13.7286	28.5580	46.5261	63.8825	87.1604
→∞	1	DQM	14.2527	30.4933	50.1823	71.5333	97.5069
		FEM	14.2930	30.5235	50.2245	71.5591	97.5375
	10	DQM	14.3169	30.7153	50.5682	72.4688	98.8460
		FEM	14.3574	30.7458	50.6110	72.4948	98.8780
	→∞	DQM	14.3241	30.7401	50.6105	72.5696	98.9829
		FEM	14.3646	30.7707	50.6532	72.5957	99.0150

Table 10. First natural frequencies $\Omega_i = \sqrt{\rho A_1(0) / EI_1(0)}\, L^2 \omega_i$ for a two-span elastically restrained rotating Timoshenko beam, with elliptical cross section and quadratic height variation along the axis. $l1/L = 1/2$ $l2/L = 1/2$, $hB1/hA1 = 1/2$, $h'B1 = 0$, $hA2/hB1 = 1/2$, $hB2/hA2 = 1/2$, $h'A2 = 0$, $KW1 = 10$, $K\psi 1 = 0.1$, $\eta = 10$.

Kψ d	KWd	Method	Ω_1	Ω_2	Ω_3	Ω_4	Ω_5
0	0	DQM	9.94295	21.1789	37.1287	54.4936	77.1422
		FEM	9.97877	21.2145	37.1668	54.5377	77.1717
	0.1	DQM	12.3497	26.6244	45.1340	63.3202	86.7544
		FEM	12.3950	26.6574	45.1757	63.3506	86.7801
	1	DQM	12.6335	27.5522	46.9619	67.3571	93.8220
		FEM	12.6802	27.5867	47.0072	67.3872	93.8529
	10	DQM	12.6651	27.6547	47.1481	67.7705	94.4732
		FEM	12.7119	27.6894	47.1937	67.8007	94.5049
	→∞	DQM	12.6686	27.6662	47.1687	67.8159	94.5425
		FEM	12.7155	27.7009	47.2144	67.8461	94.5743
	0	DQM	9.96265	21.5532	38.7860	57.0231	81.5044
		FEM	9.99874	21.5890	38.8248	57.0634	81.5315
	0.1	DQM	13.0297	27.9504	45.9575	63.5235	86.9102
		FEM	13.0763	27.9817	45.9976	63.5530	86.9362
1	1	DQM	13.5409	29.8282	49.4887	70.9743	97.0122
		FEM	13.5897	29.8623	49.5335	71.0012	97.0437
	10	DQM	13.5994	30.0410	49.8608	71.8782	98.3059
		FEM	13.6484	30.0754	49.9060	71.9052	98.3387
	→∞	DQM	13.6060	30.0648	49.9016	71.9759	98.4391
		FEM	13.6550	30.0993	49.9469	72.0030	98.4720
	0	DQM	9.96324	21.5641	38.8347	57.1056	81.6606
		FEM	9.99933	21.5999	38.8736	57.1459	81.6876
	0.1	DQM	13.0583	28.0142	46.0008	63.5342	86.9175
		FEM	13.1049	28.0454	46.0408	63.5635	86.9436
10	1	DQM	13.5802	29.9428	49.6333	71.2194	97.2490
		FEM	13.6291	29.9769	49.6781	71.2462	97.2806
	10	DQM	13.6399	30.1611	50.0148	72.1525	98.5813
		FEM	13.6890	30.1956	50.0600	72.1794	98.6142
	→∞	DQM	13.6467	30.1856	50.0566	72.2531	98.7177

		FEM	13.6958	30.2201	50.1018	72.2800	98.7508
	0	DQM	9.96331	21.5653	38.8403	57.1151	81.6785
		FEM	9.99940	21.6011	38.8792	57.1553	81.7056
	0.1	DQM	13.0616	28.0216	46.0058	63.5354	86.9184
		FEM	13.1082	28.0529	46.0459	63.5648	86.9444
→∞	1	DQM	13.5848	29.9563	49.6504	71.2486	97.2773
		FEM	13.6337	29.9904	49.6952	71.2753	97.3090
	10	DQM	13.6447	30.1752	50.0329	72.1851	98.6141
		FEM	13.6938	30.2097	50.0781	72.2120	98.6471
	→∞	DQM	13.6514	30.1997	50.0748	72.2860	98.7509
		FEM	13.7006	30.2342	50.1201	72.3129	98.7840

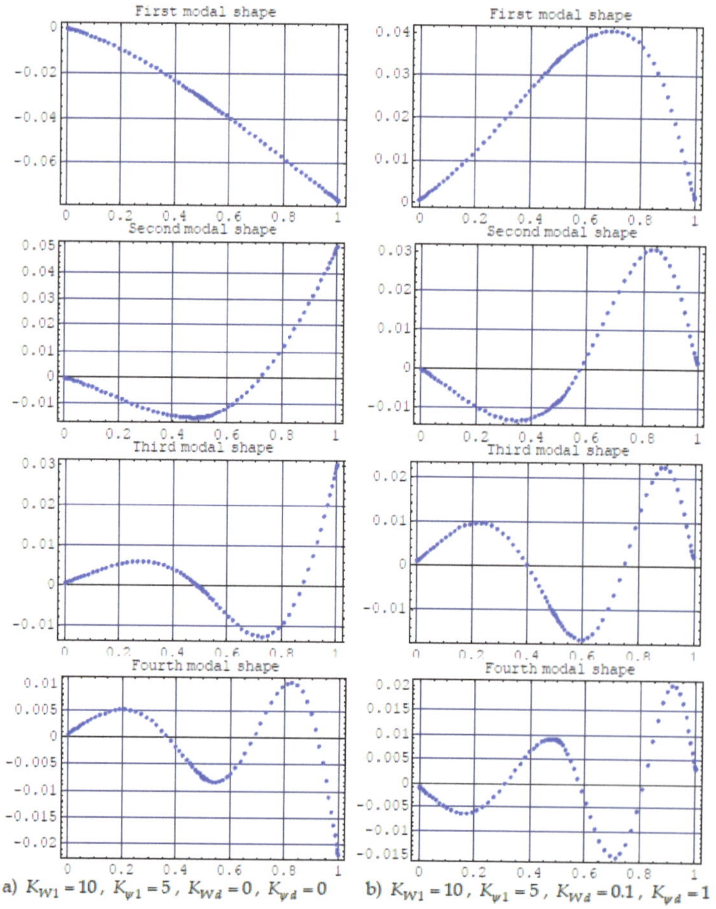

Fig. 4. Natural frequencies mode shapes for a two-span elastically restrained rotating Timoshenko beams, with elliptical cross section and quadratic height variation along the axis. $l1/L = l2/L = 1/2$; $hB1/hA1 = 1/2$; $h'B1 = 0$; $hA2/hB1 = 1/2$; $hB2/hA2 = 1/2$; $h'A2 = 0$; $\eta = 10$

Figure 4 shows the first four natural frequency mode shapes for beams, with two different kinds of boundary conditions: a) corresponds to $KW1 = 10$, $K\psi 1 = 5$, $KWd = 0$, $K\psi d = 0$, while b) corresponds to $KW1 = 10$, $K\psi 1 = 5$, $KWd = 0.1$, $K\psi d = 1$.

The next Figures, 5 and 6, present the variation of the fundament frequency parameter Ω_1 with the variation of the non-dimensional rotational speed η and the spring constant $K\psi_1$.

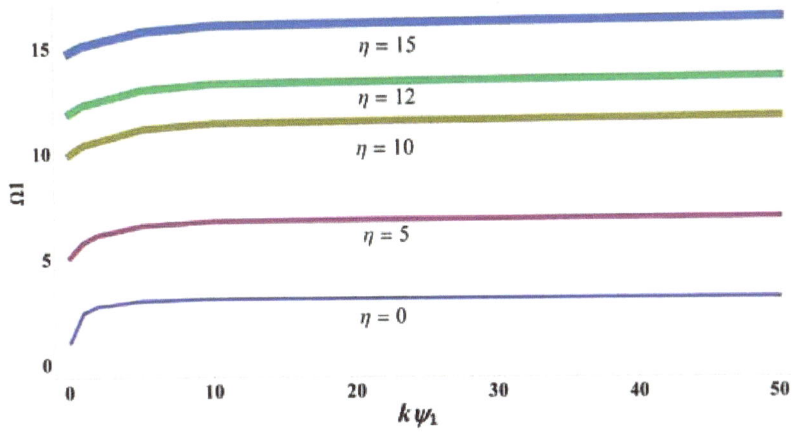

Fig. 5. The fundamental frequency coefficient Ω_1 of a one-span elastically restrained rotating Timoshenko beam versus the spring constant parameter of the rotational spring $K_{\psi 1}$, for different rotational speed parameters η. K_{w1}= 10; K_{wd}=1; $K_{\psi d}$=10

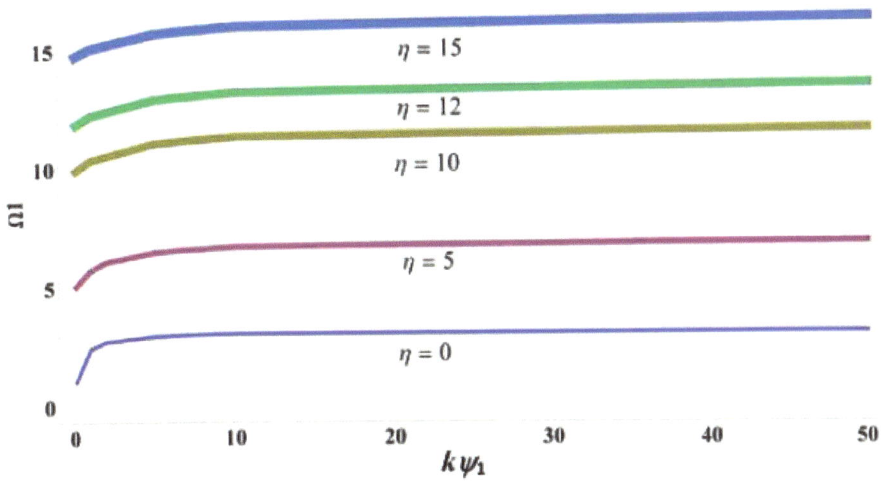

Fig. 6. The fundamental frequency coefficient Ω_1 of a two-span elastically restrained rotating Timoshenko beam versus the spring constant parameter of the rotational spring $K_{\psi 1}$, for different rotational speed parameters η. K_{w1}= 10; K_{wd}=0; $K_{\psi d}$=0

6. CONCISE SUMMARY

The differential quadrature approach is highly effective for determining the frequencies and mode shapes of natural vibrations in the rotating Timoshenko beam model. The adaptability of the suggested beam model (changing cross-section, abrupt changes in cross-section, elastic supports at both ends) enables the resolution of numerous distinct scenarios. It is noteworthy that the approach directly resolves two ordinary differential equations, hence avoiding imposing additional constraints. This phenomenon does not occur in alternative approaches, such as the dynamic stiffness method (Banerjee, 2000, 2001). The differential quadrature approach possesses the same advantages as the finite element method while requiring less computational memory than FEM. The current results indicate that the frequency coefficients exhibit greater variability when the translational spring stiffness at the beam's end, farthest from the axis of rotation Kψ, is altered.$_d$.

7. APPENDIX A

As Shu presents in his book (Shu, 2000), the differential quadrature method, DQM, is a numerical technique for solving differential equations.

In order to obtain the DQM analog equations to the governing equations of the rotating beam and its boundary conditions, the beam domain is discretized in a grid of points using the Chebyshev – Gauss - Lobato expression, (Shu & Chen, 1999):

$$x_i = \frac{1-\cos\left[(i-1)\pi/(n-1)\right]}{2} \quad ; \; i = 1,2,\ldots,n$$

where *n* is the number of discrete points or nodes and *xi* is the coordinate of node *i*

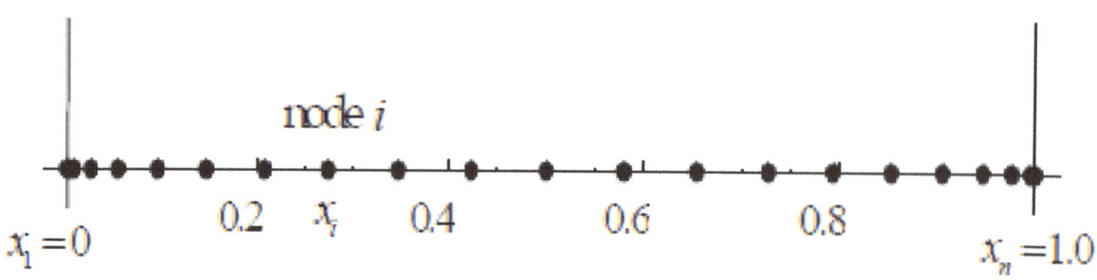

Fig. A1. Grid of *n* points

The weighting coefficients $A_{ij}^{(1)}$ and $A_{ij}^{(2)}$ which appeared in the linear algebraic equations of quadrature (28-35), were determined using the explicit expressions cited by (Bert & Malik, 1996).

The coefficients matrix of correspond $A_{ij}^{(1)}$ to first order derivatives and can be arranged in a square order *n*. (1)

The matrix elements $A_{ij}^{(1)}$ with $i \neq j$, are determined by:

$$A_{ij}^{(1)} = \frac{\Pi(x_i)}{(x_i - x_j)\,\Pi(x_j)}$$

where

$$\Pi(x_i) = \prod_{\nu=1\ \nu\neq i}^{n} (x_i - x_\nu)\ ;\ \Pi(x_j) = \prod_{\nu=1\ \nu\neq i}^{n} (x_j - x_\nu)\ ;$$

The coefficients $A^{(1)}$ with $i = j$, will tend to infinity and need to be calculated in another way.

The coefficients $A^{(2)}$ correspond to second-order derivatives and are obtained from

$$A_{ij}^{(2)} = 2\left[A_{ii}^{(1)} \cdot A_{ij}^{(1)} - \frac{A_{ij}^{(1)}}{x_i - x_j}\right]$$

with $i \neq j$ and $i, j = 1, 2, 3, \ldots, n$.

Because the sum of the weighting coefficients of a row of the matrix is zero, it is easy to calculate the diagonal terms of derivatives of any order q, using the following expression:

$$A_{ii}^{(q)} = -\sum_{j=1\ j\neq i}^{n} A_{ij}^{(q)}$$

And the equations for q equal to 1 and 2, corresponding to first and second order derivatives, are:

$$A_{ii}^{(1)} = -\sum_{j=1\ j\neq i}^{n} A_{ij}^{(1)}\ ;\ A_{ii}^{(2)} = -\sum_{j=1\ j\neq i}^{n} A_{ij}^{(2)}$$

REFERENCES

1. Bambill, D.V.; Felix, D.H. & Rossi, R. E. (2010). Vibration analysis of rotating Timoshenko beams by means of the differential quadrature method. *Structural Engineering and Mechanics*, Vol. 34, No. 2, pp. 231-245, ISSN 12254568

2. Banerjee, J. (2000). Free vibration of centrifugally stiffened uniform and tapered beams using the dynamic stiffness method. *Journal of Sound and Vibration*, Vol.233, No.5, pp. 857- 875, ISSN 0022-460X

3. Banerjee, J. (2001). Dynamic stiffness formulation and free vibration analysis of centrifugally stiffened Timoshenko beam. *Journal of Sound and Vibration*, Vol.247, pp. 97-115, ISSN 0022-460X

4. Banerjee, J.; Su, H, & Jackson, D. (2006). Free vibration of rotating tapered beams using the dynamic stiffness method. *Journal of Sound and Vibration*, Vol. 298, pp. 1034-1054, ISSN 0022-460X
5. Bellman, R. & Casti, J. (1971). Differential quadrature and long-term Integration. *J. Math. Anal.* Vol.34, pp. 235-238, ISSN 0022-247X
6. Bellman, R.E. & Roth, R.S. (1986). *Methods in approximation: techniques for mathematical modelling*, Editorial D. Reidel Publishing Company, ISBN 9-027-72188-2, Dordrecht, Holland
7. Bert, C. & Malik, M. (1996). Differential quadrature method in computational mechanics: A review. *Applied Mechanics Review* Vol.49, pp. 1-28, ISSN 0008-6900
8. Choi, S.; Wu J. & Chou Y. (2000). Dynamic analysis of a spinning Timoshenko beam by the differential quadrature method. *American Institute of Aeronautics and Astronautics* Vol.38, pp. 51-856, ISSN 0001-1452
9. Felix, D.H.; Rossi, R. E. & Bambill, D. V. (2008). Vibraciones transversales por el método de cuadratura diferencial de una viga Timoshenko rotante, escalonada y elásticamente vinculada, *Mecánica Computacional* Vol. XXVII, pp.1957-1973, ISBN 1666-6070
10. Felix, D. H.; Bambill, D. V. & Rossi, R. E. (2009). Análisis de vibración libre de una viga Timoshenko escalonada, centrífugamente rigidizada, mediante el método de cuadratura diferencial, *Revista Internacional de Métodos Numéricos para Cálculo y Diseño en Ingeniería.* Vol. 25, No. 2, pp. 111-132, ISSN 0213-1315
11. Ganesh, R and Ganguli, R. (2011). Physics based basis function for vibration analysis of high speed rotating beams. *Structural Engineering and Mechanics*, Vol.39, No.1, pp. 21-46, ISSN 1225-4568
12. Gunda, J. B. & Ganguli R. (2008). New rational interpolation functions for finite element analysis of rotating beams. *International Journal of Mechanical Sciences*; Vol. 50, pp. 578-588, ISSN 0020-7403
13. Gunda, J.B.; Singh, A.P.; Chhabra, P.S. & Ganguli, R. (2007). Free vibration analysis of rotating tapered blades using Fourier-p superelement, *Structural Engineering and Mechanics*, Vol.27, No.2, pp. 243-257, ISSN 1225-4568
14. Kumar A. & Ganguli R. (2009). Rotating Beams and Nonrotating Beams with Shared Eigenpair, *Journal of Applied Mechanics.* Vol.76. No.5, pp. 1-14, ISSN: 0021-8936
15. Hodges, D. H. & Rutkowski, M. J. (1981). Free vibration analysis of rotating beams by a variable order finite method, *American Institute of Aeronautics and Astronautics Journal.* Vol.19, No.11, pp. 1459-1466

16. Lin, S. C. & Hsiao, K. M. (2001). Vibration analysis of a rotating Timoshenko beam. *Journal of Sound and Vibration.* Vol. 240, pp. 303-322.

17. Liu, G. R. & Wu, T. Y. (2001). Vibration analysis of beams using the generalized differential quadrature rule and domain decomposition. *Journal of Sound and Vibration.* Vol.246, pp.461-481, ISSN 0022-460X

18. Naguleswaran, S. (2004). Transverse vibration and stability of an Euler–Bernoulli beam with step change in cross-section and in axial force. *Journal of Sound and Vibration.* Vol.270, pp.1045-1055, ISSN 0022-460X

19. Özdemir, Ö. & Kaya, M.O. (2006). Flapwise bending vibration analysis of a rotating tapered cantilever Bernoulli–Euler beam by differential transform method. *Journal of Sound and Vibration.* Vol.289, pp.413-420, ISSN 0022-460X

20. Ozgumus, O. & Kaya, M. O. (2010). Vibration analysis of rotating tapered Timoshenko beam using DTM. Meccanica. Vol. 45, pp. 33-42, ISSN 0025-6455

21. Rossi R.E. (2007). *Introducción al análisis de Vibraciones con el Método de Elementos Finitos.* EdiUNS, Universidad Nacional del Sur, IBSN 978-987-1171-71-2, Bahía Blanca, Argentina.

22. Rossi, R. E.; Gutiérrez R. H. & Laura P. A. A. (1991). Transverse vibrations of a Timoshenko beam of nonuniform cross section elastically restrained at one end and carrying a concentrated mass at the other. *J. Acoust. Soc. Am,* Vol.89, pp.2456-2458.

23. Seon, M. H.; Benaroya, H. & Wei, T. (1999). Dynamics of transversely vibrating beams using four engineering theories. *Journal of Sound and Vibration.* Vol.225, pp.35-988, ISSN 0022-460X

24. Singh, A.P.; Mani, V. & Ganguli, R. (2007). Genetic programming metamodel for rotating beams, *CMES - Computer modelling in Engineering and Sciences,* Vol.21. No.2, pp. 133-148.

25. Shu, C. (2000). *Differential Quadrature and Its Application in Engineering,* Springer-Verlag, ISBN 1852332093, London, England

26. Shu, C. & Chen, W. (1999). On optimal selection of interior points for applying discretized boundary conditions in DQ vibration analysis of beams and plates. *Journal of Sound and Vibration.* Vol.222, No.2, pp. 239-257, ISSN 0022-460X

27. Vinod, K. G., Gopalakrishnan, S. & Ganguli, R. (2007), Free vibration and wave propagation analysis of uniform and tapered rotating beams using spectrally formulated finite elements. *International Journal of Solids and Structures*; Vol.44, pp. 5875-5893, ISSN 0020-7683

28. Yang, J. B.; Jiang, L. J. & Chen, D. CH. (2004). Dynamic modelling and control of a rotating Euler–Bernoulli beam. *Journal of Sound and Vibration.* Vol.274, pp. 863-875, ISSN 0022-460X

14 Chapter

VIBRATION-DRIVEN DIAGNOSTICS OF STEAM TURBINES

1. INTRODUCTION: AN OVERVIEW

Among the three primary maintenance strategies—run-to-break, preventive maintenance, and predictive maintenance—the latter, known as condition-based maintenance, is increasingly acknowledged as the most successful (see e.g. Randall, 2011). To fully realize its potential, it must rely on dependable condition assessment methods and procedures. This is especially crucial for essential machinery, distinguished by elevated unit costs and significant repercussions in the event of a failure. Steam turbines serve as an exemplary illustration in this context. Technical diagnostics can be described as the assessment of technical condition by objective methodologies and metrics. Objectivity denotes that the evaluation of technical condition relies on quantifiable physical metrics. These quantities serve as sources of diagnostic symptoms. The development of technical diagnostics for any class of objects comprises four primary steps (Crocker, 2003):

- measurement,
- qualitative diagnostics,
- quantitative diagnostics,
- prognosis (forecasting).

During the measuring phase, we can quantify physical parameters pertinent to the object's technical condition. Utilizing measurement data, flaws and malfunctions are recognized and located during the qualitative diagnostics phase through an appropriate diagnostic model. Quantitative diagnostics involves assessing the extent of harm (progression), for which a reference scale is essential. Ultimately, prognosis is an assessment of the duration until action becomes necessary. Qualitative diagnostics focus on identifying hard (random) failures, whereas quantitative diagnostics seek to analyze the progression of soft (natural) faults (Martin, 1994).

Complex devices, such as steam turbines, exhibit various residual processes (including vibration, noise, heat radiation, etc.) that accompany the primary energy transformation process, resulting in multiple forms of condition symptoms. Vibration-based indicators are crucial for the technical condition assessment of all rotating equipment for at least three reasons.:

- high content of information,
- comparatively easy and non-intrusive measurement techniques,
- well-developed methods for data processing and diagnostic information extraction.

Of all vibration-based symptom types (see e.g. Morel, 1992; Orłowski, 2001), three are of particular importance for steam turbine diagnostics:

- absolute vibration spectra,
- relative vibration vectors,
- time evolution of spectral components.

These symptoms form the basis of diagnostic reasoning in both permanent (*on-line*) and intermittent (*off-line*) monitoring systems.

2. VIBRATION GENERATION AND VIBRODIAGNOSTIC SYMPTOMS

Just like all rotating machines, steam turbines generate broadband vibration, so that power density spectra typically contain a number of distinct components. Due to different vibration generation mechanisms involved, it is convenient to divide the entire frequency range under consideration (typically from a few hertz up to some 10 to 20 kilohertz) into two sub-ranges, commonly referred to as the *harmonic* (or 'low') and *blade* (or 'high') frequency ranges, respectively. Sometimes the sub-harmonic range (below the fundamental frequency f_0 resulting from rotational speed) is also distinguished. This division is shown schematically in Fig.1.

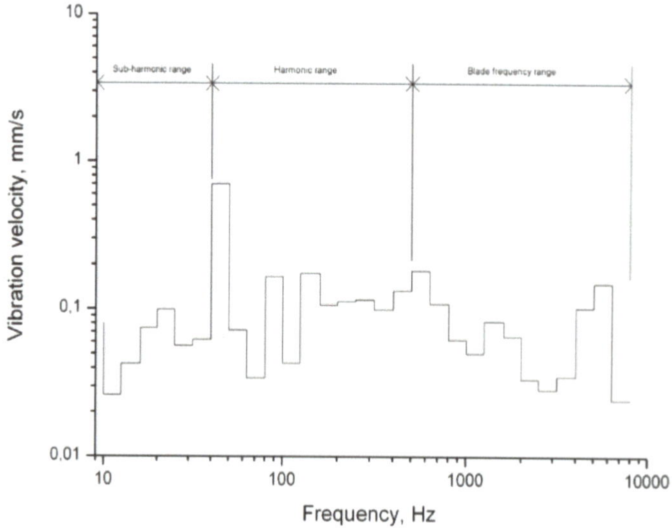

Fig. 1. Schematic representation of dividing the entire power density spectrum frequency range into sub-harmonic, harmonic and blade frequency ranges (after Gałka, 2009a).

Components from the harmonic range result directly from the rotary motion of turbine shaft and are related to malfunctions common to all rotating machines, such as:

- unbalance,
- shafts misalignment,
- bent or cracked rotors,
- magnetic phenomena in the generator.

Components that fall into the sub-harmonic range are typically determined mainly by the stability of the oil film in shaft bearings (Bently and Hatch, 2002; Kiciński, 2006). Those of very low frequencies (a few hertz) may be indicative of cracks in turbine casings and other non-rotating elements.

Individual components from the blade frequency range are produced as a result of interaction between steam flow and the fluid-flow system, and hence may be considered specific to steam turbines. There are three basic phenomena involved (Orłowski, 2001; Orłowski and Gałka, 1998), namely:

- flow disturbance caused by stationary and rotating blades edges,
- flow disturbance resulting from scatter of fluid-flow system elements dimensions.,
- flow disturbance by control valves opening.

First of these can be described in the following way: discharge edges of stationary and rotating blades introduce local interruptions of steam flow, thus reducing its thrust on a rotating blade and causing an instantaneous force of the opposite direction. Resulting force q_1 is thus periodic and can be expressed by

$$q_1 = \zeta_0 + \Sigma\zeta_k \cos k(n\omega t + \Psi_k) \quad (1)$$

where ζ_0 is time-averaged thrust, ζ_k and ψ_k are amplitude and phase of the k-th component, respectively, n is number of blades in a stage (stationary or rotating) under consideration and w denotes angular frequency. This force can thus be expressed as a series of harmonic components with frequencies equal to knw = $2nknu$, where u is the rotational speed in s-1. As for the second phenomenon, it results from the fact that manufacture of blades and their assembly into rotor stages or bladed diaphragms are not perfect, so for each blade the corresponding discharge cross-section is slightly different from the other ones. Resulting force has a form of a pulse generated once per rotation and thus may be expressed by

$$q_2 = \zeta_0 + \Sigma\zeta_k \cos k(wt + \psi_k)\ . \quad (2)$$

The third phenomenon is related to turbine control and shall be dealt with a little later. It should be mentioned, however, that – unlike the first two – the influence of control valves opening is usually limited to the vicinity of the control stage and diminishes as we move along the steam expansion path. Frequencies of basic spectral components resulting from interaction between steam flow and the fluid-flow system can be, on the basis of above considerations, expressed by

$$f_w = l \cdot u \quad (3)$$

$$f_k = b \cdot u \quad (4)$$

$$f_{(w+k)/2} = (l + b) \cdot u/2 \quad (5)$$

$$f_{(w-k)/2} = (l - b) \cdot u/2 \quad (6)$$

where l and b denote numbers of blades in rotor stages and bladed diaphragms, respectively. Components given by Eqs.(5) and (6) result from interactions between rotor stages and adjacent bladed diaphragms. Each turbine stage is thus in general characterized by as many as six individual vibration components.

Vibration signal that can be effectively measured in an accessible point of a turbine is influenced not only by relevant generation mechanisms, but also by its propagation to this point, as well as by operational parameters and interference (see e.g. Radkowski, 1995; Gałka, 2011b). In general terms it may be expressed as (Radkowski, 1995)

$$z(r,t) = h_p(r,t) * u_w(r,t) + \eta(r,t)\,, \quad (7)$$

where z denotes measured diagnostic signal, h_p is the response function for signal propagation from its origin to the measuring point and η denotes uncorrelated noise; all these quantities are functions of the spatial variable r and dynamic time t. $u_w(r,t)$ is given by

$$u_w(r,t) = \sum_{i=1}^{n} h_i(r,D_i,t) * x(t) + h(r,t) * x(t)\,, \quad (8)$$

where D_i describes development of the ith defect, h_i is the response function pertaining to this defect and h is the response function with no defect present; $x(t)$ is the input signal, generated by an elementary vibroacoustic signal source. This model is shown schematically in Fig.2.

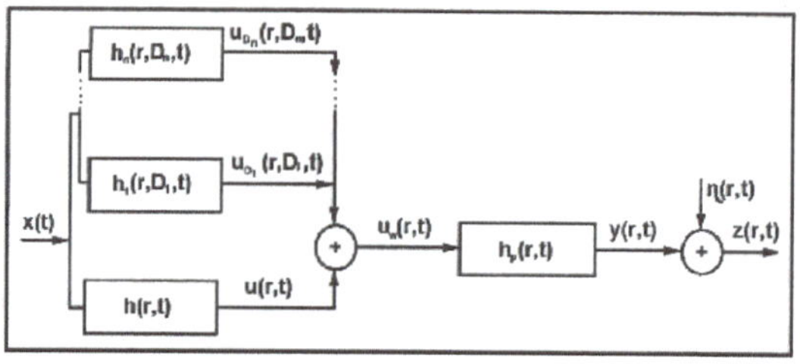

Fig. 2. Model of vibroacoustic signal generation and propagation (after Radkowski, 1995).

An alternative general relation, in a vector form, is provided by (Orłowski, 2001)

$$\mathbf{S}(\theta) = \mathbf{S}[\mathbf{X}(\theta), \mathbf{R}(\theta), \mathbf{Z}(\theta)], \quad (9)$$

where **S**, **X**, **R** and **Z** denote vectors of symptoms, condition parameters, control parameters and interference, respectively, all varying with time θ.[1] Control parameters may be defined as resulting from object operator purposeful action, aimed at obtaining demanded performance (Gałka, 2011b). In steam turbines, usually (at least in power industry) the 'demanded performance' means demanded output power; active load P_u can thus be treated as a scalar measure of the vector **R**. As for the interference, two types can be distinguished: *external interference* (the source is outside the object) and *internal interference* (the source is within the object). With some reservations, the former can be identified with measurement errors, while the latter refers to all other contributions to the uncorrelated noise $\eta(t)$ in Eq.(7).

Let us assume that the influence of interference may be reduced to a point wherein it can be neglected. As control parameters are, at any given moment, known, there is obviously a possibility of symptom normalization with respect to them, either model-based or empirical. It has to be kept in mind, however, that normalization with respect to P_u, which seems most straightforward, in practice may be only approximate. P_u can be expressed as (Traupel, 2000)

$$P_u = (dm/dt)\Delta i \eta_t, \quad (10)$$

where dm/dt denotes steam mass flow, Δi is the enthalpy drop and η_t is the turbine efficiency. Assuming that η_t remains constant (which is only an approximation), P_u may be controlled by changing Δi (qualitative control), dm/dt (quantitative control), or both. The latter method (known as group or nozzle control) is typically used in large steam turbines. Each control valve supplies steam to its own control stage section; the number of these valves in large steam turbines is usually from three to six and they are opened in a specific sequence. At the rated power the last valve is only partly open, or even almost closed, as it provides a reserve in a case of a sudden drop of steam parameters. Furthermore, Δi depends also on condenser vacuum, which for a given unit may change within certain limits depending on overall condenser condition, cooling water temperature, weather etc. Thus

$$P_u = f(r_1, r_2, \ldots, r_k, p_o), \quad (11)$$

where r_i denotes ith valve opening, k is the number of valves and p_o is the condenser pressure. In fact, r_i and p_o are the $\mathbf{R}(\theta)$ vector parameters, various combinations of which may yield the same value of P_u. In view of Eqs.(9) and (11), any $S_i(P_u)$ function ($S_i \in \mathbf{S}$) cannot thus be a single-valued one.

Some attention has been paid to developing experimental relations of the $S_i = f(P_u)$ type (see e.g. Gałka, 2001), bearing in mind that they are approximate and applicable to a given turbi- ne type only. Such relations turn out to be strongly non-linear and differences between individual symptoms are considerable. In general, within the load range given by roughly $P_u = (0.85 \div 1.0)P_n$, where P_n is rated power, variations are quite small; thus, when dealing with large sets of data, the simplest approach is to reject those acquired at extremely low or high loads. It has to be added that the fact of vibration-based symptoms dependence on control parameters and interference may serve as a basis for developing certain diagnostic procedures; this issue shall be dealt with in Section 6.

3. QUALITATIVE DIAGNOSIS

As already mentioned, qualitative diagnosis consists in determining what malfunctions or damages are present and localizing them. In this Section the influences of control and inter- ference shall be neglected, i.e. it shall be assumed that symptoms under consideration are deterministic functions of condition parameters $X_i \in \mathbf{X}$.

For obvious reasons, the following review does not claim to be exhaustive and is concentrated on issues relevant to steam turbine applications. For comprehensive and detailed treatment the reader is referred e.g. to (Morel, 1992; Bently and Hatch, 2002; Randall, 2011).

3.1 Harmonic (low) frequency range

Basically this subsection deals with absolute vibration spectral components of frequencies determined by $f = nf_0$, where f_0 results from rotational speed, and to some extent also with relative vibration vectors or orbits. In practical applications, components corresponding to $n > 4$ are seldom accounted for; this means that we are dealing with first four harmonic and sub-harmonic ($n < 1$) components. As each of these is typically influenced by a number of condition parameters, it is convenient to speak in terms of possible malfunctions and faults rather than frequencies.

3.1.1 Unbalance

Unbalance is common to all rotating elements. Primary symptom of this malfunction is the $1 \times f_0$ component of absolute vibration in a direction perpendicular to the turbine shaft line. They are, however, many other possible malfunctions (some of them quite common) that produce similar vibration patterns; additional procedures are therefore usually needed for a correct diagnosis.

In general, a 'pure' unbalance, be it static, quasi-static or dynamic, produces a $1 \times f_0$ compo- nent that remains almost constant in amplitude and phase during steady-state operation and disappears at low rotational speed. As rotor systems are non-linear, this component is typically accompanied by higher harmonics ($n > 1$), with amplitudes decreasing as n in- creases. Shaft orbits usually are quite regular and nearly circular or slightly elliptical. If such vibration pattern is present, the probability of unbalance being the root cause is high. Proper rotor balancing will usually reduce the residual unbalance to an acceptable level.

Rotor systems will always respond to balancing. Step changes of the $1 \times f_0$ component not related to any maintenance activities (but occurring mainly after turbine shutdown and sub- sequent startup) may be indicative of a loose rotor disk. Similarly, sudden and dramatic change may result from a broken rotor blade; such step changes are often big enough to enforce turbine tripping. Much slower, but continuous increase is often indicative of a permanent rotor bow (see also sub-section 3.1.3). An example is shown in Fig.3; it is easily seen that balancing results in a considerable decrease of the $1 \times f_0$ component, but the improvement is only temporary. If this component is comparatively high at low rotational speed, coupling problem (offset rotor axles) is a possible root cause, especially in turbines with rigidly coupled rotors.

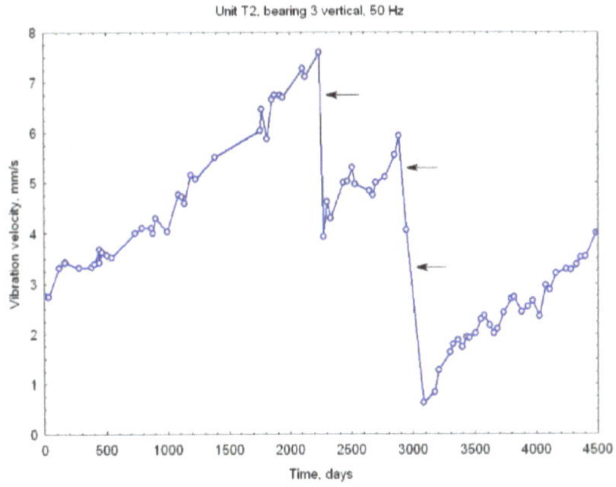

Fig. 3. Time history of the $1 \times f_0$ component with permanent rotor bow present: 230 MW unit, rear intermediate-pressure turbine bearing, vertical direction. Arrows indicate balancing sessions.

3.1.2 Misalignment

Ideally the entire turbine-generator unit shaft line (with overall length approaching 70 m in large units in nuclear power plants) should be a continuous and smooth curve; a departure from such condition is referred to as misalignment. The shape of this line is determined by shaft supports (journal bearings). As they displace during the transition from 'cold' to 'hot' condition, due to changing temperature field (this process may take even a few days to complete), at the assembly stage care has to be taken to ensure that the proper shape is maintained during normal operation. Relative vertical displacements may be even of the order of millimeters (Gałka, 2009a).

Misalignment modifies distribution of load between individual shaft bearings and therefore affects shaft orbits. With increasing misalignment magnitude they typically evolve from elongated elliptical shape through bent ('banana') and finally to highly flattened one (Bently and Hatch, 2002). High misalignment may lead to oil film instability, but in large steam turbines (especially modern ones, with only one bearing per coupling) this is a very rare occurrence. As for absolute vibration, $2 \times f_0$ component in directions perpendicular to the turbine axis is generally recognized as the basic misalignment symptom. Care, however, has to be taken when dealing with the turbine-generator coupling, as this component may be dominated by the influence of the generator (asymmetric position of rotor with respect to the stator electromagnetic field); in the latter case, dependence on the excitation current is usually conclusive. Marked misalignment is often accompanied with relatively high amplitudes of harmonic components in axial direction, but this symptom can by no means be considered specific.

3.1.3 Rotor bow

In general, three types of turbine rotor bow can be distinguished, namely:

- elastic bow, resulting from static load,
- temporary bow, caused by uneven temperature field and/or anisotropic rotor material properties, and
- permanent bow, wherein material yield strength has been exceeded (plastic defor- mation).

Permanent bow is obviously the most serious one. As it causes the center of gravity to move off from the shaft centerline, it basically produces an unbalance (cf. Fig.3). In general, rotor response vector may be expressed as (Bently and Hatch, 2002):

$$\mathrm{r} = r_e e^{j\delta} + \frac{M r_e \omega^2 e^{j\delta}}{[K - M\omega^2 + jD(1-\lambda)\omega]}, \tag{12}$$

where M denotes unbalance mass, shifted at the distance r_e in the direction determined by the angle S. K and D are stiffness and damping coefficients, respectively; w denotes rotor angular velocity and λ is the fluid circumferential velocity ratio ($\lambda = v/w$, where v denotes average fluid angular velocity). First term describes the low-speed response (which, as men- tioned earlier, is basically absent with 'plain' unbalance), while the second one refers to the dynamic synchronous response. It can be seen that for $w >> w_r$ (where w_r is the resonance angular speed), when the first and the third term in the denominator can be neglected, rotor response is close to zero. This is a feature characteristic for this malfunction (colloquially speaking, the rotor 'balances itself out'), but in large steam turbines with heavy flexible rotors the $w >> w_r$ condition is seldom fulfilled.

It has been shown (Gałka, 2009b) that permanent rotor bow causes simultaneous increase of the $1 \times f_0$ component in vertical and axial directions, so that a developing bow should result in strong correlation between these components (see also Section 6). Available data seem to confirm this conclusion, in fact based on quite simple model considerations.

3.1.4 Rotor crack

As a very serious fault with potentially catastrophic consequences, rotor crack has received considerable attention (for perhaps the most comprehensive available review, see Bach- schmid, Pennacchi and Tanzi, 2010). In general, crack reduces shaft stiffness and thus causes resonance to shift to a lower rotational speed. As a result, the $1 \times f_0$ component amplitude during steady-state operation will either increase or decrease. In large steam turbines, ope- rated above the first critical speed, the latter may be the case. This effect may be combined with that of increasing rotor bow due to reduced bending stiffness. As a result of asymmetry introduced by a crack, the $2 \times f_0$ component may also increase substantially.

It is generally recognized that considerable continuous changes of first two harmonic com- ponents amplitudes (not necessarily both increasing!) and phases during steady-state operation indicate that a shaft crack is possibly present. Rates of these changes vary within broad limits, from the order of months to days or even hours – in the latter case, a cata- strophic failure is most probably imminent. Such evolution of vibration patterns should serve as an alert. Presence of a crack may be confirmed by monitoring absolute and relative vibration during transients – typically after a turbine trip. Time histories of the $1 \times f_0$ and $2 \times f_0$ components, obtained in such manner, may be compared with reference data recorded after unit commissioning or a major overhaul. Significant reduction of critical speeds and increase of vibration amplitudes on passing through them are indicative of this malfunction, as well as is high overall relative vibration amplitude; the latter will sometimes render the startup impossible to complete, as the unit shall be tripped automatically below nominal rotational speed.

3.1.5 Bearing problems

A problem specific to shaft journal bearings is oil film instability that induces so-called self- excited vibrations. This issue has attracted considerable attention and detailed theoretical models have been developed (Bently and Hatch, 2002; Kiciński, 2006). It can be shown that threshold rotational speed for the onset of instability Ω_{th} is given by

$$\Omega_{th} = \frac{1}{\lambda}\sqrt{\frac{K}{M}}, \qquad (13)$$

where K and M denote stiffness and mass, respectively, and λ is the oil circumferential velocity ratio. It is therefore obvious that a suitable stability margin should be provided by proper design and operation, which influence all three quantities that determine Ω_{th}.

Bearing instability is nicely demonstrated with laboratory-scale model rotor systems. For large steam turbines in power industry, operated at a fixed rotational speed, this is a rare occurrence. Most frequently it results from bearing vertical displacement, due to thermal deformation and/or foundation distortion. Downward displacement reduces bearing load and causes K to decrease, so that Ω_{th} may become lower than the nominal rotational speed. In such circumstances, instability is unavoidable. Most typical symptom of this malfunction is the increase of sub-harmonic spectral components, often of the 'hump' shape centered slightly below $0.5 \times f_0$. Shaft orbits typically exhibit loops. Strong instability results in high relative vibration that leads to bearing damage. Proper adjustment of bearing positions is the primary action to be taken; sometimes reduction of the bearing size (length), in order to increase specific load, is necessary for a permanent remedy (Orłowski and Gałka, 1995).

Due to strong non-linearity, journal bearings generate higher harmonic components which may be very sensitive to bearing condition, clearances and oil pressure. An example is shown in Fig.4, in the form of a time history of the $3 \times f_0$ absolute horizontal vibration com- ponent. Initially very low, it increased dramatically following a minor bearing damage and remained at a high level, exhibiting considerable variations that suggest a resonance nature of the phenomenon. Permanent improvement was achieved only after a major overhaul. It has to be noted that such behavior is to a large extent influenced by design features; therefore care has to be taken when generalizing the results over other turbine types. In any case, sensitivity of spectral components to oil pressure is decisive.

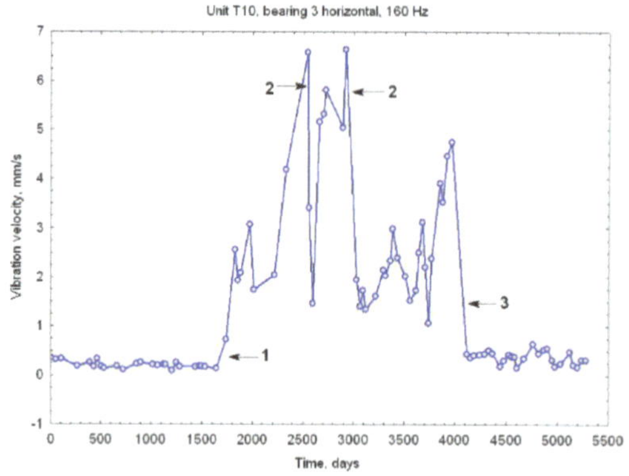

Fig. 4. Time history of the $3 \times f_0$ component: 200 MW unit, rear intermediate-pressure turbine bearing, horizontal direction. Arrows: 1, bearing damage; 2, bearing position and clearances adjustments; 3, major overhaul

Typical malfunctions which have their representations in the low frequency range and their corresponding symptoms have been listed in Table 1, which summarizes this subsection.

Table 1. Typical steam turbine malfunctions and their representation in low-frequency vibration-based symptoms.

Malfunction	Typical symptoms
Unbalance	$1 \times f_0$ component in vertical and horizontal directions, constant amplitude and phase, decreasing at low rotational speed
Misalignment	$2 \times f_0$ component in vertical and horizontal directions, 'banana-shaped' or flattened shaft orbits, high harmonic components in axial direction
Permanent rotor bow	$1 \times f_0$ component in vertical and horizontal direction (also at low rotational speed), strong correlation between $1 \times f_0$ components in vertical and axial directions,
Rotor crack	Continuous changes of $1 \times f_0$ and $2 \times f_0$ components amplitudes and phases during steady-state operation, reduction of critical speeds and increase of vibration amplitudes on passing through them
Bearing problems	Increase of sub-harmonic components (typically slightly below $0.5 \times f_0$), relative vibration increase, shaft orbits with loops, high and unstable amplitudes of higher harmonic components, sensitive to bearing oil pressure

3.2 Blade (high) frequency range

So-called blade spectral components, with frequencies given by Eqs.(3) to (6), are usually low in amplitude. Typically they fall into the frequency range from a few hundred hertz to about 10÷20 kilohertz. In vibration displacement spectra they are undistinguishable, so velocity or acceleration spectra have to be employed. Constant-percentage bandwidth (CPB) analysis is the most convenient tool; 23% CPB yields satisfactory results.

Technical condition of the individual fluid-flow system components, i.e. rotor stages and bladed diaphragms, influences the ζ_k coefficients in Eqs.(1) and (2) and hence the vibration amplitudes in relevant frequency bands. Blade components are, however, highly sensitive to control and interference. Influence of control may be seen as a competition between two mechanisms. First, with nozzle control typical for large steam turbines, there is an asymmetry of steam pressure distribution over the turbine cross-section that depends on the control valve opening. This asymmetry affects forces resulting from the steam flow thrust, again via the ζ_k coefficients. As turbine load increases and consecutive valves are opening, pressure distribution becomes more uniform. Second, with increasing turbine load and steam mass flow, the ζ_0 coefficient also increases. As already mentioned, it may be expected that the former mechanism shall influence vibration patterns at points close to the control stage, as the asymmetry decreases as we move downstream the steam expansion path. The latter should be noticeable for last low-pressure turbine stages, with long blades and large cross-section area. In practice, influence of steam flow asymmetry on blade components is quite strong in points located at the high-pressure turbine; operation at extremely low loads2 may cause them to increase even by a few times. Steam mass flow influence is usually much weaker.

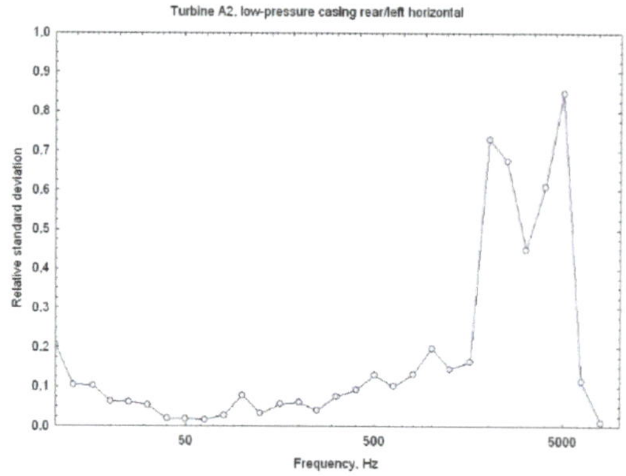

Fig. 5. Relative standard deviation vs. frequency: results for a 120 MW unit, low-pressure turbine casing rear/left side, horizontal direction; data obtained from 90 consecutive measu- rements (after Gałka, 2011b).

Fig.5 shows relative standard deviation (o/$\hat{S}$, where $\hat{S}$ denotes mean value) plotted against mid-frequency of 23% CPB spectrum bands, determined for a 120 MW steam turbine. It is immediately seen that for the harmonic range o /$\hat{S}$ is below 0.1, while in the blade range it may be as high as about 0.6 to 0.8. Similar analysis for other turbine types has yielded quan- titatively comparable results (Gałka, 2011b). In such circumstances, a time history of a blade spectral component has to be considered a monotonic curve with large fluctuations imposed; an example is shown in Fig.6.

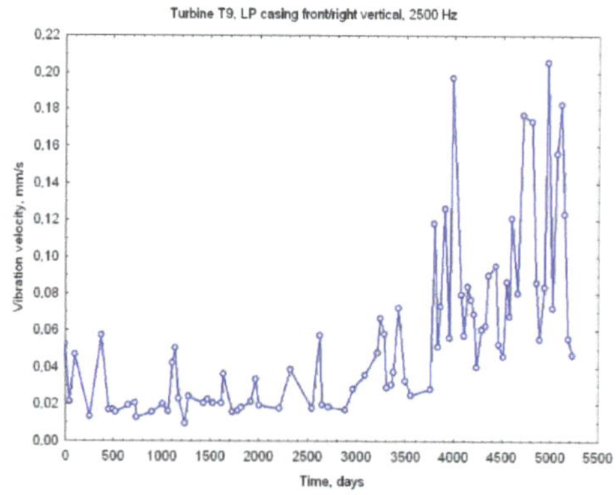

Fig. 6. Time history of the 2500 Hz component: 200 MW unit, low-pressure turbine casing front/right side, vertical direction

Therefore the very occurrence of a high amplitude cannot be unanimously considered as indicative of a fluid-flow system failure. From the point of view of measurement data processing, values heavily influenced by control and/or interference have to be treated as outliers.

It has to be noted that in steam turbines there are sources other than the fluid-flow system that generate vibration components with frequencies in the same range. Typically this is the case with oil pump and governor, driven from turbine shaft via gears. If unexpectedly high amplitudes are encountered, additional narrow-band analysis provides conclusive data, as frequencies of these components may be easily calculated.

4. QUANTITATIVE DIAGNOSIS

In short, qualitative diagnosis provides an answer to the question 'what', while quantitative diagnosis is expected to tell 'how much'. This problem is becoming particularly important when a turbine is operated beyond its design lifetime, which is by no means uncommon. It has to be kept in mind that many turbines still in operation had been designed a few dozen years ago, with much less knowledge of lifetime consumption mechanisms and therefore larger safety margins. Quantitative diagnosis is obviously mandatory if condition-based maintenance is to be introduced.

By necessity, for a quantitative condition assessment a reference scale of some kind has to be used. Such scale may be provided by three values: basic, limit and admissible. Basic value S_b corresponds to a new object with no malfunctions or faults present. Limit value S_l may be considered as determining the 'normal' operation range: if $S > S_l$, further operation is still possible, but the machine cannot fulfill all requirements (concerning e.g. reliability, economy, output, environmental impact etc.). Admissible value S_a is determined from safety considerations: $S > S_a$ indicates high possibility of imminent breakdown and should result in machine tripping.

As S_a is in practice irrelevant to technical diagnostics and S_b may be determined in a rather straightforward manner, the S_l estimation is fundamental for quantitative diagnostics. A complex machine is characterized by a large number of symptoms, and obviously each of them may be assigned its specific limit value. An approach to this estimation is provided by the Energy Processor model and the concept of symptom reliability (for a comprehensive and detailed treatment, see Natke and Cempel, 1997). This approach is based on the fact that any energy-transforming object is a source of residual processes, such as vibration, noise, thermal radiation etc. The power of these processes V can be shown to depend on the object condition. In the simplest case the relation is given by

$$V = V_0\left(1 - \frac{\theta}{\theta_b}\right)^{-1}, \qquad (14)$$

where $V_0 = V(Ø = 0)$ and $Ø_b$ denotes time to breakdown, determined by the time-invariant properties of the object. As Eq.(14) has been derived with quite restrictive assumptions, several modifications have been proposed, applicable for various types of diagnostic objects (see e.g. Gałka and Tabaszewski, 2011); they inevitably result in considerable complication of the mathematical description.

In practice V is usually non-measurable and accessible only via measurable symptoms. A symptom is related to V by so-called *symptom operator* Ø. Several types of symptom operators have been proposed (see e.g. Natke and Cempel, 1997). In steam turbine applications, Weibull and Fréchet operators have been found particularly appropriate; it also has to be added that they conform to all relevant requirements (in particular, vertical asymptote at $Ø = Ø_b$), while some other operators (e.g. Pareto or exponential) are valid only for small values of $Ø / Ø_b$. Weibull operator results in the following expression for a symptom as a function of Ø:

$$S(\theta) = S_0\left(\ln\frac{1}{1-\theta/\theta_b}\right)^{1/\gamma}, \tag{15}$$

while Fréchet operator yields:

$$S(\theta) = S_0(-\ln\theta/\theta_b)^{-1/\gamma}. \tag{16}$$

In both cases, γ is the shape factor to be determined empirically and $S_0 = S(Ø = 0)$.

In order to determine S_l, the concept of symptom reliability is introduced. Symptom reliability $R(S)$ is defined (Cempel, Natke and Yao, 2000) as the probability that a machine classified as being in good condition ($S < S_l$) will remain in operation with the symptom value $S < S_{br}$, where S_{br} denotes value corresponding to breakdown. This may be written as

$$R(S) \equiv P(S_{br} > S \mid S < S_l). \tag{17}$$

Analytically this may be expressed as

$$R(S) = \int_S^\infty p(S^*)dS^* \tag{18}$$

where $p(S)$ denotes the symptom probability density function. Determination of the limit value must involve some measure of acceptable operational risk. This may be accomplished by using the Neyman-Pearson rule, known from statistical decision theory (Neyman and Pearson, 1933). In this particular case, it yields

$$R(S_l)\cdot G = G\int_{S_l}^\infty p(S)dS = A\ , \tag{19}$$

where G denotes the availability of the machine (or group of machines) and A is the accep- table probability of erroneous condition classification as 'faulty', i.e. performing an un- necessary repair.

For a given symptom operator, $p(S)$ may be estimated from experimental data, providing that the available database is sufficiently large. In practice (Gałka, 1999) about 100 indivi- dual data points will allow for a reasonable estimation. Weibull and Fréchet operators usually yield S_l values differing just by a few percent.

A set of limit values should be considered specific for a given turbine example; experience has shown that generalization of results over the entire type should be avoided. It has to be kept in mind that an overhaul often results in a considerable modification of vibration characteristics. This refers mainly to harmonic components, which are sensitive even to minor repairs or adjustments, while blade components are typically influenced only by major overhauls that involve opening of turbine casings. Formally such overhaul is equivalent to creating a new object. Normalization of the influence of overhauls (which determine machine life cycles) is quite straightforward if S_0 values are available, which is usually the case.

5. EVOLUTIONARY SYMPTOMS

Insofar attention has been focused on vibration characteristics recorded at some given moment Ø. Diagnostic information is obviously also contained in symptom time histories. Although state-of-the-art vibration monitoring systems facilitate so-called trending, i.e. plotting of S against Ø, this is seldom used for diagnostic purposes. It has to be mentioned that this refers to steady-state operation data, not transients (startups or shutdowns). In general, any quantity pertaining to the S(Ø) time history may be evaluated in terms of diagnostic reasoning and treated as a symptom itself.

Time histories of vibration components, especially in the blade frequency range, are usually quite irregular. As already mentioned, symptom time history may be considered a monotonic trend with superimposed fluctuations resulting from control and interference (cf. Eq.(9)). If a fault develops fast and strongly influences vibration patterns, this trend is clearly visible (see Fig.3). On the other hand, if condition evolution is slow, it may be suppressed by control and interference to a point where it is barely distinguishable. The latter is often the case for the blade frequency range. Various data smoothing procedures have been proposed to extract the monotonic trend, including three-point averaging, wherein kth symptom reading $S(Ø_k)$ is replaced with $S'(Ø_k)$ given by:

$$S_i{}'(\theta_k) = \frac{1}{3}[S_i(\theta_{k-1}) + S_i(\theta_k) + S_i(\theta_{k+1})] \,. \tag{20}$$

Another option is peak trimming, which in fact consists in eliminating isolated outliers. This method is based on the assumption that if

$$S(\text{Ø}_k)/S(\text{Ø}_{k-1}) > c \text{ and } S(\text{Ø}_k)/S(\text{Ø}_{k+1}) > c\,, \tag{21}$$

then the $S(\text{Ø}_k)$ value is suspicious and treated as an outlier; in such cases, $S(\text{Ø}_k)$ is replaced by $S'(\text{Ø}_k) = [S(\text{Ø}_{k-1}) + S(\text{Ø}_{k+1})]/2$. For steam turbines $c = 1.5$ is reasonable.

Six basic types of vibration evolution can be distinguished for rotating machines in general (Morel, 1992), namely:

- simple evolution (linear or nearly linear),
- complex evolution (usually variations or fluctuations superimposed on a monotonically increasing curve),
- stepwise changes (discontinuous evolution),
- exponential increase,
- cyclic or nearly cyclic variations,
- rapid random variations.

Moreover, each type is characterized by a 'timescale' ranging within broad limits, from seconds to years. Both evolution type and timescale depend on the malfunction or damage type and on the turbine element involved, so the primary idea was to employ this approach in qualitative diagnostics. General guidelines for steam turbines are given in Table 2 (after Orłowski, 2001).

Vibration evolution assessment is, however, far more important for a quantitative diagnosis. If we limit our attention to Weibull and Fréchet operators, we may expand relevant ex- pressions for $S(\text{Ø})$ into Taylor series around $\text{Ø} / \text{Ø}_b = a$, wherein $0 < a << 1$ (for mathematical reasons, $a = 0$ is unacceptable). Truncating higher-order terms, we obtain $S(\text{Ø})$ in the form of

$$S(\theta) \cong S_0\left(1 + A\frac{\theta}{\theta_b}\right), \tag{22}$$

which is valid for $\text{Ø} << \text{Ø}_b$. The constant A depends on the symptom operator and is given by

$$A = \frac{1}{\gamma(1-a)}\left(\ln\frac{1}{1-a}\right)^{1/\gamma-1} \tag{23}$$

Table 2. Failures and damages of steam turbines revealed in vibration evolution parameters (after Orłowski, 2001)

Evolution type	Timescale	Failure
Simple (linear or nearly linear increase)	over 24 h	deformation of casings
	a few minutes to a few hours	deformation of rotors, thermal unbalance (temporary)

Complex (usually fluctuations superimposed on an increasing trend)	a few minutes to a few hours	thermal unbalance (temporary)
	a few hours to a few days	variations of natural frequencies
	a few days to a few months	deformation of casings and/or foundations
Stepwise (discontinuous)	a few seconds	damage of blades, cracks of rotor elements
Exponential or nearly exponential	a few to a few dozen minutes	rubbing in labyrinth seals
	a few hours to a few weeks	material creep effects
Cyclic or nearly cyclic	variable	soft rubbing in seals
	a period of a few seconds	flutter, problems with control
Rapid random		bearing instability, steam flow instability

for the Weibull operator and

$$A = \frac{1}{\gamma a}(-\ln a)^{-(1/\gamma+1)} \tag{24}$$

for the Fréchet operator. We may thus infer that, if lifetime consumption (given by θ/θ_b) is small, symptom time history will be well approximate d by a straight line and its slope will not change substantially with time. In fact this is compatible with the main mechanisms of lifetime consumption, i.e. fatigue and creep, for which linear approximations for $\text{Ø} << \text{Ø}_b$ are also valid. Such case is illustrated by an example shown in Fig.7a.

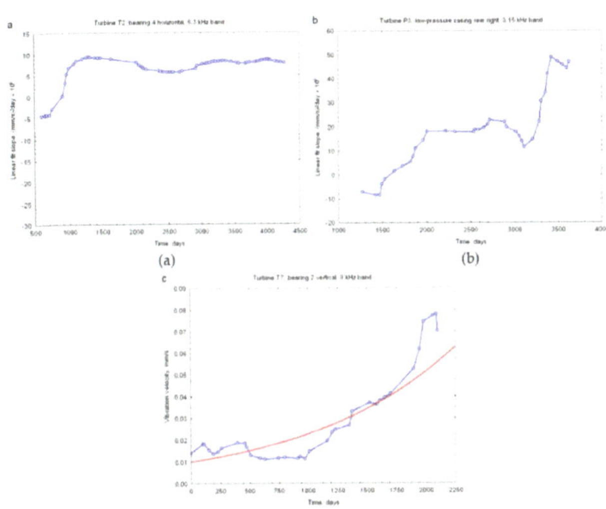

Fig. 7. (a) Linear fit slope vs. time: 230 MW unit, front low-pressure turbine bearing, horizontal direction, 6.3 kHz band; (b) the same, 200 MW unit, low-

pressure turbine casing rear/ right side, horizontal direction, 3.15 kHz band; (c) vibration velocity vs. time: 200 MW unit, high pressure/intermediate pressure bearing, vertical direction, 8 kHz band. Data smoothing: peak trimming at $c = 1.5$ followed by three-point averaging. Red line in (c) represents exponential fit.

On the other hand, for a considerable lifetime consumption, the slope will initially increase with time (Fig.7b) to a point wherein linear approximation is no longer acceptable and $S(Ø)$ resembles an exponen- tial curve. For $Ø$ close to $Ø_b$ even exponential fit fails (Fig.7c).

Although quantitative assessment results should not be generalized over different turbine types, it has been estimated that, for components from the blade frequency range, vibration velocity vs. time plots with linear slope values below about (10 ÷ 20) × 10-6 (mm/s)/day are typical for normal lifetime consumption (natural damage) with $Ø$ substantially lower than θ_b. For the harmonic frequency range, the value of 10-4 (mm/s)/day may be accepted as a very rough estimate. Accelerated damage may result in a value higher by an order of magnitude (cf. Fig.3). It has to be kept in mind, however, that in the harmonic range normalization of life cycles is mandatory.

6. STATISTICAL SYMPTOMS

6.1 Dispersion measures

Up to this point, the deterministic approach has been employed. It may be argued that this is not compliant with the statistical nature of vibration generation mechanisms. What is more important, however, is the fact that statistical approach allows for eliminating prob- lems resulting from the influences of control and interference. The basic idea may be summed up as 'if we cannot get rid of it, then try to make use of it'.

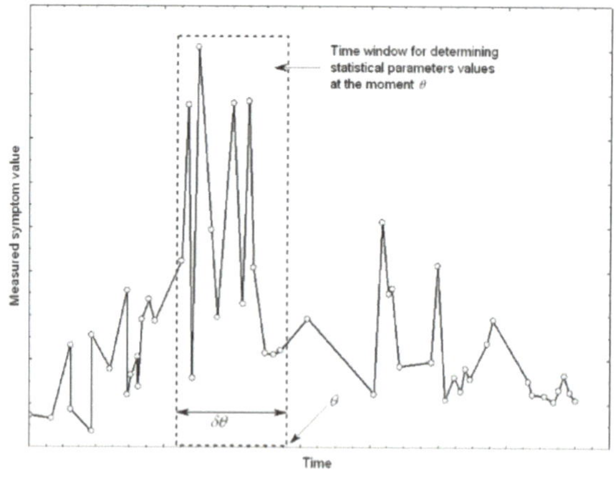

Fig. 8. The idea of statistical symptom determination: parameters pertaining to measured symptom value distribution are determined within a time window $\delta\theta$.

The main assumption in the statistical approach is that a symptom is a random variable rather than a deterministic function of machine condition parameters. Parameters of this random variable also depend on object condition and thus may be themselves accepted as symptoms (sometimes they are referred to as meta-symptoms, in order to stress that they are not directly measurable physical quantities). The idea of determining such symptoms is shown schematically in Fig.8.

Obviously elements of the control and interference vectors are also random variables. More- over, for a given turbine at some fixed location, it is reasonable to assume that statistical parameters of these random variables do not change with time, so each of them is characterized by a time-invariant probability distribution. Now, let us assume that we deter- mine probability distribution of a vibration-based symptom S (say, vibration velocity level in a given frequency band, measured in a given point) in a manner shown in Fig.8. If it can be shown that

$$\partial S/\partial R_i = f(\mathbf{X}) \text{ and/or } \partial S/\partial Z_i = f(\mathbf{X}) \,, \tag{25}$$

then the distribution of S will obviously change with condition parameters. Intuitively we may expect that with deteriorating technical condition both $\partial S/\partial R_i$ and $\partial S/\partial Z_i$ will increase as $\varnothing \rightarrow \varnothing_b$, i.e. S shall be more and more sensitive to control and interference parameters. It can be shown on the basis of a suitably modified Energy Processor model that this is exactly the case (Gałka and Tabaszewski, 2011), providing that measurement errors are excluded. Thus, a measure of vibration level dispersion may be accepted as a diagnostic symptom.

Standard deviation is the most commonly used dispersion measure. From the point of view of this application, it has two main deficiencies. First, normal distribution is tacitly assumed, which in general is not the case. Second, standard deviation is very sensitive to outliers, and obviously no data smoothing can be employed in estimating dispersion. Although standard deviation has yielded basically encouraging results (Gałka, 2008b), robust measures are far superior. Mean absolute difference between two consecutive measurement results Δ was first proposed (Gałka, 2008b) and $\Delta(\varnothing)$ time histories have been found much more regular and easier to interpret than those of $\partial(\varnothing)$. Other possibilities include median absolute deviation about the median m, defined as

$$m = \mathrm{Med}[S - \mathrm{Med}(S)] \,, \tag{26}$$

which in fact consists in centering the data around median rather than mean value, and interquartile range given by:

$$q = Q_3(S) - Q_1(S), \tag{27}$$

where Q_i is the ith quartile:

$$Q_1 = F\text{-}1(0.25),\, Q_2 = F\text{-}1(0.5),\, Q_3 = F\text{-}1(0.75) \,, \tag{28}$$

F being the cumulative distribution function. Obviously, for a symmetrical distribution these two approaches are equivalent, but with a heavy-tailed distribution this is not the case. For the normal distribution, both m and q are constant multiples of ∂.

Time window width ∂ is obviously a compromise. Larger ∂ yields better estimation of dispersion but it has to be kept in mind that the approach schematically shown in Fig.8 is in fact based on the assumption that

$$\wedge Xi\,(\emptyset + \partial\,\emptyset\,) \approx Xi\,(\emptyset\,)\,. \tag{29}$$

If this condition is not fulfilled, centering data around any value 'averaged' over the entire time window becomes groundless. This is certainly the case when Ø is close to $\emptyset_b$. In analyzing time series one should speak in terms of deviations from the trend rather than from some mean value corresponding to the entire time window. This in fact explains why

$\Delta(\emptyset)$ yields better results, as differences between consecutive measurement results better represent such deviations. Another symptom may be thus proposed, in the form of

$$\varepsilon = \frac{\sum_{i=1}^{n}\left[\left|S(\theta_i) - S_t(\theta_i)\right|\right]}{n}, \tag{30}$$

where i is the number of data points contained in the time window, $S(\emptyset_i)$ are consecutive symptom value readings and $S_t(\emptyset)$ represents symptom trend estimated for the entire period under consideration.

Comparison of these five dispersion measures is shown in the example presented in Fig.9, which refers to a natural damage (last measurement was performed shortly before rotor replacement). It is easily seen that they all increase with Ø, almost monotonically, but $\partial(\emptyset)$ is obviously influenced by outliers and hence is of 'step-like' form. Both $m(\emptyset)$ and $q(\emptyset)$ are more regular, but certainly $\Delta(\emptyset)$ and $E(\emptyset)$ are superior; in particular, the latter is most regular and almost perfectly monotonic. The 'dynamics' of these symptoms is also noteworthy: during the period covered by observation they increase roughly by one order of magnitude. They may be thus considered highly sensitive to lifetime consumption. It may also be noted that a marked increasing tendency starts well before rotor replacement (about four years). Symptoms of this type may thus provide an 'early warning', with a lead long enough e.g. to re-schedule maintenance or purchase replacement parts.

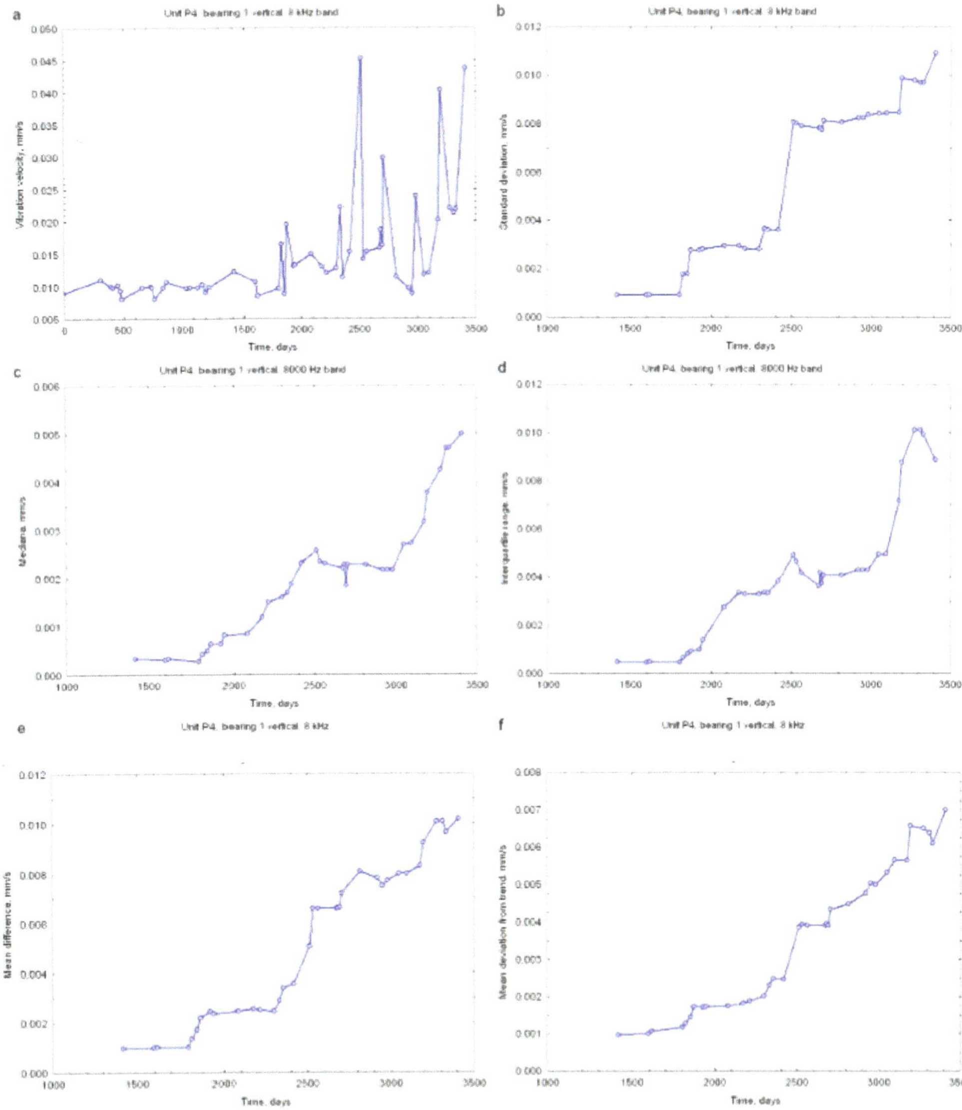

Fig. 9. K-200 unit, front high-pressure turbine bearing, vertical direction, 8 kHz band. (a) symptom time history, (b) $\partial(\emptyset)$, (c) $m(\emptyset)$, (d) $q(\emptyset)$, (e) $\Delta(\emptyset)$, (f) $\varepsilon(\emptyset)$; data window containing 20 measurements.

6.2 Correlation measures

The use of a correlation measure in vibration-based condition assessment is twofold. First, we may check the very existence of correlation or, more precisely, determine if it is 'weak' or 'strong'. This may be very useful, because – as already noted – in steam turbines several possible malfunctions sometimes produce similar changes of vibration characteristics. Such approach is thus applicable in qualitative diagnostics. Second, we may study how a correla- tion measure changes with time and utilize the results for a quantitative diagnosis.

The most commonly used measure of correlation is the Pearson product-moment correlation coefficient *r*, given by the normalized covariance

$$r = \frac{E\{(S_1 - \eta_1)(S_2 - \eta_2)\}}{\sqrt{E\{(S_1 - \eta_1)^2\}E\{(S_2 - \eta_2)^2\}}}, \tag{31}$$

where *E* denotes expected value and

$$\eta_1 = E(S_1),\ \eta_2 = E(S_2)\ . \tag{32}$$

This measure is very sensitive to outliers (see e.g. Maronna, Martin and Yohai, 2006), but is often sufficient for a qualitative diagnosis. The basic idea stems from the fact that if two symptoms can be shown to be correlated, we may infer that they are dependent, i.e. that their changes have been caused by the same condition parameter. The reverse is not true: if two random variables are not correlated, this does not imply that they are independent.

Fig.10 shows two vibration time histories recorded with a 200 MW turbine that suffered an intermediate-pressure rotor failure and secondary fracture of steam guiding fences. Manifes- tation of this failure in vibration patterns was quite complex. It may be noted, however, that before repair both these components tended to increase simultaneously. Several other components from the blade frequency range – up to about 2 kHz – behaved in a similar manner. We may thus suspect that comparatively high level of the $4 \times f_0$ component was a result of the fluid-flow system failure. This is corroborated by correlation analysis; for 23% CPB spectra bands from 800 Hz to 2 kHz coefficients of correlation with the $4 \times f_0$ compo- nent ranged from $r = 0.689$ to $r = 0.912$, while for two other turbines of the same type $|r|$ was below 0.2 (in several cases negative). Shortly after the repair the $4 \times f_0$ component increased again, eventually reaching even substantially higher level; this time, however, there is virtually no correlation with the blade components, *r* being about –0.1. The root cause was thus different.

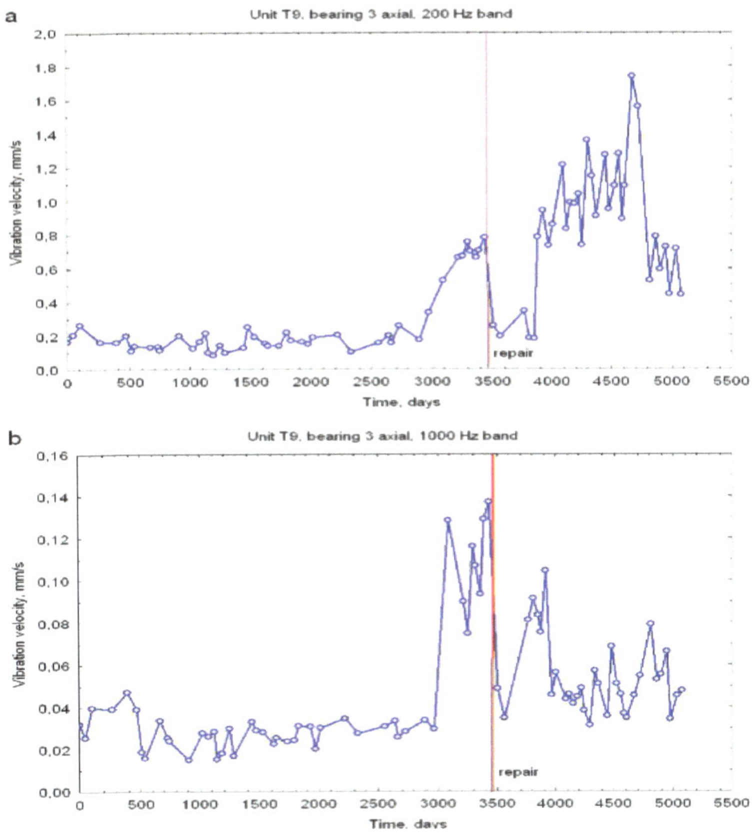

Fig. 10. Time histories of vibration amplitudes: K-200 unit, rear intermediate-pressure turbine bearing, axial direction, 4×f_0 (a) and 1000 Hz (b) components. Vertical lines indicate repair.

In addition to being sensitive to outliers, the Pearson correlation coefficient is deficient in that a linear relation is assumed. Both above-mentioned disadvantages may be eliminated or at least alleviated to some extent by using non-linear and more robust measures of correlation. These include the Kendall rank correlation coefficient *r*, given by

$$\tau = 1 - \frac{2d_\Delta}{N(N-1)}, \qquad (33)$$

and Spearman rank correlation coefficient *p*, given by

$$\rho = 1 - \frac{6\sum d_i^2}{N(N^2-1)}. \qquad (34)$$

In these formulae, *N* is the number of scores (elements) in two data samples, d_Δ denotes symmetric difference distance and d_i are differences between individual ranks. It should be noted here that both *r* and *p* are calculated on the basis of ranks rather than standard deviations and therefore more suitable for analyzing time series (Salkind, 2007). They are thus more appropriate when dealing with correlation as a function of Ø.

As mentioned earlier, for a rapidly developing damage one should speak in terms of a deviation from the trend than of some mean or expected value. We may therefore, as suggested in (Gałka, 2011a), introduce yet another correlation measure, tentatively termed 'modified Parsons coefficient' *r*'. Using the notation from Eqs.(30) and (31), *r*' is given by

$$r' = \frac{E\{(S_1 - S_{t1})(S_2 - S_{t2})\}}{\sqrt{E\{(S_1 - S_{t1})^2 (S_2 - S_{t2})^2\}}} . \tag{35}$$

Let us assume that lifetime consumption degree $D = Ø / Ø_b$ is the only condition parameter that is taken into account. Then Eq.(9) for a given symptom *S* may be rewritten as

$$S = f(D, R_1, R_2, ..., R_k, Z_1, Z_2, ..., Z_m) . \tag{36}$$

Within the framework of the Energy Processor model, the influence of *D* on *S* is purely deterministic and *S*(*D*) is a monotonically increasing function. As *D* approaches unity, both *S* and *dS*/*dD* tend to infinity (cf. Eq(14)), so equal increments of *D* will result in increasing increments of *S*:

$$D \to 1 \Rightarrow \Delta S = S(D + \Delta D) - S(D) \to \infty \ (\Delta D = \text{const.}) , \tag{37}$$

and this will hold for all symptoms. Correlation is thus expected to increase with *D*, as for any two symptoms S_j and S_k both will, to a growing extent, be dominated by *D* rather than other factors and thus become more deterministic with respect to *D*. Speaking in a descriptive manner, Eq.(35) may be viewed as revealing a competition between the random (represented by R_i and Z_i) and the deterministic (represented by *D*). The above argumentation suggests that for $D \to 1$ the latter should prevail and consequently a measure of correlation should increase in value. This phenomenon has been termed the 'Old Man Syndrome'.

Fig.11 shows comparison of the above four correlation measures plotted against time for the same unit as in Fig.9 (albeit for different frequency bands). All plots exhibit a more or less pronounced 'saddle', which was found to have resulted from an overhaul which 'de- correlated' the symptoms to a certain extent. The terminal increasing section is, however, evident.

Due to a large number of vibration-based symptoms generated by a typical multi-stage steam turbine, the number of pairs to be analyzed in terms of correlation is large, of the order of a few dozen or more. It is, however, possible to select those with the highest content of diagnostic information. Such selection may employ the Singular Value Decomposition method, known from linear algebra (see e.g. Cempel, 2003). This approach has been applied to a number of steam turbines and results have been found very encouraging (Gałka, 2011a). In general, components generated by rotor stages are more informative in this respect than those generated by bladed diaphragms. Best results have been obtained with high-pressure turbines; this is not particularly surprising, due to high temperature and pressure, which contribute to accelerated lifetime consumption.

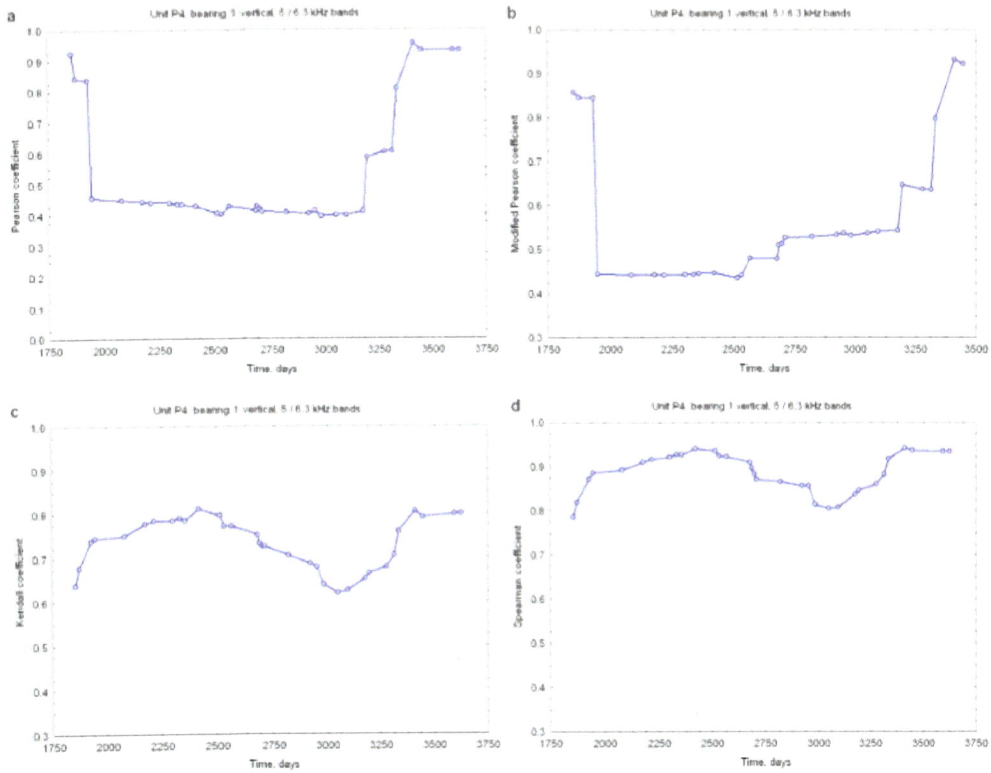

Fig. 11. Plots of Pearson (a), modified Pearson (b), Kendall (c) and Spearman (d) correlation coefficients: 200 MW unit, front high-pressure turbine bearing, vertical direction, 5 kHz and kHz bands. Data window containing 25 measurements.

7. CONCISE SUMMARY

Steam turbines, essential to any economy, have consistently remained at the forefront of advancements in technical diagnosis. A range of vibration monitoring systems is commercially available, typically customized to specific requirements. Some are termed 'diagnostic systems', however this designation is not always accurate, as many just supply data for diagnostic reasoning.

Qualitative diagnostics is predominantly founded on recognized techniques and guidelines, particularly with harmonic components. Quantitative condition evaluation appears to be in a nascent stage of development, particularly for practical applications. Its significance is acknowledged, particularly with the implementation of predictive maintenance. A turbine model appropriate for the theoretical establishment of quantitative diagnostic relationships has yet to be produced, hence much of the research in this domain relies on empirical data. It is reasonable to assert that accurate forecasting of the evolution of technical conditions is now the primary issue confronting specialists in this domain globally.

FOOTNOTES

1. The reason for using t and θ symbols for denoting time is that the former refers to the 'dynamic' time (e.g. that enters equations of motion), while the latter is for the 'operational' time – the argument of equations pertaining to technical condition evolution.
2. Load minimum is usually imposed by the steam generator (boiler or nuclear reactor) stable operation considerations.
3. Detailed case study may be found in (Gałka, 2008a)
4. To the author's best knowledge, this term has been first used in the context of technical diagnostics by Cempel (see Cempel, 1991).

REFERENCES

1. Bachschmid, N., Pennacchi, P. and Tanzi, E. (2010). *Cracked Rotors. A Survey on Static and Dynamic Behaviour Including Modelling and Diagnosis.* Springer, ISBN 978-3-642- 01484-0, Berlin-Heidelberg, Germany
2. Bently, D.E. and Hatch, C.T. (2002). *Fundamentals of Rotating Machinery Diagnostics*, Bently Pressurized Bearings Press, ISBN 0-9714081-0-6, Minden, USA
3. Cempel, C. (1991). *Vibroacoustic Condition Monitoring*, Ellis Horwood, ISBN 0-13-931718-X, New York, USA
4. Cempel, C., Natke, H.G. and Yao J.T.P. (2000). Symptom reliability and hazard for systems condition monitoring, *Mechanical Systems and Signal Processing*, vol.14, No.3 (2000) pp. 495-505, ISSN 0888-3270
5. Cempel, C. (2003). Multidimensional Condition Monitoring of Mechanical Systems in Ope- ration. *Mechanical Systems and Signal Processing*, vol.17, No.6 (2003) pp. 1291-1303, ISSN 0888-3270
6. Crocker, J. (2003). Prognostics in Aero-Engines, *Proceedings of the 16th International Congress COMADEM 2003*, pp. 145-154, ISBN 91-7636-376-7, Växjö, Sweden, August 27-29, 2003
7. Gałka, T. (1999). Application of energy processor model for diagnostic symptom limit value determination in steam turbines. *Mechanical Systems and Signal Processing*, vol.13, No.5 (1999) pp.757-764, ISSN 0888-3270

8. Gałka, T. (2001). Influence of Turbine Load on Vibration Patterns and Symptom Limit Value Determination Procedures, *Proceedings of the 14th International Conference COMADEM 2001*, pp. 967-976, ISBN 0 08 0440363, Manchester, UK, September 4-6, 2001

9. Gałka, T. (2008a). Correlation-Based Symptoms in Rotating Machines Diagnostics, *Proceedings of the 21st International Congress COMADEM 2008*, pp. 213-226, ISBN 978- 80-254-2276-2, Praha, Czech Republic, June 11-13, 2008

10. Gałka, T. (2008b). Statistical Vibration-Based Symptoms in Rotating Machinery Diagnostics. *Diagnostyka*, vol. 2(46)/2008, pp. 25-32, ISSN 1641-6414

11. Gałka, T. (2009a). Large Rotating Machines, In: *Encyclopedia of Structural Health Monitoring*, C.Boller, F.Chang and Y.Fujino (Ed.), 2443-2456, Wiley, ISBN 978-0-47-006162-6, Chichester, UK

12. Gałka, T. (2009b). Rotor Bow in a 230 MW Steam Turbine: A Case Study, *Proceedings of the 6th International Conference on Condition Monitoring and Machine Failure Prevention Tech- nologies*, pp. 1053-1063, Dublin, Ireland, June 23-25, 2009 (CD-ROM edition)

13. Gałka, T. and Tabaszewski, M. (2011). An Application of Statistical Symptoms in Machine Condition Diagnostics, *Mechanical Systems and Signal Processing*, vol.25, No.1 (2011) pp. 253-265, ISSN 0888-3270

14. Gałka, T. (2011a). The 'Old Man Syndrome' in Machine Lifetime Consumption Assessment, *Proceedings of the 8th International Conference on Condition Monitoring and Machine Failure Prevention Technologies*, paper No. 108, Cardiff, UK, June 20-22, 2011 (CD- ROM edition)

15. Gałka, T. (2011b). Influence of Load and Interference in Vibration-Based Diagnostics of Rotating Machines, *Advances and Applications in Mechanical Engineering and Tech- nology*, vol. 3, No. 1/2 (2011), pp. 1-19, ISSN 0976-142X, available from http://scientificadvances.co.in/index.php?cmd=artical&j=7&su=66

16. Kiciński, J. (2006). *Rotor Dynamics*, IFFM Publishers, ISBN 83-7204-542-9, Gdańsk, Poland Martin, K.F. (1994). A Review by Discussion of Condition Monitoring and Fault Diagnosis in

17. Machine Tools. *Int. Journal of Machine Tools and Manufacture*, vol. 34 (1994), pp. 527- 551, ISSN 0890-6955

18. Maronna, R.A., Martin, R.D. and Yohai, V.J. (2006) *Robust Statistics. Theory and Methods*, John Wiley & Sons, ISBN 0-470-01092-4, Chichester, UK

19. Morel, J. (1992). *Vibration des Machines et Diagnostic de Leur État Mécanique*, Eyrolles, ISBN 0399-4198, Paris, France
20. Natke, H.G. and Cempel, C. (1997). *Model-Aided Diagnosis of Mechanical Systems*, Springer, ISBN 978-3540610656, Berlin-Heidelberg, Germany
21. Neyman, J. and Pearson E.S. (1933). On the problem of the most efficient tests of statistical hypotheses. *Philosophical Transactions of the Royal Society of London*, Ser. A, 231, pp. 289-337
22. Orłowski, Z. and Gałka, T. (1995). Excessive Vibration of a Small Steam Turbine: Diagnosis and Remedy, *Proceedings of the Inter-Noise'95 International Congress*, pp. 1133-1137, ISBN 0-931784-32-8, Newport Beach, USA, July 10-12, 1995
23. Orłowski, Z. and Gałka, T. (1997). Determination of Diagnostic Symptom Limit Values for Steam Turbines, *Proceedings of the Condition Monitoring'97 International Conference*, pp. 247-253, ISBN 7-118-01719-1, Xi'an, China, March 24-26, 1997
24. Orłowski, Z. and Gałka, T. (1998). Vibrodiagnostics of Steam Turbines in the Blade Frequency Range, *Proceedings of the COMADEM'98 International Conference*, pp. 683- 692, ISBN 0-7326-2027-9, Monash University, Australia, December 8-11, 1998
25. Orłowski, Z. (2001). *Diagnostyka w życiu turbin parowych*, WNT, ISBN 83-204-2642-1, Warsaw, Poland
26. Radkowski, S. (1995). Low-Energy Components of Vibroacoustic Signal as the Basis for Diagnosis of Defect Formation. *Machine Dynamics Problems*, vol. 12 (1995), ISSN 0239-7730
27. Randall, R.B. (2011). *Vibration-Based Condition Monitoring*, Wiley, ISBN 978-0-470-74785-8, Chichester, UK
28. Salkind, N.J. (Ed.) (2007). *Encyclopedia of Measurements and Statistics*, SAGE Publications, ISBN 978-1-412-91611-0, Thousand Oaks, USA
29. Traupel, W. (2000). *Thermische Turbomaschinen. Zweiter Band: Geanderte Betriebsbedingungen, Regelung, mechanische Probleme, Temperaturprobleme*, ISBN 978-3-540-67376-7, Sprin- ger, Berlin

15 Chapter

REGARDING THE MECHANICAL COMPLIANCE OF TECHNICAL SYSTEMS

1. INTRODUCTION: AN OVERVIEW

In safe physical human-machine interaction, the compliance of technological systems is a fundamental necessity (Zinn et al., 2004; Bicchi & Tonietti, 2004). The physical compliance of technical systems can be achieved through the installation of control functions and/or by optimizing structural configuration and material qualities (Beder & Suzumori, 1996; Wang et al., 1998). The latter offers advantages due to enhanced reliability and the overall simplicity of the design and manufacturing processes (Beder & Suzumori, 1996; Ham et al., 2009). We will concentrate on mechanical systems characterized by inherent mechanical compliance.

The deformability of structures is primarily characterized by their stiffness. Stiffness quantifies a structure's capacity to withstand deformation caused by external forces (IFToMM Terminology, 2010). Compliant mechanisms are devices whose operation relies on their capacity to deform. The mobility of these systems arises from their predominantly elastic or plastic deformability (as defined by Bögelsack, 1995; Howell, 2001; Christen & Pfefferkorn, 1998). It is recommended to utilize compliance rather than stiffness for the description of these systems. Compliance is the inverse of stiffness and is defined as the capacity of a structure to undergo deformation in response to external forces (IFToMM Terminology, 2010). The objective of each engineer is to build methods that establish compliance based on the intended use. Compliance is contingent upon various elements. The optimal design of these mechanisms can only be achieved with accurate understanding of the influencing parameters and potential types of compliance.

2. INFLUENCE FACTORS OF COMPLIANCE

First, the factors will be considered that determine the compliance of mechanisms generally, without a specific application.

The compliance of a mechanism is determined with respect to a displacement of a specific selected reference point or area of the mechanism as a result of an external force. That approach is necessary because the deformation of a mechanism is usually associated with varying displacements for differing areas of a mechanism. Accordingly, the compliance of a mechanism depends on the location of the reference point. At the same mechanism for the same load and boundary conditions, the evaluation of compliance with respect to different points leads to different results. For the reference usually the force application point or area is chosen. Depending on the application, the amount of the displacement vector of the reference point or its components are used when specifying the compliance.

The compliance is not a pure structure-related property, defined only by the initial geometric configuration and initial material properties. Geometric boundary conditions (location, type), the loading situation (location, type, magnitude, direction and loading history) and environmental conditions (e.g. thermal, chemical) must be also considered in order to formulate the compliance of a mechanism. On the material side the compliance is influenced by actual and previous environmental conditions and by the loading history (elastic or plastic behaviour). The geometric configuration for a given load is dependent on the material properties and geometric boundary conditions. Therefore the compliance of the mechanism depends on its actual geometric configuration and actual material properties and is valid only for the considered reference point by the given actual load and boundary conditions (Figure 1).

In practical applications, the boundary conditions, load levels and the reference point are given. In this case, the adjustment of the compliance can be achieved by appropriate design and material selection. The effort for the design depends on the variety of possible future applications of the mechanism.

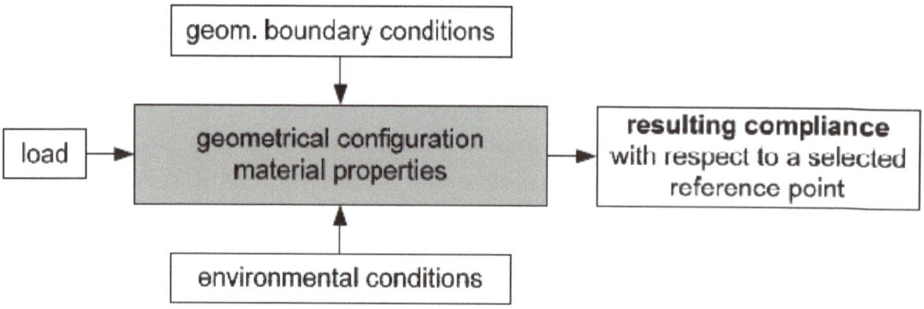

Fig. 1. Influence factors on the mechanical compliance of mechanisms

2.1 Variability of compliance

Generally, the compliance can be either constant or variable. The constant compliance is impossible in the nature. However, we can use the theoretical models with constant compliance, for example, in the linear theory of small bending of beams. In this case the force is linear proportional to the displacement. Table 1 shows deferent compliance for a compliant quarter-circle shaped beam with radius R=20 mm. For this problem we use the Castigliano's theorem for describing the displacement of the end of beam with the geometric linear theory:

$$u = \sqrt{u_1^2 + u_2^2} = \frac{FR^3}{2EI_3}\sqrt{\frac{\pi^2}{4} + 1} \tag{1}$$

The compliance applied of the end of beam can be given by:

$$\frac{u}{F} = \frac{R^3\sqrt{\pi^2 + 4}}{4EI_3} \tag{2}$$

The compliance would be demonstrated by means of one point for constant bending stiffness in one dimension domain. Consequently, the constant compliance is the compliance of the zero degree.

The mathematical model for the large displacements is based on the theory of curved beams. The following equations are nonlinear equations of equilibrium and constitutive equations of a curved beam (Zentner, 2003).

$$\begin{aligned} &Q_1' - \kappa Q_2 = 0 \\ &Q_2' + \kappa Q_1 = 0 \\ &EI_3\kappa' + Q_2 = 0 \\ &\theta' = \kappa - \frac{1}{R} \\ &u_1' = \cos\theta - 1 \\ &u_2' = \sin\theta \end{aligned} \tag{3}$$

Where κ - curvature of the loaded beam, Q_i – internal forces, EI_3 - bending stiffness, u_i – displacements on the directions of x_1 and x_2, θ - angle between the tangent and the axis x_1. All these parameters depend on the beam coordinate s (0≤s≤L). This system of nonlinear equations was solved with the program MATHEMATICA with boundary conditions:

$$\begin{aligned} &Q_1(L) = 0 \\ &Q_2(L) = F \\ &\kappa(L) = 0 \\ &\theta(0) = 0 \\ &u_1(0) = 0 \\ &u_2(0) = 0 \end{aligned} \tag{4}$$

As for the large displacements, we have different compliances depending on the particular position. To each position, a different force dF corresponds, which displaces a beam point on du. This situation is presented in Table 1 as a curve in 2D domain. The different compliance is characterized by ∂u/∂F dependent on u. By means of changing the prestressing of a compliant structure its compliance can be changed. Another example of this is a structure in the ring shape as a prestressed spring. While changing the clamping with the help of the parameter h, the compliance of the spring can be purposefully set onto the point P. Such compliance can be called the compliance of the first degree.

In addition to this, if we also take the temperature into consideration concerning a mechanism made of a temperature sensitive material, the compliance will depend on the two parameters, namely the displacement position and the temperature. In such a case we have compliance of the second degree. The surface F=F(u, T) would reflect the compliance depending on the two parameters, as a value for compliance ∂u/∂F for a definite temperature T and a particular displacement u.

The list of such parameters, which influence the compliance, can continue to be developed. If the compliance depends on the N parameter, we deal with the case of the compliance of the N degree.

Table 1. Compliance of three degrees, from zero till two, for a compliant quarter-circle shaped beam (EI_3=100 Nmm2, F=1N)

Degree of compliance	Modelling	Figure	Compliance
N=0	Linear theory	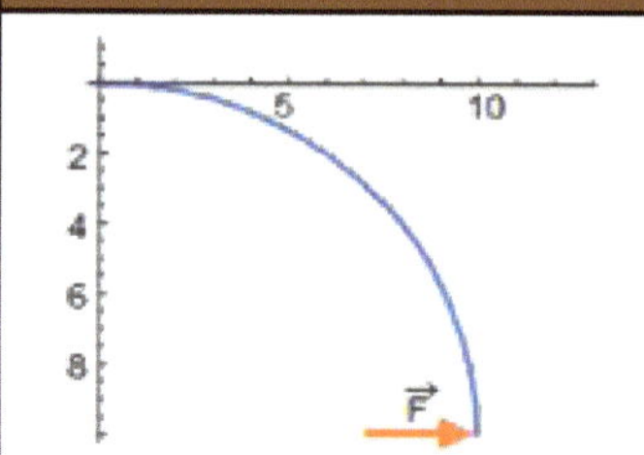	$\frac{u}{F}=\frac{R^3\sqrt{\pi^2+4}}{4EI_3}$
N=1	Non-linear theory (large deformations)	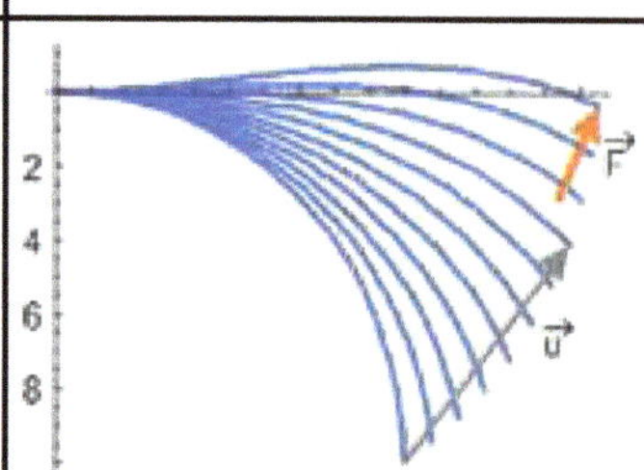	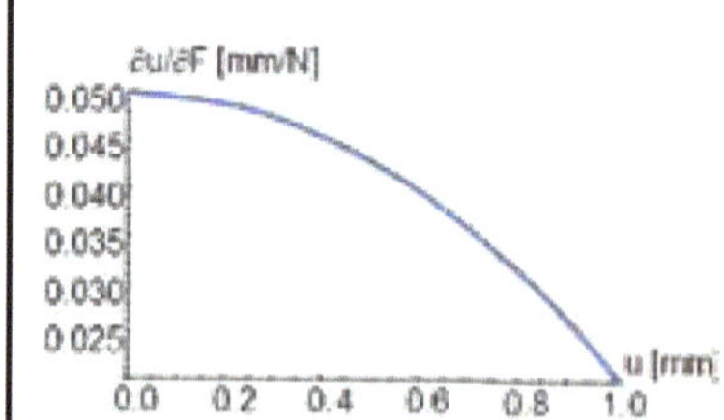
N=2	Non-linear theory and dependence of the temperature, e.g. E(T)	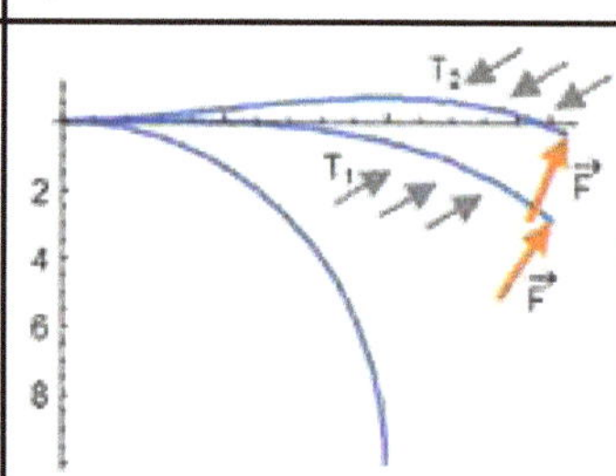	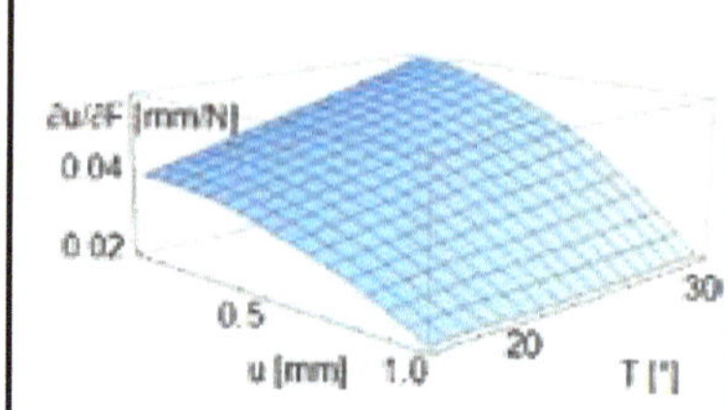

2.2 Distribution of compliance

The mostly deformable parts of a compliant mechanism are called as compliant joints. Compliant joints can be classified by their distribution of their compliance.

The joints with concentrated, local compliance have a small deformable area with the reference to the dimension of a mechanism. In contrast, compliance joints with distributed compliance include a large area of the deformable part. The decision whether a deformable area is small or large, depends on the purpose of the modelling. For example, the installation of a substitute rigid body model for a compliant mechanism, a great role is played by extension of the joint.

In case of a joint with a local compliance, the rigid body joint is introduced in the most cases into the middle of the compliant part. With a compliant joint possessing the distributed compliance, it is important that the position of the rigid body joint is determined for a substitute model. It can be admitted that for the compliant joints the following conditions are available: if the extension of the joint is 10 or more times smaller than the biggest dimension of the whole mechanism, it is classified as a joint with a distributed compliance.

Mechanisms with concentrated compliance behave like classic rigid link mechanisms, where kinematic joints are replaced with flexible hinges, and in consequence methods conceived to design rigid body mechanisms can be modified and applied successfully in this case (Albanesi et al., 2010). Mechanisms with distributed compliance are treated as a continuum flexible mechanism, and Continuum Mechanics design methods are used instead of rigid body kinematics (Albanesi et al., 2010). An overview of calculation methods of compliant mechanisms is indicated in (Albanesi et al., 2010; Shuib et al., 2007).

3. CLASSIFICATION OF COMPLIANT MECHANISMS CONCERNING THE DEFORMATION

In case the deformation-behaviour is chosen as a criterion for classifying compliant mechanisms, two subgroups – dynamic and static deformation – can be distinguished (Zentner & Böhm, 2009).

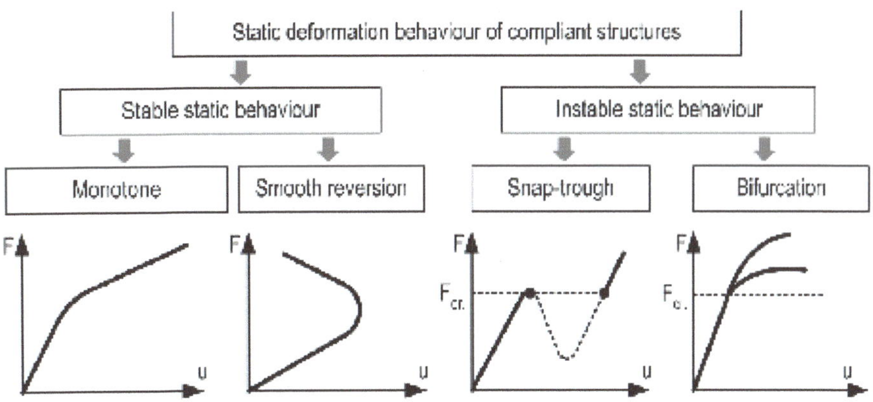

Fig. 2. Classification of the static deformation of compliant mechanisms.

Furthermore the static deformation behaviour of compliant mechanisms having a fixed compliance are considered, whereat influences caused by inertia are neglected. The static deformation behaviour is divided into stable and instable behaviour (Figure 2). Stable deformation behaviour is characterised by a surjective mapping of a particular load F on the deformation u. Thereby one can differentiate between a monotonic behaviour and the behaviour with a singular smooth reversion. In case of instable behaviour of compliant mechanisms snap-through (deformation-behaviour with jump-discontinuities) and bifurcation (local bifurcation of the behaviour) are possible.

3.1 Stable deformation-behaviour of compliant mechanisms: monotonic deformation

Figure 3 shows an example of the monotonic deformation behaviour of a pneumatically driven compliant mechanism. By increasing the load (here: internal pressure) the characteristic deformation parameters such as the angle between the longitudinal-axis of the rigid structural parts also increases. This mechanism is used as a finger of a gripper.

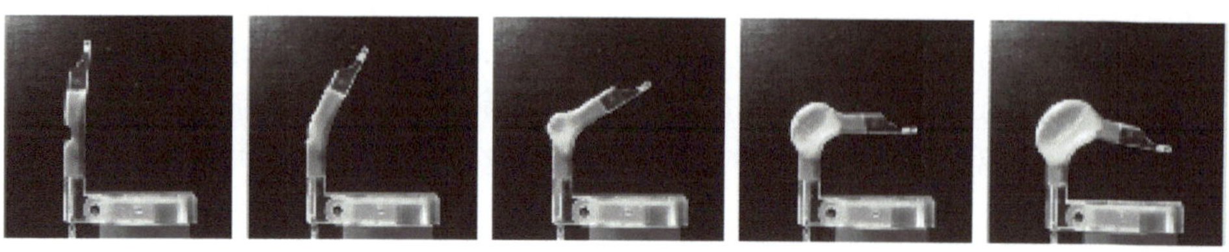

Fig. 3. Monotonic deformation behaviour of a pneumatically driven compliant mechanism made of silicone rubber

Another example concerns a compliant fluid driven structure, which is applied as a medical probe. The cross-section diameter of probe changes from at the fixed end (3 mm) to the free end (1 mm) linearly. In the probe model there is a hollow with constant diameter of 0.2 mm. An unstretchable thin fibre is embedded in the wall with constant distance h from the symmetry axes of probe-beam. Under inner pressure p in the hollow, with the cross-section of A, the probe structure will bend towards the embedded fibre. Linear material law is supposed. The equation for displacement of probe is calculated analytically in order to examine the possibilities to obtain the required bending.

$$\begin{aligned} &Q_1' - \kappa Q_2 = 0 \\ &Q_2' + \kappa(Q_1 - pA) = 0 \\ &EI_3\kappa' + Q_2 = 0 \\ &\theta' = \kappa \\ &u_1' = \cos\theta - 1 \\ &u_2' = \sin\theta \end{aligned} \tag{5}$$

Corresponding boundary conditions are:

$$\begin{aligned} &Q_1(L) = pA(L) \\ &Q_2(L) = 0 \\ &\kappa(L) = \frac{pA(L)h}{EI_3} \\ &\theta(0) = 0 \\ &u_1(0) = 0 \\ &u_2(0) = 0 \end{aligned} \tag{6}$$

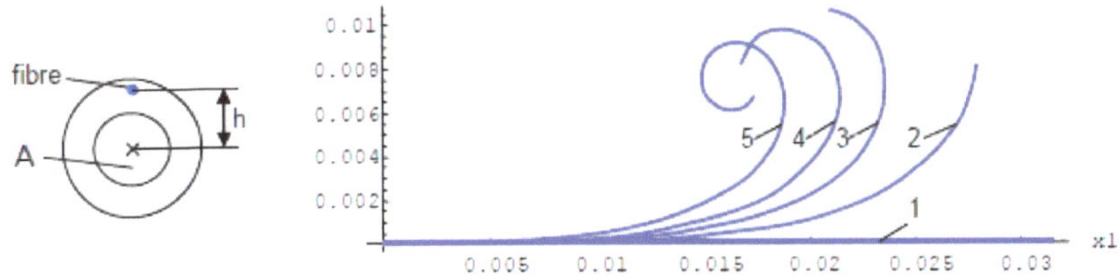

Fig. 4. Displacement behaviour of a compliant fluid driven structure by increasing internal pressure (1-5)

The characteristic parameter for such structure is the angle between the tangent of beam-end and the x_1-axe. This angle is strictly monotonic increasing, when the pressure rises.

3.2 Stable deformation-behaviour of compliant mechanisms: Behaviour with direction reversion

The deformation behaviour with direction reversion of a pneumatically driven compliant mechanism is shown in Figure 5. The geometry is optimised, to achieve a reversion by loading the internal pressure. The horizontal displacement of the working element was chosen as one characteristic parameter, whereat the vertical displacement and the distance of the working element to the clamping are also possible characteristic parameters which can be taken into account. In the present case the increasing load initially yields to an extension of the horizontal displacement of the working element and finally causes a smaller displacement.

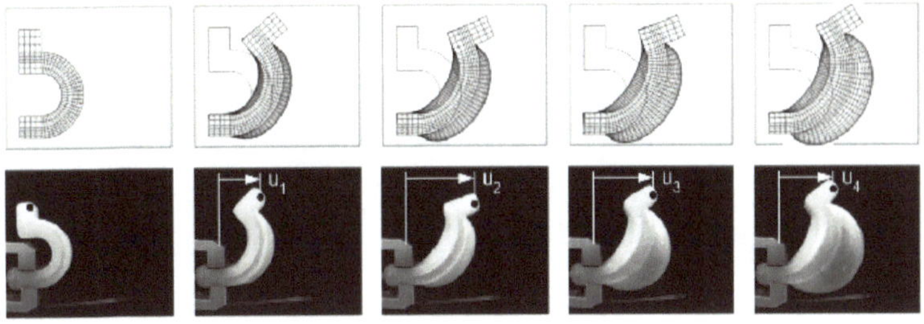

Fig. 5. Static deformation behaviour with direction reversion of a compliant mechanism: at the top – FEM-calculations; at the bottom – compliant mechanism made of silicone rubber with characteristic parameters u_i, i=1,..,3 (u_2>u_3>u_1)

Figure 6 a graphically presents the change of the characteristic parameters mentioned above. Therein the displacement of the working element u_3 equals the reversal point of the structure.

The considered deformation effect, referenced here as reversal effect, consists in reversion of movement, which the working element of the mechanism performs during unidirectional change of the internal pressure (Zentner et al., 2009). Thus movements in two opposite directions depend on the magnitude of the uniformly increasing pressure, and two opposite working directions follow one another. One of the important consequences of this effect is that the movement range of the mechanisms in the first working direction is not limited by the maximum pressure but by the material properties and geometric forming, whereby the sensors utilization can be minimized in grasping applications for example. The main characteristic criteria of the mechanisms considered are: (1) fulfilling movements in two opposite directions, where both movements are caused by unidirectional pressure-activating of the actuator; (2) the movement range in the first working direction being, independently from the size of the pressure load, bounded; (3) a given position of the working element of the mechanism is to be reached at two different pressure loads and therefore with different compliances. The reversal effect can be generally realized in two ways: (1) through cascading of several conventional structures; or (2) with a non-conventional actuator. In both cases such an effect can be achieved by both geometrical properties (geometrically asymmetrical actuators), and material properties (actuators with variation of the material properties). A combination of materials of different elasticity and/or anisotropic materials can fulfil this characteristic, too. The numerical calculations approved, that the mechanism behaviour is influenceable geometrically and materially. Hence it can be adjusted to specific tasks.

Compared to other mechanisms, more complex motion trajectories can be easily provided with unidirectional pressure change with the help of these mechanisms.

One application for a mechanism having the property of a direction reversion is using it as gripping fingers. Through model based optimisation a novel dependency of the load on the displacement could be achieved (Figure 6 a, broken line).

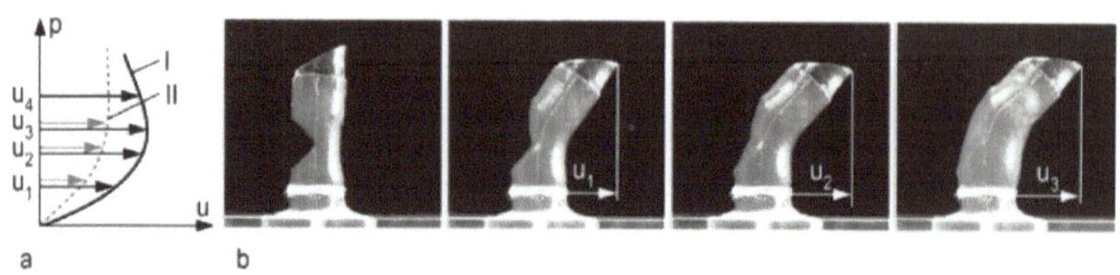

Fig. 6. a: I – Dependency of displacement u on internal pressure p of the mechanism introduced in Figure 4, II – p(u) for gripping-fingers with defined gripping-force; b: gripping-fingers with defined gripping-force made of silicone rubber

Therein the characteristic displacement will not increase at a defined value of load irrespective of further increase of load. This property puts aside the sensory effort to monitor the gripping force. This force is already defined by the structure's mechanical properties. Such a structure is shown in Figure 6 b.

3.3 Instable deformation behaviour of compliant mechanisms: Snap-through

In contrast to the stable deformation behaviour of compliant mechanisms, the instable case has more than one equilibrium position for a particular load. The instable deformation behaviour shows snap-through or bifurcation.

In case of snap-though a sudden transition from one equilibrium position to another happens. Thereby a given load corresponds to several equilibrium positions. In Figure 7, a rotational structure is shown having a half-toric curve around the spherical curve in the origin state. One characteristic feature of such a mechanism is the potential bistable deformation behaviour, which can be enforced by the specified geometric parameters (shape, wall thickness, etc.).

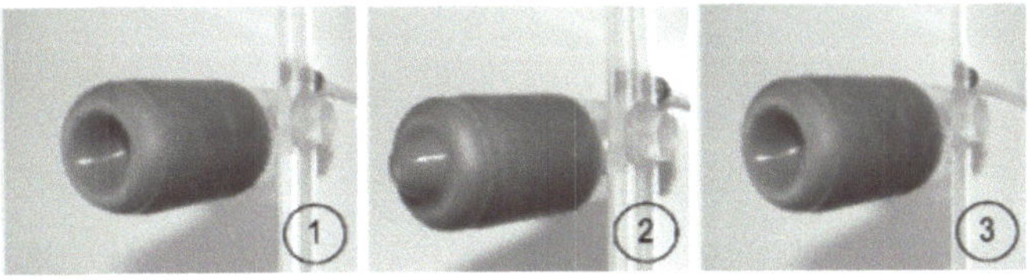

Fig. 7. Snap-through of a curved mechanism having a monostable deformation behaviour

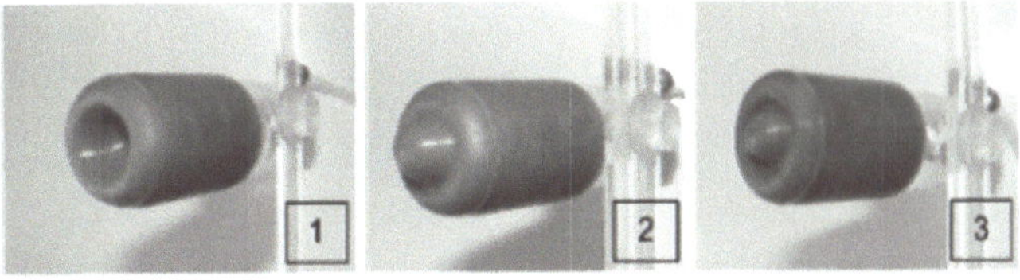

Fig. 8. Snap-through of a curved mechanism having a bistable deformation behaviour

Two different mechanisms with a big and small wall thickness are presented in Figure 7 and 8, respectively. By increasing the load, the angular point (centre point) of the median curvature of both mechanisms moves outwards up to the critical load (Figure 9 c). Herein the value of the critical load is different to the named structures. An arbitrary small rise of load causes a huge displacement, as soon as the critical load is reached. In this process the median curvature penetrates completely (state 2 in Figure 9). Removing the load causes the first mechanism to reverse to the original position (monostable deformation). This is demonstrated by state 3 in Figure 9 a. The second mechanism switches to another equilibrium position (bistable deformation), named in Figure 9 b with state 3. A sketch of both characteristics and the calculated positions by means of FEM is shown in Figure 9 c.

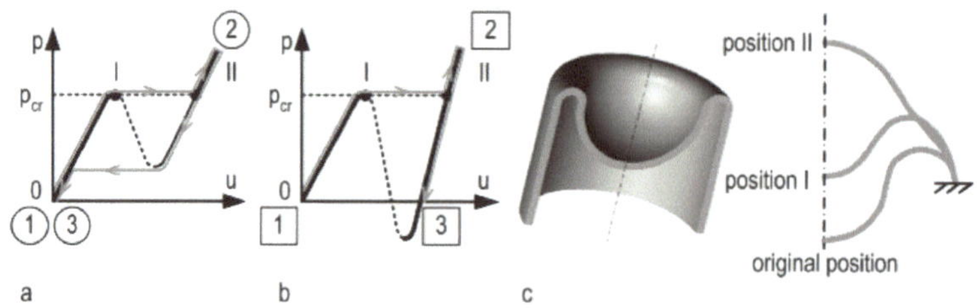

Fig. 9. Sketch of Snap-through behaviour of a curved mechanism: mechanism with monostable deformation (l.), mechanism with bistable deformation (r.)

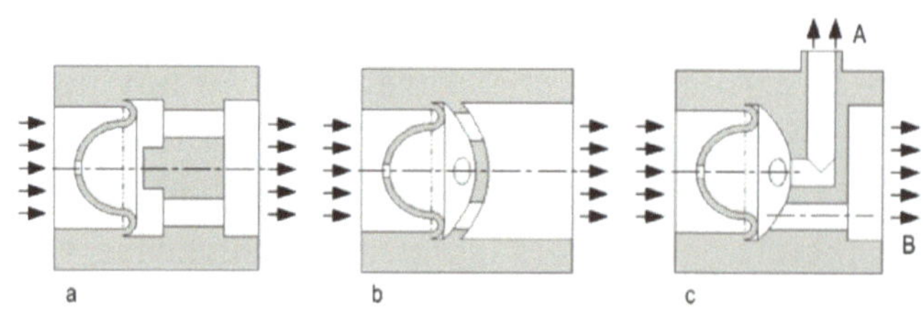

Fig. 10. Applications as mechanical valves demonstrated for double-curved rotational mechanisms: a, b – pipe with one output is disabled for $p=p_{cr}$, c – pipe with two outputs A and B, output B is closed if critical load is reached

Some applications of these mechanisms used as mechanical valves are shown in Figure 10. To generate bistable deformation (Figure 10 a) an opening is inserted in the centre point of the mechanism. Because of the critical dynamical pressure the mechanism is deformed and the flow is interrupted. In this case the current position guarantees the closure of the pipe. Low-pressure on the compliant part of the valve makes the flow possible again. The next Figure 10 b shows that the effect of the critical pressure yields to a deformation of the monostable mechanism, so the pipe is completely closed. If the pressure falls under the critical level, the original position is recaptured and the flow rate is reconstituted. The last example in Figure 10 c illustrates a valve installed in a pipe with two outputs. Output B is disabled, as soon as the critical pressure is reached. Decreasing the pressure enables this output.

3.4 Instable deformation behaviour of compliant mechanisms: Bifurcation

Situations with bifurcation of structures are avoided systematically in engineering. The following theoretical analyses reveal some opportunities to apply this behaviour profitably to a technical system.

The best known examples for the loss of stability under static loads are the Eulerian cases of stability. For loads under the critical level, the equilibrium is determinate, whereas at the critical level of loads, bifurcations in the solutions occur to state equations. The solutions are no longer bijective, one load situation may lead to more than one possible geometric configurations of the system. Such structures are shown in Figure 11. Herein, the load is generated by the attraction force of the filaments e.g. SMA-wires or by the low-pressure in cavities. If the wires are uniformly pulled or the cavities possess the same low-pressure, the classical Euler stability problem (bifurcation) is regarded as replacing the named rotationally drive configuration by an axial acting force.

The following statement explains how the bifurcation effect can be used profitably. The response of a systematically designed system with bifurcation behaviour (deformation or displacement in several directions) on external (e.g. temperature change) or on user-defined conditions leads to one preferred direction. The deformation direction is selected "autonomously" whereas the drive regime for each process always remains the same. The sensory and control effort are minimised enormously. The control of the system is partly adopted by "intelligent" mechanics.

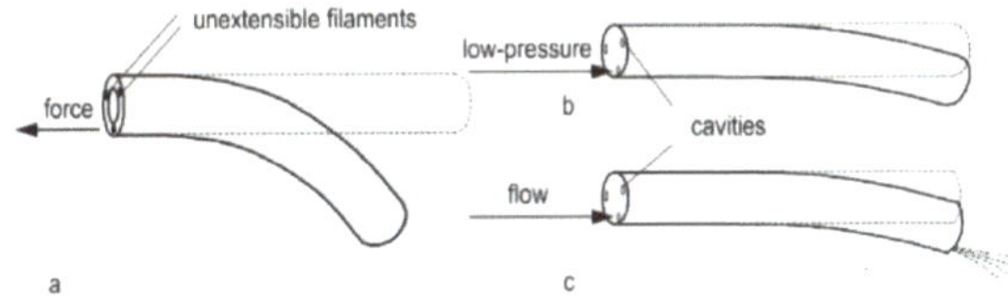

Fig. 11. Structures among the influence of an axial load which is generated by the attraction force of the filaments or Shape-Memory-Alloy-wires (a) and by the low-pressure in cavities (b, c)

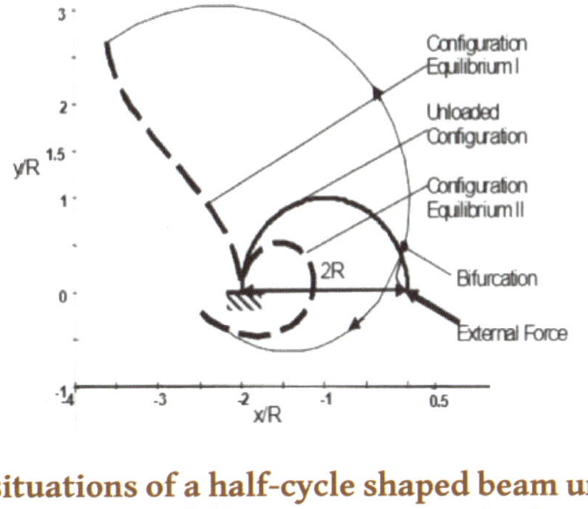

Fig. 12. Equilibrium situations of a half-cycle shaped beam under external load; A force constant in amount and directions traces the free end of the beam

Figure 12 exemplifies this phenomenon in the case of a half-cycle shaped bending beam subject to loading by a single force with constant direction (conservative force) but with a 2D-free floating location of the site of application under load. Two possible trajectories of this point and two realisations of the equilibrium are illustrated. Solutions have been determined numerically, a current application is the design of compliant grasping devices.

4. CONCISE SUMMARY

The proposed classification that accounts for the deformation of compliant mechanisms aims to further their development and simplify their integration into rigid body systems or the functional substitution of specific components of the rigid body. The significant utilization of compliant mechanisms, particularly those with unstable static behavior, presents substantial developmental potential. The function of the sensor system may be partially or wholly assumed by "intelligent" mechanics. The implementation of compliant mechanisms and structural components, exhibiting unstable static behavior and transitioning between states based on external conditions, can alter fundamental characteristics of the system (Risto et al., 2008; Linß et al., 2008; Risto et al., 2010; Griebel et al., 2010). Consequently, such systems will independently and directly adjust to the operational conditions.

Regarding functionally dominant compliance qualities, many application-oriented tasks, such as gripping fingers with specific attributes and medical structural components and systems, are feasible.

REFERENCES

1. Albanesi, A. E., Fachinotti, V. D., Pucheta, M. A. (2010). A review on design methods for compliant mechanisms. *Mecánica Computacional*, Vol.29, E. Dvorkin, M. Goldschmit, M. Storti (Eds.), Buenos Aires, (2010), pp. 59-72.
2. Beder, S., Suzumori, K. (1996) Elastic materials producing compliant robots. *Robotics and Autonomous Systems*, Vol.18, No.1, (July 1996), pp. 135-140, ISSN 0921-8890
3. Bicchi, A., Tonietti A. (2004) Fast and "soft-arm" tactics. *IEEE Robotics and Automation Magazine*, Vol. 11, No. 2, (June 2004), pp. 22-33, ISSN 1070-9932
4. Bögelsack, G. (1995). Nachgiebige Mechanismen in minaturisierten Bewegungssystemen, *Proceedings of the 9th World Congress on Theory of Mach. and Mech.*, Milano, August- September 1995
5. Christen, G., Pfefferkorn, H. (1998). Nachgiebige Mechanismen, *VDI Berichte Nr. 1423*, 1998 Griebel, S., Fiedler, P., Streng, A., Haueisen, J., Zentner L. (2010). Medical sensor placement with a screw motion. *Proceedings of Actuator 10 / International Conference on New Actuators*, ISBN 978-3-933339-12-6, Bremen, June 2010

6. Ham, R., Sugar, T., Vanderborght, B., Hollander, K., Lefeber, D. (2009) Compliant actuator designs. IEEE Robotics and Automation Magazine, Vol.16, No.3, (September 2009), pp. 81-94, ISSN 1070-9932

7. Howell, L. L. (2001). *Compliant Mechanisms*, John Wiley & Sons, ISBN 978-0471384786, New York IFToMM Terminology, Version 2.3 – April. 2010, http://www.iftomm.3me.tudelft.nl/2057/frames.html

8. Linß, S., Zentner, L., Schilling, C., Voges, D., Griebel, S. (2008). Biological inspired development of suction cups. *Proceedings of 53. Internationales Wissenschaftliches Kolloquium, IWK. Technische Universität Ilmenau*, ISBN 978-3-938843-37-6, Ilmenau, September 2008

9. Risto, U., Zentner, L., Uhlig, R. (2008). Elastic structures with snap-through characteristic for closing devices. *Proceedings of 53. Internationales Wissenschaftliches Kolloquium, IWK. Technische Universität Ilmenau*, ISBN 978-3-938843-37-6, Ilmenau, September 2008

10. Risto, U., Uhlig, R., Zentner, L. (2010). Thermal controlled expansion actuator for valve applications. *Proceedings of Actuator 10 / International Conference on New Actuators*, ISBN 978-3-933339-12-6, Bremen, June 2010

11. Shuib, S., Ridzwan M. I. Z., Kadarman, H. (2007). Methodology of Compliant Mechanisms and its Current Developments in Applications: A Review. *American Journal of Applied Sciences*, Vol.4, No.3, (March 2007), pp. 160-167, ISSN 1546-9239

12. Wang, W., Loh, R. N. K., Gu, E. Y. (1998) Passive compliance versus active compliance in robot based automated assembly systems. *Industrial Robot*, Vol.25, No.1, (1998), pp. 48-57, ISSN: 0143-991X

13. Zentner, L. (2003). Untersuchung und Entwicklung nachgiebiger Strukturen basierend auf innendruckbelasteten Röhren mit stoffschlüssigen Gelenken, *Ilmenau ISLE Verlag*, 2003, ISBN 3-932633-77-6

14. Zentner, L., Böhm, V., Minchenya, V. (2009). On the new reversal effect in monolithic compliant bending mechanisms with fluid driven actuators. *Mechanism and Machine Theory*, Vol.44, No.5, (May 2009), pp. 1009–1018, ISSN 0094-114X

15. Zentner, L., Böhm, V. (2009). On the classification of compliant mechanisms. *Proceedings of EUCOMES 08 The Second European Conference on Mechanism Science*, ISBN 978-1- 4020-8914-5, Cassino, September 2009

16. Zinn, M. Khatib, O., Roth, B., Salisbury, J. K. (2004). A new actuation approach for human friendly robot design. *The International Journal of Robotics Research*, Vol. 23, No.4-5, (2004), pp. 379 – 398, ISSN 0278-3649

INDEX

S

T

V

www.ingramcontent.com/pod-product-compliance
Lightning Source LLC
LaVergne TN
LVHW070849220926
848486LV00032B/358